# The Future Past of Tourism

THE FUTURE OF TOURISM
*Series Editors*: Ian Yeoman, *Victoria University of Wellington, New Zealand* and Una McMahon-Beattie, *Ulster University, Northern Ireland, UK*

Some would say that the only certainties are birth and death; everything else that happens in between is uncertain. Uncertainty stems from risk, a lack of understanding or a lack of familiarity. Whether it is political instability, autonomous transport, hypersonic travel or peak oil, the future of tourism is full of uncertainty but it can be explained or imagined through trend analysis, economic forecasting or scenario planning.

This new book series, The Future of Tourism, sets out to address the challenges and unexplained futures of tourism, events and hospitality. By addressing the big questions of change, examining new theories and frameworks or critical issues pertaining to research or industry, the series will stretch your understanding and generate dialogue about the future. By adopting a multidisciplinary perspective, be it through science fiction or computer-generated equilibrium modelling of tourism economies, the series will explain and structure the future – to help researchers, managers and students understand how futures could occur. The series welcomes proposals on emerging trends and critical issues across the tourism industry and research. All proposals must emphasize the future and be embedded in research.

All books in this series are externally peer-reviewed.

Full details of all the books in this series and of all our other publications can be found on http://www.channelviewpublications.com, or by writing to Channel View Publications, St Nicholas House, 31–34 High Street, Bristol BS1 2AW, UK.

THE FUTURE OF TOURISM: 2

# The Future Past of Tourism

Historical Perspectives and Future Evolutions

Edited by
**Ian Yeoman and
Una McMahon-Beattie**

**CHANNEL VIEW PUBLICATIONS**
Bristol • Blue Ridge Summit

DOI https://doi.org/10.21832/YEOMAN7079
**Library of Congress Cataloging in Publication Data**
A catalog record for – book is available from the Library of Congress.
Names: Yeoman, Ian, editor. | McMahon-Beattie, Una, editor.
Title: The Future Past of Tourism: Historical Perspectives and Future
    Evolutions/Edited by Ian Yeoman and Una McMahon-Beattie.
Description: Bristol, UK; Blue Ridge Summit, PA: Channel View
    Publications, 2020. | Series: The Future of Tourism: 2 | Includes
    bibliographical references and index. | Summary: "This book offers a
    critical account of the historical evolution of tourism through the
    identification and discussion of key turning points. Based on these
    considerations, future turning points are identified and evaluated. The
    core findings of the book provide the first perspective on how the
    history of tourism will shape its future"—Provided by publisher.
Identifiers: LCCN 2019029537 (print) | LCCN 2019029538 (ebook) |
    ISBN 9781845417062 (paperback) | ISBN 9781845417079 (hardback) |
    ISBN 9781845417086 (pdf) | ISBN 9781845417093 (epub) | ISBN 9781845417109
    (kindle edition)
Subjects: LCSH: Tourism—History. | Tourism—Forecasting. | Hospitality
    industry—History. | Hospitality industry—Forecasting.
Classification: LCC G156.F87 2020 (print) | LCC G156 (ebook) | DDC
    338.4/791—dc23
LC record available at https://lccn.loc.gov/2019029537 LC
ebook record available at https://lccn.loc.gov/2019029538

**British Library Cataloguing in Publication Data**
A catalogue entry for this book is available from the British Library.

ISBN-13: 978-1-84541-707-9 (hbk)
ISBN-13: 978-1-84541-706-2 (pbk)

**Channel View Publications**
UK: St Nicholas House, 31–34 High Street, Bristol, BS1 2AW, UK.
USA: NBN, Blue Ridge Summit, PA, USA.

Website: www.channelviewpublications.com
Twitter: Channel_View
Facebook: https://www.facebook.com/channelviewpublications
Blog: www.channelviewpublications.wordpress.com

Copyright © 2020 Ian Yeoman, Una McMahon-Beattie and the authors of
individual chapters.

All rights reserved. No part of this work may be reproduced in any form or by any
means without permission in writing from the publisher.

The policy of Multilingual Matters/Channel View Publications is to use papers that
are natural, renewable and recyclable products, made from wood grown in
sustainable forests. In the manufacturing process of our books, and to further
support our policy, preference is given to printers that have FSC and PEFC Chain of
Custody certification. The FSC and/or PEFC logos will appear on those books
where full certification has been granted to the printer concerned.

Typeset by Nova Techset Private Limited, Bengaluru and Chennai, India.
Printed and bound in the UK by Short Run Press Ltd.
Printed and bound in the US by NBN.

# Contents

|  | Figures, Tables, Exhibits | vii |
|---|---|---|
|  | Alastair J. Durie (1946–2017): The Historian who was a Futurist<br>*Ian Yeoman* | xi |
|  | Contributors | xv |
| 1 | Introduction: Does the Past Shape the Future?<br>*Ian Yeoman and Una McMahon-Beattie* | 1 |

**Part 1: Globalization**

| 2 | History of the Future of Tourism<br>*Bertus van der Tuuk* | 11 |
|---|---|---|
| 3 | The Development of Mass Tourism<br>*Jim Butcher* | 20 |

**Part 2: The Development of Destinations**

| 4 | The Historical Future of Tourism: The Case of Malta's Policies<br>*Marie-Louise Mangion* | 39 |
|---|---|---|
| 5 | Jules Verne as a Key to Understanding Irish Tourism<br>*Klaus Pfatschbacher* | 53 |
| 6 | The Growth, Decline and Resurgence of the City-State<br>*Brian Hay* | 65 |
| 7 | Geohistorical Analysis of Coastal Tourism in China (1841–2017)<br>*Benjamin Taunay* | 78 |

**Part 3: Mobility**

| 8 | The Future Past of Aircraft Technology and its Impact on Stopover Destinations<br>*Rafael Castro, Gui Lohmann, Bojana Spasojevic, Carla Fraga and Thiago Allis* | 93 |
|---|---|---|

| 9 | Forever Young and New: Cruise Tourism<br>*Wendy London and Wallace Farias* | 105 |
|---|---|---|
| 10 | From Muscles to Electrons: A Technological Look at the Futures of Energy, Transport and Tourism<br>*Jonathan Hui* | 118 |

### Part 4: The Hotel

| 11 | Hotel History<br>*Kevin James* | 133 |
|---|---|---|
| 12 | Historical Employment Relations in the New Zealand Tourism Hotel Sector: From a Collective Past to an Individual Future<br>*David Williamson* | 146 |

### Part 5: Diversification into Niche Tourism

| 13 | Film Tourism through the Ages: From Lumière to Virtual Reality<br>*Peter Bolan and Mihaela Ghisoiu* | 161 |
|---|---|---|
| 14 | The Evolution of the Grand Tour in the Digital Society<br>*Sabrina Seeler* | 174 |
| 15 | Shopping on the Edge: Identifying Factors Contributing to Tourist Retail Development in Heritage Villages<br>*Gianna Moscardo, Laurie Murphy, Karen Hughes and Pierre Benckendorff* | 188 |
| 16 | Tourism and Religion: Pilgrims, Tourists and Travellers – Past, Present and Future<br>*Richard Butler and Wantanee Suntikul* | 201 |
| 17 | The History and Future of Mountaineering Tourism<br>*Ghazali Musa and Md Moniruzzaman Sarker* | 215 |
| 18 | Sustainability, Ecotourism and Scotland: Concerns, Complaints, Conflicts and Conservation<br>*Alastair Durie, Ian Yeoman and Una McMahon-Beattie* | 229 |

### Part 6: Evolution

| 19 | Does the Past Shape the Future of Tourism? A Cognitive Map(s) Perspective<br>*Ian Yeoman and Una McMahon-Beattie* | 243 |
|---|---|---|
| | Index | 308 |

# Figures, Tables, Exhibits

**Figures**

| | | |
|---|---|---|
| **Figure 4.1** | Malta's tourism volumes and guest nights, 1959–2017 | 40 |
| **Figure 6.1** | The continuous cyclic city-state model | 76 |
| **Figure 19.1** | History of the future | 247 |
| **Figure 19.2** | The development of mass tourism | 249 |
| **Figure 19.3** | The historical future of tourism: The case of Malta's policies | 251 |
| **Figure 19.4** | Jules Verne as a key to understanding Irish tourism | 254 |
| **Figure 19.5** | The growth, decline and resurgence of the city-state | 256 |
| **Figure 19.6** | Geohistorical analysis of coastal tourism in China | 258 |
| **Figure 19.7** | The future past of aircraft technology and its impact on stopover destinations | 260 |
| **Figure 19.8** | Forever young and new: Cruise tourism | 262 |
| **Figure 19.9** | From muscles to electrons: A technological look at the futures of energy, transport and tourism | 265 |
| **Figure 19.10** | Hotel history | 267 |
| **Figure 19.11** | Historical employment relations in the New Zealand tourism hotel sector: From a collective past to an individual future? | 269 |
| **Figure 19.12** | Film tourism through the ages: From Lumière to virtual reality | 271 |
| **Figure 19.13** | The evolution of the Grand Tour in the digital society | 274 |

| Figure 19.14 | Shopping on the edge: Identifying factors contributing to tourist retail development in heritage villages | 276 |
|---|---|---|
| Figure 19.15 | Tourism and religion: Pilgrims, tourists and travellers – past, present and future | 278 |
| Figure 19.16 | History and future of mountaineering tourism | 280 |
| Figure 19.17 | Sustainability, ecotourism and Scotland: Concerns, complaints, conflicts and conservation | 282 |
| Figure 19.18 | Mindfulness | 286 |
| Figure 19.19 | Mobility | 288 |
| Figure 19.20 | Step changes determining mass tourism | 290 |
| Figure 19.21 | The leisure class of consumption | 293 |
| Figure 19.22 | Future past | 295 |
| Figure 19.23 | A conceptualization of tourism's future | 296 |

## Tables

| Table 7.1 | Foreign coastal concessions and ports open to foreign powers between the Nanking Treaty and the end of WWII (1842–1946) | 81 |
|---|---|---|
| Table 8.1 | Qantas' Kangaroo Route over time | 95 |
| Table 8.2 | The longest hauls | 100 |
| Table 12.1 | Key voices in the narrative | 147 |
| Table 16.1 | Significant events of a religious nature | 202 |
| Table 17.1 | Remarkable events in the history of modern mountaineering | 222 |
| Table 19.1 | Chapter 2: Historical and future turning points | 246 |
| Table 19.2 | Chapter 3: Historical and future turning points | 248 |
| Table 19.3 | Chapter 4: Historical and future turning points | 250 |
| Table 19.4 | Chapter 5: Historical and future turning points | 253 |
| Table 19.5 | Chapter 6: Historical and future turning points | 255 |
| Table 19.6 | Chapter 7: Historical and future turning points | 257 |
| Table 19.7 | Chapter 8: Historical and future turning points | 259 |

Table 19.8   Chapter 9: Historical and future turning points    261
Table 19.9   Chapter 10: Historical and future turning points   264
Table 19.10  Chapter 11: Historical and future turning points   266
Table 19.11  Chapter 12: Historical and future turning points   268
Table 19.12  Chapter 13: Historical and future turning points   270
Table 19.13  Chapter 14: Historical and future turning points   273
Table 19.14  Chapter 15: Historical and future turning points   275
Table 19.15  Chapter 16: Historical and future turning points   277
Table 19.16  Chapter 17: Historical and future turning points   279
Table 19.17  Chapter 18: Historical and future turning points   281

## Exhibits

Exhibit 19.1   Central command    283
Exhibit 19.2   Domain command     284

# Alastair J. Durie (1946–2017): The Historian who was a Futurist

Ian Yeoman

When I was the Scenario Planner for VisitScotland, there was one person in particular who inspired me to write about the future. That was Dr Alastair Durie. Alastair possessed a great mind and he knew 'everything' about Scottish tourism. He always had a story or tale to tell. When I talked to Alastair about the future and the scenarios VisitScotland were preparing, he could always tell me a story about the same scenario from a historical perspective. This book is based upon the historian Hobsbawm's (1995) observation that the past is a permanent component of human consciousness and the patterns of the future of human society. Each generation learns from the previous one and, in the knowledge transfer process, copies, improves and reproduces its predecessor as far as possible. As Alastair would recall, if many 'big new ideas' in business are examined critically, it will be found that these 'new ideas' have been present in Scottish tourism for a long time. Alastair's studies of the historical development of Scottish tourism (Durie, 2003, 2006, 2013, 2017) are a representation of Scotland today and in the future. Simply, he was a historian who understood the future and he would often say that the future is just the past but with different actors.

We published two articles together (Durie et al., 2006; Yeoman et al., 2005) which helped me understand how the future is a continuum from the past. Chapter 18 in this book, 'Sustainability, Ecotourism and Scotland: Concerns, Complaints, Conflicts and Conservation', was a paper we started together when I was at VisitScotland but never finished. He started it, but I have had to finish it. Alastair passed away in 2017.

Alastair Durie was educated at Edinburgh Academy where he distinguished himself academically. He enjoyed sport, and completed an MA (Hons) at the University of Edinburgh and a PhD about the Scottish linen industry (1707–1775). Tourism was his passion. I first came across Alastair's writing when the National Library of Scotland was hosting an exhibition on the history of Scottish tourism with Alastair's (2003) book

*Scotland for the Holidays: Tourism in Scotland c1780–1939* as the prominent focus. (Brodie, 2017) noted:

> Although notionally restricted to one country, his breadth of knowledge and wide reading was clear from the first page onwards. The book [Scotland for the Holidays, Tourism in Scotland 1780–1939] opens with the sentence: 'If statistics are to be believed, tourism is now the most important economic activity in the world, ahead even of oil, with scarcely any part of the globe unaffected.' In the same paragraph, he noted both the positive and negative impacts of tourism money, stimulating the creation of resorts, but also contributing to the corruption and corrosion of traditional societies. What is also apparent from the first paragraph is the author's sense of humour. He described how Scots promote Scotland 'as a land of heather, the kilt and whisky', a message that he amusingly characterised as 'a dash of truth, a splash of history and a good deal of manufacture and manipulation!' (Brodie, 2017: 274)

This was my first insight into the mind of this wonderful man. Meeting Alastair for a 'lemonade', I would always just listen, as I had so much to learn. He was a man with a sense of humour, a typical Scot who could spin a yarn.

*Scotland for the Holidays: Tourism in Scotland c1780–1939* (Durie, 2003) also contained a chapter on the search for health. Alastair described how Scotland embraced hydrotherapy, a subject which he obviously felt required more than 20 pages of discussion (Brodie, 2017). Hydrotherapy is not about spa treatments or beauty therapy as we know them today, but rather a very Scottish affair of very cold water bathing to cure a range of ailments. This was the subject of Alastair's next project. With the support of VisitScotland he published *Water is Best: The Hydros and Health Tourism in Scotland 1840–1940* (Durie, 2006). As Brodie (2017) said:

> *Water is Best* combined carefully chosen, often amusing insights with telling detailed extracts from visitors books and balance sheets. In the chapter discussing the interwar years, a table drawn from the annual accounts of the Creiff Hydro vividly tell the institution's story of initial expansion after World War I, followed by a lengthy decline, a significant drop in visitor numbers and modest profit margins. (Brodie, 2017: 276)

In 2017 Alastair published *Scotland and Tourism: The Long View, 1700–2015*, which covered three centuries of Scottish tourism in 130 pages. Here he examined everything from golf, fishing, shooting and even the Broons' cartoon strip. The final sentence of the book states that tourism 'is an industry of the past, for the present and for the future' (Durie, 2017: 121). My belief is that Alastair was an historian whom I trusted with the future; therefore this book is dedicated to him. It was his inspiration and words that helped my co-editor, Professor Una McMahon-Beattie, and me to bring about this collection of chapters.

He is sadly missed by many.

## References

Brodie, A. (2017) Alastair J. Durie (1946–2017). *Journal of Tourism History* 9 (2–3), 274–276. doi:10.1080/1755182X.2017.1415744

Durie, A.J. (2003) *Scotland for the Holidays: Tourism in Scotland c1780–1939*. Edinburgh: Tuckwell Press.

Durie, A.J. (2006) *Water is Best: The Hydros and Health Tourism in Scotland 1840–1940* Edinburgh: John Donald.

Durie, A.J. (2013) Sporting tourism flowers – the development from c. 1780 of grouse and golf as visitor attractions in Scotland and Ireland. *Journal of Tourism History* 5 (2), 131–145. doi:10.1080/1755182X.2013.828783

Durie, A.J. (2017) *Scotland and Tourism: The Long View, 1700–2015*. London: Routledge.

Durie, A., Yeoman, I.S. and McMahon-Beattie, U. (2006) How the history of Scotland creates a sense of place. *Place Branding* 2 (1), 43–52. doi:10.1057/palgrave.pb.5990044

Hobsbawm, E.J. (1995) *The Age of Extremes: The Short Twentieth Century, 1914–1991*. London: Abacus.

Yeoman, I., Durie, A., McMahon-Beattie, U. and Palmer, A. (2005) Capturing the essence of a brand from its history: The case of Scottish tourism marketing. *Journal of Brand Management* 13 (2), 134–147. doi:10.1057/palgrave.bm.2540253

# Contributors

**Editors**

**Ian Yeoman** is an Associate Professor of Tourism Futures at Victoria University of Wellington and Visiting Professor at the European Tourism Futures Institute and Ulster University. Ian is co-editor of the *Journal of Tourism Futures* and co-editor of Channel View's The Future of Tourism series. He is the author and editor of more than 20 books, including the forthcoming titles, *Science Fiction, Disruption and Tourism* and *Global Scenarios for World Tourism*. Outside the future, Ian is New Zealand's number one Sunderland AFC fan.

**Una McMahon-Beattie** is Professor and Head of Department for Hospitality and Tourism Management at Ulster University (UK). Her research interests include tourism futures, tourism and event marketing and revenue management. Una is co-editor of Channel View's The Future of Tourism series and sits on the editorial board of the *Journal of Tourism Futures*. She is the author/editor of a number of books, including the forthcoming book entitled *Science Fiction, Disruption and Tourism*.

**Chapter Authors**

**Thiago Allis** is Assistant Professor in the School of Arts, Sciences and Humanities at the University of São Paulo (Brazil). He holds a Masters in Latin American studies and a PhD in urban planning. Thiago's research interests include urban tourism, tourism mobilities and, more recently, post-conflict tourism.

**Pierre Benckendorff** is an award-winning educator and researcher specializing in visitor behaviour, technology and tourism. He has authored over 90 publications in these areas and is a regular speaker at tourism research conferences. Pierre serves on the editorial boards of several leading tourism journals.

**Peter Bolan** is a Senior Lecturer and Director for International Travel and Tourism Management at the Ulster University Business School. His research interests and consultancy specialisms include film/screen tourism, digital tourism, golf tourism and food tourism. Peter also writes regularly for a number of business and hospitality trade publications.

**Jim Butcher** is a Reader in the Geography of Tourism at Canterbury Christ Church University in Kent (UK). He is the author of three books and numerous articles looking at the social and political debates surrounding modern tourism. He blogs at https://politicsoftourism.blogspot.com/ and tweets at @jimbutcher2.

**Richard Butler** is Emeritus Professor at Strathclyde University. A geographer, he has taught in Canada, the UK, Holland and Italy. He has published more than 20 books and 100 papers on tourism. A past president of the International Academy for the Study of Tourism, Richard was UNWTO Ulysses Laureate 2016.

**Rafael Castro** is Assistant Professor in Tourism and Transport in the Federal Center for Technological Education Celso Suckow da Fonseca's Tourism Department (Rio de Janeiro, Brazil). Rafael is a research member of the Transport and Tourism Research Group (GPTT/UNIRIO) and the Strategic Planning of Transport and Tourism Centre (PLANETT/UFRJ).

**Wallace Farias** is a Professor in Tourism, Hospitality and Leisure at the Federal Institute of Brasilia. He has a BA in hospitality (Hons) from the Federal Fluminense University and a Masters in tourism from the University of São Paulo. Wallace's Masters research focused on cruise infrastructure development. He has worked as a tourism guide at Rio de Janeiro's Cruise Terminal.

**Carla Fraga** is Assistant Professor in Tourism in the Department of Tourism and Heritage at the Federal University of the State of Rio de Janeiro (UNIRIO). Carla is leader of the Transport and Tourism Research Group (GPTT/UNIRIO).

**Mihaela Ghisoiu** is a doctoral researcher in film tourism at Ulster University. Originally from Romania, she came to Northern Ireland to study for a tourism degree and fell in love with the country and its people so much that she continued her postgraduate studies there. Mihaela has a passion for nature, books and art in all its forms.

**Brian Hay** was a town planner in England and for 20 years was Head of Research at VisitScotland. He is currently an Honorary Professor at Heriot-Watt University. Brian's qualifications range from a BSc in town and country planning from Heriot-Watt University to a PhD in recreation resources development from Texas A&M University.

**Karen Hughes** researches and teaches in the area of sustainable tourism, tourist behaviour and visitor management. Her research focuses on the areas of interpretation, environmental education and using stories to connect with new and emerging visitor markets. Karen is currently investigating the impact of interpretation on visitors' long-term environmental knowledge, attitudes and behaviour.

**Jonathan Hui** is a futures researcher and consultant working at the intersection of energy, climate and digital infrastructure issues. Jonathan is a Master's graduate of the Hawai'i Research Center for Futures Studies and an incoming PhD in global governance at the Balsillie School for International Affairs.

**Kevin J. James** is Professor of History at the University of Guelph, Canada, and founder of the Tourism History Working Group. Kevin's publications include *Tourism, Land and Landscape in Ireland: The Commodification of Culture* (Routledge, 2014) and *Histories, Meanings and Representations of the Modern Hotel* (Channel View Publications, 2018).

**Gui Lohmann** is Associate Professor in Aviation Management and Head of the Aviation Discipline in Griffith University's Aviation Department (Australia). Gui has authored several books and peer-reviewed journal articles on air transport studies, particularly in the interface with tourism, transport geography and transport economics.

**Wendy London** is an Adjunct Research Fellow at Griffith University's Cities Research Institute and a cruise sector consultant. She has a special interest in cruise infrastructure development. Prior to following her passion in cruise tourism, Wendy was an IT/IP lawyer who was an internationally recognized expert in technology law and the use of computers in the law.

**Marie-Louise Mangion** lectures in the Department of Public Policy at the University of Malta, and previously served for 20 years in the public sector, managing desks for tourism strategy and European affairs and policy development in the fields of tourism, environment and culture. Her current research focuses on evidence-based policy making.

**Gianna Moscardo**'s research interests include evaluating tourism as a sustainable development strategy, understanding the relationships between the characteristics of tourism development and the dimensions of destination community wellbeing. Gianna is the current Chair BEST EN, an international organization committed to the creation and dissemination of knowledge to support sustainable tourism.

**Laurie Murphy**'s research interests focus on improving tourism's contribution to regional communities with an emphasis on tourism marketing, including backpackers, destination image and choice, destination branding and tourist shopping villages. Laurie serves on the Tourism Development and Marketing Strategic Advisory Committee for Townsville Enterprise Ltd.

**Ghazali Musa** is a Professor and a medical doctor and holds a PhD in tourism. He is the Head of Strategy and Business Policy Department in the Faculty of Business and Accountancy, University of Malaya, Kuala Lumpur. Ghazali has a wide interest in tourism research which includes scuba diving tourism, mountaineering tourism, backpacking tourism, medical tourism and international second homes.

**Klaus Pfatschbacher** is a Lecturer in French at the University of Applied Sciences IMC- FH-Krems, Austria. Klaus has authored a number of publications on French literature and the link between literature and tourism.

**Md Moniruzzaman Sarker** is a PhD candidate in the Department of Marketing, Faculty of Business and Accountancy, University of Malaya. His research interests evolve around services branding, consumer behaviour, travel and tourism marketing, transportation services and adventure tourism research.

**Sabrina Seeler** is a Postdoctoral Research Fellow in Experienced-based Tourism at Nord University Business School (Bodø, Norway) and is part of the research group for Marketing, Management and Innovation of Experiences (MMIE). Having experience in qualitative and quantitative research methods, Sabrina's research interests are consumer behaviour in tourism, tourist experiences, destination branding and strategic destination management.

**Bojana Spasojevic** is a Lecturer in Aviation Management and a PhD candidate at Griffith University, Australia, within the field of air transport and tourism. Her research focuses on stakeholder engagement, governance and leadership during the process of air route development. She is also a member of the Griffith Institute for Tourism.

**Wantanee Suntikul**'s recent books include *Tourism and Political Change*, *Tourism and Political Change* (2nd edn), *Tourism and War* and *Tourism and Religion: Issues and Implications*. Wantanee is joint Editor-in-Chief of the journal *Tourism, Culture & Communication*. Her other interests include gastronomy and tourism.

**Benjamin Taunay** is an Associate Professor, currently working as Academic Attaché at the French Ministry of Foreign Affairs. He has written several academic papers on Chinese domestic and outbound tourism. Benjamin's current research focus is on the political issues of tourism and the spread of Chinese tourism around the world.

**Bertus van der Tuuk** is a researcher and consultant in leisure and tourism. He has an educational background in leisure, tourism, sociology, research methods, statistics and demography. Since 1990 Bertus has been the Director of the research and consultancy firm, Bureau Vandertuuk (www.vandertuuk.nl).

**David Williamson** is a Senior Lecturer in the School of Hospitality and Tourism at the Auckland University of Technology. He previously worked for 18 years in the hospitality industry as a bar and hotel manager and restaurateur. David's research interests include work, employment and labour market issues in hospitality and tourism.

# 1 Introduction: Does the Past Shape the Future?

Ian Yeoman and Una McMahon-Beattie

## The Concept

In 1950, 25 million tourists undertook an international holiday whereas today international tourism has reached 1.3 billion arrivals (UNWTO, 2016). Why has this exponential growth occurred? Simply put, mobility increases wealth, the development of economies and the expansion of the experience economy. Historically, the evolution of tourism can be depicted through several key phases. For example, travelling in the Middle Ages increased due to the number of people going on pilgrimage, while travelling in the Romantic period, with its emphasis on emotion and individualism as well as the glorification of the past and nature, led to the development of the Grand Tour and the modern concept of the 'tourist'. From another perspective, technological advancement has been a driver of tourism development. For example, in the past, travellers to New Zealand by sailing ship from England undertook a journey lasting up to six months. Steam ships cut this journey time to six weeks (McClure, 2004). Today, the same journey by jet aircraft is only 24 hours. Looking to the future, Yeoman (2012) predicts that international tourism arrivals will reach 4.2 billion by 2050 with an economic value of US$4.7 trillion. But what is the connection between the past and the future? The historian Hobsbawn (1995) believed that the future is a replication of the past. If this is the case, it should be possible to analyse the past in a scientific manner in order to inform the future, thus answering this chapter's question: Does the past shape the future?

Thomas Kuhn (1962), in his book *The Structure of Scientific Revolutions*, coined the concept of a paradigm for scientific discussions. Kuhn's account of a paradigm describes it as a set of achievements which a scientific community 'acknowledges as supplying the foundation of its further practice'. They are achievements that 'attract an enduring group of adherents away from competing modes of scientific activity' and are 'sufficiently open-ended to leave all sorts of problems for the redefined group of practitioners to resolve'. In futures research, change is the constant from the past to the future. Mannermaa (1991) has stated:

> The concepts of evolution and development are used as synonyms [in this article]. The idea of development or evolution was perhaps the most important single feature in the cultural physiognomy of the 19th century, and it has maintained its importance until our times. At least as much as the roots of this idea belong to the domain of biology, they can be sought in an awakening sense of historicity. Historicity meant a new interest in the origin of things and a view of history, not as a haphazard flow of disjointed events, but as development or evolution, which meant a directed and ordered succession of 'genetically' related stages. Central to the idea of development is that the succeeding states of the system are dependent on each other, and that this process has a direction. (Mannermaa, 1991: 349)

Descriptive futures research is based on what Karl Popper (Popper & Miller, 1983) has called the historicism idea of the general nature of societal development. Miles (1978) stated that:

> Historicism refers to the attempt to predict the future, the basis of supposed laws of historical evolution, whose operation may be projected forward to provide a view of the future. Historicism is manifest in the extrapolation of trends or cyclical regularities, the application of macro sociological theories that deduce the inevitable progression of societies from one to another form, and the use of deterministic accounts (technological, economic, or ecological) of social change. Human choices and praxis are assigned an epiphenomenal or mechanical role in making past and future history. (Miles, 1978: 68)

Thus, the future can be predicted based upon the past (Yeoman & McMahon-Beattie, 2016). One of the roles of futures research is to model the development of society, looking for signs, social movements, technological advancement and signs of change at the point of evolution, or what Gladwell (2002) calls 'tipping points'. The identification of tipping points is key to the evolutionary paradigm from futures studies (Mannermaa, 1991). Turning points encompass the notion of era analysis. From a historical perspective, eras are incremental and continuous phases in the development of tourism with relative stability and coherence. Thus, a specific era or turning point can be identified from one era to the next (Li & Petrick, 2008; Reed & Gill, 1997). The same principles can also be used from a futures perspective (Mackay, 1999; Millard, 2010; Virginia, 2017).

## The Structure of the Book

The book is structured into a number of sections which overview the historical development of tourism. The first section, **Globalization**, is about the birth and development of tourism. With the establishment of mass tourism, the next section, **The Development of Destinations**, examines tourism in specific countries, followed by the section on **Mobility** as a driver of growth. The section on **The Hotel** looks at the largest sector within tourism, and the section on **Diversification into Niche Tourism** provides an analysis of tourism experiences and products. The final section, **Evolution**, brings the book together as a conceptual framework based on a series of cognitive maps of the contributing authors' commentaries.

## Globalization

In Chapter 2, *History of the Future of Tourism*, Van der Tuuk explores the evolution of tourism through identifying a number of turning points that have impacted on the history of tourism and the motivation to travel. These include the move from travelling as a necessity to travelling for fun, the shift from 'verge-tourism' to mass tourism, religious pilgrimages in the Middle Ages, travel as a source of knowledge and education (17th and 18th centuries), Grand Tour travel (18th and 19th centuries), and the emergence of travel and tourism as an 'experience'. Van der Tuuk uses the Netherlands as an exemplar of tourism's European evolution. Looking to the future, a number of turning points based upon demography, family structures, the global economy and cultural values are identified as important.

In Chapter 3, Butcher builds upon the previous chapter, focusing on *The Development of Mass Tourism*. Butcher argues that mass tourism emerges with mass modern society and is hence distinct in important respects from previous forms of leisure travel. Key societal and technological developments are highlighted, such as rail travel, the motor car, the jet engine and, more recently, the technological/informational 'revolution'. The chapter has a global remit and considers relative geographical shifts over time in the generation and receipt of tourists, such as the growth in the relative importance of China and other Asian economies. It allows the reader to quickly establish the key turning points in the history of mass tourism and likely future turning points, focusing on Asia and Africa's potential.

## The Development of Destinations

In Chapter 4, *The Historical Future of Tourism: The Case of Malta's Policies*, Mangion recognizes that the evolutionary pattern in any system of tourism depends on its environment. The chapter examines how this applies to Malta, a small island nation and the smallest member state within the European Union. Key turning points in Malta's tourism activity are identified, taking a historical perspective on the country's tourism policy and the private sector's response. The development of Malta's tourism activity, spanning the period from 1958 to the present day, is of interest due to the amplified challenges of accessibility, increased vulnerability, relatively negligible promotional resources, and competing demands which need to be met by minimal resources. Within this context, the chapter outlines what transformations took place in the sector, what triggered change and what provoked policy responses and considerations governing the threshold below which a decline would not be acceptable to policy makers.

Chapter 5, *Jules Verne as a Key to Understanding Irish Tourism*, considers how Ireland is portrayed as a touristic nation. It begins by looking at past images of the Emerald Isle, with descriptions intertwined with the notion of poverty. The French traveller and eminent sociologist Tocqueville, for example, attracts the reader's attention to the less than modest dwellings of the rural Irish population in the first half of the 19th century, thus presenting numerous obstacles on the way to touristic prosperity. Pfatschbacher analyses Verne's novel *Foundling Mick*, which describes Irish idiosyncrasies outstandingly well and thus offers a rich source of information at the disposal of the Irish agency of tourism. Moving forward, Pfatschbacher notes the importance of hotels and other tourist accommodation as offering narrative and Ireland's cultural history as a shaper of the future.

In Chapter 6, *The Growth, Decline and Resurgence of the City-State*, Hay demonstrates that the growth of tourism is closely associated with the development of the city-state. From the early Sumerian cities to the modern nation state, he notes that over the last two millennia the government of cities has been subject to continuous change. Until two centuries ago, the city-state in some form was the most common system of government, but with improvements in transport and the centralization of power their numbers have declined, so today there are only three city-states (Singapore, Monaco and Vatican City). Hay concludes by looking to the future in which an illiberal democracy dominates, placing priority on an efficient and effective government. City-states will restrict tourism through laws to govern the interaction/behaviours between citizens and tourists. Finally, Hay suggests that the term tourism will become redundant as tourists will be viewed as 'non-citizens'.

It is acknowledged that China will be the largest inbound and outbound tourism destination in the future (Yeoman, 2008). In Chapter 7, Taunay writes about the *Geohistorical Analysis of Coastal Tourism in China (1841–2017)*, highlighting the role of colonial powers and the Communist Party. The chapter thus contributes to deconstructing the idea that sea-bathing in China is a recent practice. The filiations between current and other older practices in China or elsewhere in the world are longstanding, emphasizing that the study of tourism in each place and period should be explored in a global historical context.

## Mobility

In Chapter 8, *The Future Past of Aircraft Technology and its Impact on Stopover Destinations*, Castro and colleagues trace the development of aircraft technology on the Kangaroo Route between Australia and Europe. Plane technology has markedly shaped the global tourism industry, shrinking the world by making travel fast and affordable. In particular, long-haul travel has significantly increased over the last 100 years, with new aeroplane technology expanding the range of flights and the comfort of passengers. This chapter highlights turning points in the historical development of aircraft technology and their effect on stopover destinations, from the Lockheed Constellation era with many stopovers, to the emergence of the Boeing 707 and 747, as well as the Airbus A-380 and the rise of global stopover destinations such as Singapore and Dubai. From a destination management perspective, some of these stopover hubs have not only developed their aviation infrastructure but have also invested in marketing and branding strategies and events management, and have built attractions, accommodation and other leisure/tourism features. With the likelihood of aircraft technology improving further, the stopovers that we know today will not be required in the future.

London and Farias in Chapter 9, *Forever Young and New: Cruise Tourism*, begin by stating that cruising continues to experience exceptional growth. The patterns and rate of that growth have been affected by a series of episodes or events. Leisure cruise travel began in 1844, with the first dedicated cruise ship launched in 1900. However, war interrupted leisure cruising as ships were appropriated for troop transport. When leisure travel by sea was resumed, it was again supplanted in 1958 as air transport became accessible. However, modern dedicated cruise ships were launched in the 1970s, marking the beginning of cruising as it is known today. Looking to the future, climate change, technology and conflict will dominate the discourses of cruise tourism.

In Chapter 10, *From Muscles to Electrons: A Technological Look at the Futures of Energy, Transport and Tourism*, Hui traces the historical development of energy and how it has shaped the social and experiential life of tourism. Specifically, he highlights the five key prime movers driving transportation and tourism, starting with the human body and the domestication of horses, before moving on to three inventions since the Industrial Revolution: the steam engine, the internal combustion engine and the gas turbine. The thrust of the chapter is the nexus of energy and transportation and how their co-evolution reshapes the types of tourism that become possible.

## The Hotel

In Chapter 11, *Hotel History*, James explores the history of the hotel, and also surveys, from a critical-analytic perspective, key debates in the field, evaluating their influence over scholarship. These debates range from the modern hotel's disputed origins in antiquity to the hidden history of hotel labour. Overall, this chapter argues that the material form of the hotel is highly protean and adaptive, and that hotel history must account for that feature, as well as for the dominance and cultural salience of hotel ideas and ideals in understanding this complex institution.

As in many countries, tourism has become a major economic driver in New Zealand, recently overtaking the dairy sector to become the nation's largest earner of export dollars. Yet despite its economic importance, labour in this sector demonstrates all of the challenges commonly associated with the international tourism workforce: low wages, high turnover, high levels of casualization, skills shortages and a dependence on migrant workers with the hotel at the forefront of tourism. Williamson, in Chapter 12, *Historical Employment Relations in the New Zealand Tourism Hotel Sector: From a Collective Past to an Individual Future*, looks at the growing concern that labour issues may be a major limiter on future tourism growth and development. This chapter addresses the following central question: How did we get here and what does this mean for the future? The chapter concludes by suggesting two dramatically different scenarios for the future of the tourist hotel workforce.

### Diversification into Niche Tourism

Bolan and Ghisoiu, in Chapter 13, *Film Tourism through the Ages: From Lumière to Virtual Reality*, explore how film and subsequently television and digital streaming services have increasingly influenced people's interests and decision-making processes, including their travel choices and tourism experiences. This chapter charts and examines key turning points, from the Lumière brothers' first public screening in the 1800s through to the influence of *Crocodile Dundee* on tourists to Australia in the 1980s and the enormous phenomenon of *Game of Thrones* and its impact on tourist visitation to Northern Ireland. As part of post-modernist discourse it is clear that the future of film tourism is linked to advances in technology.

In Chapter 14, *The Evolution of the Grand Tour in the Digital Society*, Seeler explores the development of tourism since the times of the Grand Tour in Europe in the 17th and 18th centuries. She describes the early explorers and their travel motives, explains the spatial and temporal developments of the Grand Tour and addresses the declining significance of the Grand Tour as a result of technological advancements during the Industrial Revolution and the emergence of a leisured society in the early 19th century. Thereby the chapter brings the development of mass tourism and the changing motivation of tourists into focus. It sheds light onto the digitized travellers and their continuous desire for knowledge enhancement and self-development. As such, the contribution of the chapter lies in the exploration of the relationship between early forms of travelling when educational purposes, identity formation and personal growth were the dominant motivation to travel, and today's increasingly experienced tourists who seek similar outcomes when engaging in leisure travel.

In Chapter 15, *Shopping on the Edge: Identifying Factors Contributing to Tourist Retail Development in Heritage Villages*, Moscardo and colleagues report on the historical case studies of three Tourism Shopping Villages (TSVs): Hahndorf in Australia, St Jacobs in Canada and Cheddar in England. These villages are well-established tourist destinations with diverse tourism development histories. The chapter's central contribution to the study of tourism futures lies in demonstrating that a historical or evolutionary approach uncovers a wider range of underlying processes than cross-sectional analyses. Decisions made by tourism governance organizations that encouraged growth without consideration of impacts, a shift in attention from locals to tourists in retail and services, and supported external investment to support growth, all contributed to the turning points for the villages.

Butler and Suntikul, in Chapter 16, *Tourism and Religion: Pilgrims, Tourists and Travellers – Past, Present and Future*, clarify the roles that the world's organized religions have played in motivating tourism and influencing its past and current patterns and practices. This is done by first identifying key events in the past related to the major religions in the world which have had major impacts upon tourism, both in past times and currently. A number of the world's major faiths are examined in terms of their roles in and influences on tourism with specific reference to major historic events and the continued relevance of these for present and future tourism.

In Chapter 17, *The History and Future of Mountaineering Tourism*, Musa and Sarker explore mountaineering tourism's historical development in order to understand its future by identifying a number of turning points including the early era, the emergence of the golden era (when mountaineering emerged as an adventurous activity), mountaineering's post golden era (when the sport went global), the climbing of the last frontiers and the modern tourist era of mountaineering. Moving forward, the authors forecast that mountaineering tourism activity will remain resilient and will benefit from a range of entrepreneurial adaptations.

In Chapter 18, *Sustainability, Ecotourism and Scotland: Concerns, Complaints, Conflicts and Conservation*, Durie and colleagues use a historical analysis and archive search to identity turning points and incidents associated with Scottish tourism from an environmental and sustainable perspective. These include the abuse and degradation of Scotland's landscapes by early tourists and the impact of tourism on the physical, natural and cultural environment. Looking to the future, the authors focus on sustainability and climate change.

## Evolution

In the final chapter, Yeoman and McMahon-Beattie explore the question: Does the past shape the future of tourism? Using a conceptual framework based on a series of cognitive maps developed from the preceding chapters, the editors of the book identify the following historically significant turning points in the evolution of tourism, namely, *Mindfulness, Mobility, Step Changes Determining Mass Tourism* and *The Leisure Class of Consumption*. Further analysis identifies four key future turning points, which are *Fluid Identity, Sustainable Futures, Mass Maturity* and *Ubiquitous Future*. Two scenarios are then proposed: *Degradation – If Only We Had Listened to the Past* and *A Balanced Future – Learning from the Past*. The importance of the chapter lies in its analysis of how the past shapes the future, linking history to the future as a continuum.

### References

Gladwell, M. (2002) *The Tipping Point: How Little Things Can Make a Big Difference*. Boston, MA: Back Bay Books.
Hobsbawm, E.J. (1995) *The Age of Extremes: The Short Twentieth Century, 1914–1991*. London: Abacus.
Kuhn, T.S. (1962) *The Structure of Scientific Revolutions*. Chicago, IL: University of Chicago Press.
Li, X. (Robert) and Petrick, J.F. (2008) Tourism marketing in an era of paradigm shift. *Journal of Travel Research* 46 (3), 235–244.
Mackay, H. (1999) *Turning Point: Australians Choosing Their Future*. Sydney: Pan Macmillan.
Mannermaa, M. (1991) In search of an evolutionary paradigm for futures research. *Futures* 23 (4), 349–372. doi:10.1016/0016-3287(91)90111-E
McClure, M. (2004) *The Wonder Country: Making New Zealand Tourism*. Auckland: Auckland University Press.
Miles, I. (1978) The ideologies of futurists. In J. Fowles (ed.) *Handbook of Futures Research* (pp. 68–69). London: Greenwood Press.
Millard, B. (2010) *Future Trends from Past Cycles: Identifying Share Price Trends and Turning Points through Cycle, Channel and Probability Analysis*. London: Harriman House.
Popper, K.R. and Miller, D. (1983) A proof of the impossibility of inductive probability. *Nature* 302 (5910), 687. doi:10.1038/302687a0
Reed, M.G. and Gill, A.M. (1997) Tourism, recreational, and amenity values in land allocation: An analysis of institutional arrangements in the postproductivist era. *Environment and Planning A: Economy and Space* 29 (11), 2019–2040.
UNWTO (2016) *UNWTO Tourism Highlights, 2016 Edn*. Madrid: UNWTO.
Virginia, G. (2017) Turning point: Uncertain future. *Nature* 541 (7636), 249–249.
Yeoman, I. (2008) *Tomorrow Tourist: Scenarios and Trends*. London: Elsevier Science.
Yeoman, I. (2012) *2050: Tomorrow's Tourism*. Bristol: Channel View Publications.
Yeoman, I. and McMahon-Beattie, U. (2016) An ontological classification of tourism futures. Paper presented at *CAUTHE 2016: The Changing Landscape of Tourism and Hospitality: The Impact of Emerging Markets and Emerging Destinations*. Sydney: CAUTHE.

# Part 1

# Globalization

# 2 History of the Future of Tourism

Bertus van der Tuuk

### History

Scarcely 15,000 years ago, humankind consisted of a mere few thousand individuals. Scattered across the globe, these nomads migrated in search of food, led by changes in the climate and the seasons. Four million years earlier, their distant ancestors did not do things any differently. So, it is safe to assume that nomadic instincts are deeply rooted in *Homo sapiens*' genes. It is an urge for 'elsewhere' which, in recorded world history, has been translated into a powerful need to conquer and into inexplicable mass exoduses (van Hesteren, 2011).

The physical needs of hunters and gatherers, who set off again and again in search of the next place to find food and shelter, have ended up as the luxury needs of settled humanity who, armed with their payment cards, travel guides and sun cream, go hunting for sights a couple of times a year. Travelling out of necessity has evolved into travelling for pleasure, a self-chosen 'adventure' or a well-earned rest (Lansink, 2002).

This chapter mainly describes the situation in the Netherlands – a situation that is largely exemplary for Western Europe and partly for the Western world. What do we see in the Netherlands in mid-July? We see cars with white vehicles attached to the back of them, backed-up roads, traffic jams in the direction of the south of Europe, whining children, sunburnt skin and people laughing and arguing, and eating and drinking too much. What are these people looking for? What is driving these people? Adventure? Sunshine? Contact? Relaxation? Other cultures? Unspoilt nature? Authenticity? Excesses of pleasure?

Tourism in its current mass form started only relatively recently. At the end of the 1950s and beginning of the 1960s, Dutch people enjoyed a kind of tourism known as 'highway tourism'. This form of leisure activity, enjoying watching the cars driving past, created an illusion of travel. Within only a few decades, this illusion or virtual travel had transformed into opportunities for mass travel and crossing great distances. Around

1900, an average adult travelled approximately 2000 km per year (Lansink, 2002); modern-day people achieve this distance with a single flight to a Spanish costa or Greek island.

The increased urge to travel has not just been the result of growth in technical and economic possibilities. Values, norms and motives have also changed. In the Middle Ages, pilgrims travelled for religious motives – not for the journey's sake but rather in the hope of meeting God. From the Renaissance onwards, people started to turn towards the profane world. These new travellers were scientists, merchants and explorers, looking for knowledge and for economic and political expansion (Lansink, 2002).

In the 17th and 18th centuries, travelling was a serious business, a way to acquire knowledge and to develop. A journey brought people into contact with other cultures and ways of life, other ideas and habits. In one of his essays, Montaigne wrote, 'I know no better method of shaping someone than by confronting them with the diversity of ways of life of other people, of their morals and customs, of their natures and their opinions' (cited in Lansink, 2002: 2). Travel was viewed as a source of knowledge and a way to finish their upbringing (Maczak, 1998).

### Turning point 1: Grand Tours of Europe

From the 18th to the start of the 20th century, it was the custom for young English adults of a higher social status to make a Grand Tour of the European continent. Large numbers of young Englishmen (and some Englishwomen) visited France and Italy. In cities like Paris, Rome, Venice, Florence and Naples they experienced their rites of passage from childhood to adulthood. These young men not only had the opportunity to know and value the classical arts, but they also got to sow their wild oats and learned to practise the most classical art of them all: they had their first sexual and romantic experiences, and these prepared them for marriage.

In this way the Grand Tour reflected a new phenomenon – a more or less hedonistic approach to consuming. For the members of the aristocracy it was an expression of wealth and taste. Italy had the cultural reputation of the Renaissance; it was truly 'classic', an easily consumable expression of refinement (Zuelow, 2015). Although the purpose of the Grand Tour was self-improvement with knowledge of foreign languages, customs, politics and culture, the reality was often different. 'Italy was long a leading center of sexual promiscuity. ... Parisian wives had the reputation of being accommodating with their sexual favors, performing every imaginable act with travelers, especially when it was common for young men to lavish gifts of £1000 on them for services rendered' (Zuelow, 2015: 25–26).

From the Romantic era onwards (at the end of the 18th century), the modern tourist appeared – someone for whom travel was a means unto itself. Romantic travellers wanted to find themselves in authentic places, untouched by bourgeois and industrial society. Dark woods, the coast and the wide sea, quiet lakes, deep valleys and tall mountain peaks: these became the new 'holy' places. Nature had an intrinsic value, a sovereign beauty which elevated her beyond her usefulness to man (Lansink, 2002). The Romantic traveller took the road less travelled, just as in the poem *The Road Not Taken*:

> Two roads diverged in a wood, and I –
> I took the one less traveled by,
> And that has made all the difference. (Frost, 1916)

He wanted to be alone and eye to eye with overwhelming, mysterious nature. Nature was supposed to evoke emotions, to touch you and even to be terrifying and frightening. To put it in modern terms, 'you need to experience nature'.

## Turning point 2: Mass tourism

From the 1960s, increasing prosperity, the longing for freedom of the baby boomers who were growing up at the time, an explosion in increased mobility and a shift from a work-oriented to a leisure-oriented society caused tourism to accelerate. In the Netherlands, participation in holidays grew from 50% to over 80% in 2016 and the number of holidays increased from 0.55 to approximately 2.08 per year. The ratio of domestic to foreign holidays made a radical flip: currently there is one domestic holiday to 1.13 foreign holidays; at the end of the 1960s the ratio was 1:0.73 (NRIT Media *et al.*, 2017). You have to have very good reasons to stay at home nowadays, to deny yourself one of the basic human needs – the right to a holiday, preferably abroad in the sun.

What has all this led to? In 1950, David Riesman wrote a book entitled *The Lonely Crowd*. He predicted that modern society would have to deal with a new middle class of apathetic and fragmented individuals. These individuals would move in a modern vacuum, focusing on the transient, on the exterior world and on others (other-directed), rather than focusing on the eternal, tradition and the inner world (inner-directed) (Mous, 2010).

## Turning point 3: Modern tourism

We can recognize this in modern tourism. Our holiday plans show who we are and where we belong, just like our cars, the way we decorate our houses, our clothes and our taste in music. The French sociologist Pierre Bourdieu has wonderful examples to reinforce this 'distinction' theory (Bourdieu, 1984). Our holiday plans are determined by the norms and values in our social environment. According to the British sociologist John Urry, our wanting to visit a certain tourist destination is not down to the good weather, getting to know another culture, the good food or the beautiful natural surroundings. Instead, it is a result of cultural norms and learned tastes. Our environment teaches us that certain popular destinations are worth visiting: holiday destinations are more human constructions than places of intrinsic attractiveness. Our holiday choices allow us to show which social group we (would like to) belong to; they are a direct result of the 'other-directedness' of our society (Urry, 1995).

So how does this relate to all those people who choose holidays 'off the beaten track': a trip to Antarctica, a trek through Tibet, a characteristic little hotel in a rustic Greek street 'where no tourist has been before'? Indeed, in all kinds of advertisements, tourist areas claim to be 'still undiscovered by mass tourism' or 'untouched by mass tourism'. Apparently, undiscovered and untouched places need to be safeguarded from hordes of tourists. The elite, no longer aristocratic but still wealthy, higher educated or intellectual, make desperate attempts to distinguish themselves.

## Turning point 4: Nature

Alas, even the best kept secrets never remain secret for long. If after a few years you return to that authentic bench by that undiscovered little lake, you will probably find a hotel with numerous picnic tables, boat hire and kite surfers. Modern man is trapped in a 'prisoner's dilemma'. In actual fact, you should not tell anyone about your holiday experiences, but how else can you show your environment that you are 'not a tourist'?

Is the massive nature of tourism digging its own grave? Will nature's popularity cause its destruction? We manfully work at protecting and conserving nature and even at creating 'new nature'. Even so, we are seeing an unprecedented attack on that very same nature. We want to enjoy the silence of the savannahs, the setting sun in the uplands of Peru and the unspoilt beauty of Antarctica. But to do so we must first produce noise, pollute the air and use up fossil fuels to get there.

There is different kind of loneliness for the modern tourist from the kind that the Romantic traveller was looking for. This is the loneliness of 'other-directedness' (Riesman *et al.*, 1950), the search for safety among the like-minded. Away from work, the city or their living environment, this *lonely mass* inhabits overfilled beaches and terraces. Silently they cruise to the toilet block in enormous child-friendly 'glamping sites', are entertained by special animation teams and enjoy the view from specially built 'view points'. This is the paradox of tourism: if you are looking for freedom, you need to adapt and conform, to follow other cars over the same roads, to see what others see, to take the same photos at the same sightseeing spots and to look at the same chipped lions in the same national parks. Campers huddle together at special stopover spots, Dutch colonies in Marbella drink Dutch beer, and young people – triggered (or not) by greedy media and ditto television viewers – colour in their rites of passage with drink, amusement, sex and adventure(s). Modern tourists regress to exactly that existence they are trying to escape from.

### Turning point 5: Pilgrimage

Just as the Bible kept pilgrims on the path of righteousness, so it is with tourists clinging to their travel guides telling them what they *have to* see. They are not so much looking for an authentic confrontation with the strange and unknown, as a confirmation of the images and descriptions given by tourists who have gone before. This is even true for the adventurous *backpacker*, the 'romantic traveller' of the 21st century who, having just arrived from the other side of the world, lets himself be guided by the tips in his well-eared *Lonely Planet* and so mainly meets more people just like himself. The modern tourist travels through a world which has already been completely mapped out. His destination has been reproduced multiple times in travel guides and holiday brochures and his arrival has been prepared for with extensive accommodation, planned walks and local summer programmes (Lansink, 2002).

Tourism has become an important economic product; for several countries it has even become their greatest source of income. A traveller is not much more than a consumer of pre-cooked experiences from videos, photos and postcards which prove where he has been, of expensive consumptions and of the rays of sun which he has been missing out on. He has become part of the (economic) system from which he so wanted to break free (Lansink, 2002).

## The Future

In 2035, demographic ageing will have almost reached its peak. The baby boomers will have become very elderly or will have passed away. Care costs will have risen dramatically and those in work will have to do their best to carry the now heavy collective load. We will need to work long and hard; retirement age will have increased to 70. Companies will be network organizations, work will have become independent of time and place, and new ways of working will have become commonplace.

Society will be sustained by the current 10–20 age group. In the Netherlands we call them 'Generation Einstein' (also known as Generation Z). They will be 35–45 in 2035. These are people who have grown up in the information age, who are faster, smarter and more sociable than previous generations, less individualistic, more collectivistic, who live more in networks, both online and offline, and who are more oriented towards values such as authenticity, respect, self-development and honour (Boschma & Groen, 2007).

Holidays will be no less important in 2035. They are and will remain a primary need, certainly for the hard-working generation of 2035, who will really need their holidays. Almost half of all Europeans state that having a holiday has become an indispensable part of life. This has become self-evident even in times of economic recession. Having holidays has a higher value (as a luxury good) than houses, fast cars, expensive perfumes or designer clothing: 'After a good job, a good relationship or a good education, leisure activities and holidays rank as top priorities for today's Europeans' (Yeoman, 2010).

New ways of working, the increased pressures of work and the disappearance of the boundaries between work and one's private life will lead to more stress and, as a result, an increasing need to 'get away from it all'. It would not be at all surprising if long holidays were to be replaced by various forms of 'short stays'. The working generation of 2035 will find city breaks, weekends away, attraction parks, multi-day events, resorts and health or wellness centres indispensable for keeping going, both physically and mentally.

### Future turning point 1: The global economy of tourism

The global economy will have an open character, even more so than now. There will be more competitors and there will be scarcely any countries not presenting themselves as attractive tourist destinations. China, India and the USA will be the leading global economies.

The entire world will be within reach because of technological progress, availability and increasing options from the internet. Countries such as Russia and China, where just a few years ago the inhabitants could only dream of foreign holidays, will be recognized by the inbound tour operators and turned into important new tourist countries of origin.

Thanks to the global expansion of available tourist destinations, people will increasingly come into contact with all kinds of events. Whereas a few year ago the Palio di Siena was an event on only a regional or national scale, nowadays it has become a significant international tourist attraction. Tourists in 2035 will be spoilt for choice, throughout the world. Their choice of destination will be made in a much more ad hoc way than is now the case and will be influenced by all kinds of sudden events such as extremes of weather, terrorist attacks, natural disasters, etc.

### Future turning point 2: Cultural capital

In a society where working people come across each other less and less in the workplace, where virtual networks of groups of friends will largely determine mutual contacts and where work will have become independent of time and place, free time and tourism will become more important domains to distinguish oneself from others. The consumer, even more so than today, will use tourist destinations to be able to say: 'This is who I am.' A destination's cultural capital, in particular, will play a central role in this – the cuisine, the people, the surroundings, the cultural heritage and the buildings. In addition, consumers will not fit as neatly into the standard marketers' pigeonholes; they will become more elusive, 'fluid': 'They will preach the virtues of the environment and take holidays in Africa with an eco-friendly flavour but at the same time will be found on a stag weekend in some dodgy lap dancing establishment' (Yeoman, 2010).

The 2035 tourist will want to experience things, even more so than at present, and to have a tale to tell, preferably something authentic: new, real and rich experiences. The tourist industry will focus even more on selling the story behind the holiday. Exotic and adventurous holidays will be fascinating. Countries such as Zambia, Botswana, Laos and Cambodia will grow in popularity. Just as Vietnam is growing as a tourist destination at the moment, in 2035 it will be Afghanistan and Iraq capturing your imagination – naturally depending on the political and security situation.

In 2035, global warming will result in Dutch summers lasting longer as well as being warmer and drier. This might make it more attractive to take holidays in one's own country and reduce the need to go abroad on holiday (Hamilton, 2003; Hamilton & Tol, 2004). According to Lise and Tol (2002), the demand for domestic tourism in the Netherlands will increase by 5% for each degree of temperature rise. Something to consider with this is that people will often alternate a domestic holiday in one year with a foreign holidays in the next. The Dutch coast will profit from this rise in temperature. Mediterranean countries will increasingly encounter periods of extreme heat and drought. If this region becomes less attractive for tourism in the coming years and if winter sports increasingly suffer from climate change, one's own country will become an attractive alternative. The need for foreign holidays will not cease to exist, however, even if the weather in one's own country improves.

### Future turning point 3: Family structure

Age and life phase are important variables influencing the demand for tourism. In 2035, 40% of the population will be over 50. Most will continue to work until they are 70. People will remain healthy for longer and will be better informed about their own health. Healthy ageing will be even more in the spotlight than now. Achievements in biotechnology will also contribute to this.

We expect to see a restructuring of families towards a more vertical structure. That means fewer children and more contact with grandparents, uncles and aunts. For the tourism sector, this means that it will be important to react to the need for multigenerational holidays, from (great)-grandparents to (great)-grandchildren.

## Concluding Scenario: A Family in 2035

Taking all this into consideration, it is quite possible that an average family in an average suburb in 2035 will look back and wonder: Where did it all go wrong? What happened to our good intentions and ideals for creating a 100% sustainable society? What could we have done differently?

Sustainability means more than technical measures to prevent $CO_2$ emissions, more than cradle to cradle, more than new fuels, more than solar panels, wind turbines or anaerobic digesters, more than the three Ps in People, Planet and Profit. Sustainability is a way of life, a philosophy for experiencing life, a way of being able to enjoy the moment, honest food, warm social connections, a way of sharing knowledge and skills. Sustainability is about awareness, the awareness that you should pass your life on earth on to the following generations in a responsible manner. And we should not forget that sustainability is also pleasure: it is about People, Planet, Profit *and Pleasure*, certainly if you are talking about free time and tourism.

During a scarce moment of reflection, the average 2035 family will realize that they have failed to achieve their ideals of doing things differently from their parents. Life will slip out of their hands on an almost daily basis. To keep their heads above water, in addition to their hard work, they will lose themselves in instant pleasure. Their children will be looked after by their grandmothers and grandfathers. They will have no lack of material goods, but there will be an almost permanent risk of burnout. People will have extensive networks and be online with friends on a daily basis; they will be able to visit virtual shops in Berlin and Rio de Janeiro, go on a virtual trek through the African wilderness, using a chip so they can experience the danger of wild animals for real. But what will have happened to the real awareness about sustainability?

It will take an '*Umwertung aller Werte*' (revaluing all your values) to escape from this extremely tiring, apathetic, excessively 'other-directed' society. This is the same way forward that Friedrich Nietzsche saw for the nihilism of his time. We need to escape from the emptiness of modern times, from the loneliness of the drifting masses of tourists who are served pre-cooked dishes of authenticity and instant pleasure for a few weeks a year. Our hope is pinned on the young people who now populate our primary and secondary schools. A new generation is arising, creating new and other values, creating its own values, separate from the existing conventions of '1.0 thinking'. This is a generation which is focused on *sharing*, which realizes that you know more together than alone, which is more sociable than ever before, which attaches importance to people, environment and self-development. They no longer need to search for all those things which strengthen their 'own selves'; they are not looking for themselves because they are already themselves, everywhere and in every situation (Boschma & Groen, 2007).

In this way, 2035 could become a year of hope, hope of an era fundamentally different from all eras that have gone before – no new Romanticism, no new Age of Enlightenment, no 'dawning of the age of Aquarius'. It will be an era that will exceed all other eras, where ethnic and religious discrepancies will be things of the past, where you will work in network companies that will place people and results at the forefront, not systems or processes. There will be no more managers, only real leaders who will give people a free rein instead of checking up on them, who trust instead of distrusting, who coach instead of commanding and who focus on what you can do best instead of what you can't do.

## References

Boschma, J. and Groen, I. (2007) *Generatie Einstein, Slimmer Sneller en Socialer* (2nd edn). Benelux: Pearson Education.
Bourdieu, P. (1984) *Distinction: A Social Critique of the Judgement of Taste*. Cambridge, MA: Harvard University Press.
Frost, R. (1916) The road not taken. See https://www.poetryfoundation.org/poems/44272/the-road-not-taken (accessed 17 March 2019).
Hamilton, J.M. (2003) *Climate and the Destination Choice of German Tourists*. Hamburg: Centre for Marine and Climate Research, University of Hamburg.
Hamilton, J.M. (2005) *Tourism, Climate Change and the Coastal Zone*. Hamburg: University of Hamburg.
Hamilton, J.M. and Tol, R.S.J. (2004) The impact of climate change on tourism and recreation. Working Paper No. FNU-52. See https://www.researchgate.net/publication/24130163_The_impact_of_climate_change_on_tourism_and_recreation.
Lansink, C. (2002) De Reiziger en de Toerist, Intermediair. See https://translate.google.com/translate?hl=en&sl=nl&u=https://taalschap.nl/examens/havo/2005/1/toa_2.pdf&prev=search (accessed 5 September 2019).
Lize, W. and Tol, R.S. (2002) Impact of climate on tourist demand. *Climatic Change* 55, 429–449.
Maczak, A. (1998) De ontdekking van het reizen. Utrecht: Het Spectrum.
Mous, H. (2010) *Verlangen Naar een Betere Wereld*. See http://www.huubmous.nl (accessed 17 March 2019).
NRIT Media, CELTH, CBS, NBTC (2017) *Trendrapport Toerisme, Recreatie en Vrije Tijd*. See https://www.cbs.nl/nl-nl/publicatie/2017/47/trendrapport-toerisme-recreatie-en-vrije-tijd-2017 (accessed 17 March 2017).
Riesman, D., Glazer, N. and Denney, R. (1950) *The Lonely Crowd: A Study of Changing American Character*. New Haven, CT and London: Yale University Press.
Urry, J. (1995) *Consuming Places*. London and New York: Routledge.
Van Hesteren, G. (2011) Toerisme themarapport, Our Common Future 2.0. See http://docplayer.nl/7928459-Toerisme-themarapport.html (accessed 12 September 2019).
Yeoman, I. (2010) *2050 – Tomorrow's Tourism*. See http://www.tomorrowtourist.com/trends.php (accessed 17 March 2019).
Zuelow, E. (2015) *A History of Modern Tourism*. London: Palgrave.

# 3 The Development of Mass Tourism

Jim Butcher

**Considering Mass Tourism**

An understanding of the past can help us to put the present into perspective and to consider what the future holds; that is the spirit of this chapter. It will sketch out the development of tourism as a mass phenomenon, and describe how key developments were viewed in their time. The chapter will also situate mass tourism in the context of the development of mass industrial society in general – often the positive and negative perceptions associated with the growth of tourism are part of wider celebration of or disillusionment with modernity itself. Key societal and technological developments will be highlighted, such as rail travel, the motor car, the jet engine and, more recently, the technological/informational 'revolution'. The chapter takes a variety of examples from different places to illustrate developments, but seeks to establish themes that are evident in all developed 'mass' societies in different ways. Finally, the chapter will comment on the contradiction between tourism's rapid technologically inspired growth on the one hand, and the attendant sense of limits to mass tourism, both environmental and cultural, on the other.

**The Birth of Mass Tourism in Historical Context**

Today mass tourism constitutes the world's largest industry, comprising some 9% of GDP, one in 11 jobs and 6% of world exports. But how did we get here? To situate and consider what is distinctive about modern mass tourism, it is useful to consider what came before.

The Grand Tour was the European tour undertaken by the sons of the landed aristocracy from Britain and elsewhere, with its heyday in the 18th century. These wealthy tourists travelled to Naples, Paris, Rome and other cities associated with the highest, or most 'civilized', cultural achievements in the arts, humanities, music and architecture. The travelling was arduous and lengthy; the infrastructure of tourism we think of today did not exist.

The tour served as a means to gain a cultural knowledge that served as a marker of one's status, one's right to rule (Feiffer, 1985; Withey, 1997).

The pre-eminence of the aristocracy was overturned in Europe in the late 1700s and 1800s through the rise of capitalism as a social system, through the modern ideas of rights and statehood associated with the French and American Revolutions, and by the Industrial Revolution. The new bourgeoisie – the capitalist class – were becoming the driving force in society and also the new elite, displacing the landowning aristocracy. The century witnessed rapid industrialization and urbanization. For increasing numbers of people, especially in Europe and North America, the factory, wage-labour and the clock dictated the rhythms of one's life, rather than the weather, harvests and landed authority as under the previous feudal society. These were 'the masses' of the new mass society.

So just as the Grand Tour reflected its times – the pre-eminence of an elite born to rule – mass tourism reflected the new bourgeois society. It was the new industrial capitalist class who employed the new industrial working class to build the tracks and the trains that ran on them. Business pioneers such as Thomas Cook in the UK (a name still synonymous with mass tourism) packaged and supplied commercial holidays, and the growing number of consumers, with their share of the new wealth gained through struggle and more enlightened laws and employment practices, bought them (Zuelow, 2015).

So mass tourism as we know it today is premised on the social and economic changes associated with the advent of a capitalist, mass society. It is distinctive to modern society culturally, too: the need to escape from the city to the liminal zone of the beach, perhaps; to spend one's wages earned from an employer on consumption beyond necessities; to utilize free time (away from wage labour) and holiday entitlement. It comprised a commercial, later to be global, industry, and catered for the expanding desires of the masses and their growing ability to afford indulging these desires. In some respects, its growth is a good metaphor for modern development itself: the growth of the technology, infrastructure, personal freedoms and mobility central to tourism have been intrinsic to modern society.

## Turning point 1: Tourism as a mass phenomenon in the 19th century

Western societies generally went through a process of profound economic transformation in the 1800s. While it was in Britain that this revolution was initially most advanced and most striking, the same patterns are reflected in the experience of other countries that have undergone modern development.

The Industrial Revolution was powered by steam. The 18th century technology of horse-drawn vehicles and sailing boats was never going cater for mass travel, and made it arduous even for those with the resources to undertake the Grand Tour. Steam-powered industrial processes, including the train and the steam ship, were both key to the growth of mass tourism.

With the advent of the railways, in the words of Dr Thomas Arnold, Headmaster of Rugby School, 'feudality had gone forever'. In many ways the railways ushered in the modern era and modern leisure travel. The first public railway was the Stockton to Darlington line in the UK, opened in 1825. While built to carry coal, it pioneered the technology that was to underpin the British seaside holiday. In 1830 the first regular scheduled service began, running from Manchester to Liverpool. Thereafter the railways developed rapidly, as did the number of people availing themselves of it as tourists. The coast was increasingly linked to the town by rail, and the seaside holiday that was to become iconic of European mass tourism was born (Feiffer, 1985; Withey, 1997; Zuelow, 2015).

While the UK was the pre-eminent economic power of the century, the developments there were evident – in different ways and at a different pace – throughout the industrializing world. For example, 1862 saw the completion of the first railroad to Coney Island in the United States, and it quickly became a popular resort for day visitors from New York.

The American engineer Robert Fulton built the first steamboat, the *North River Steamboat*, on the Hudson river in 1807. In 1811 he designed and built *The New Orleans* in Pittsburgh, then sailed down the Ohio and Mississippi rivers to New Orleans. Others followed, carrying both freight and travellers, some travelling for leisure. Steamboats transported tourists elsewhere too. Many visitors to the UK resort of Margate took a steam boat from London up the Thames to get to this once fashionable resort. Glaswegians took the first steamer down the Clyde to Greenock in 1812, and by 1900 300 steamers were taking holidaymakers 'doon the watter' on Scotland's west coast (Zuelow, 2015).

Steamboats operated internationally, too. The Peninsular and Oriental Steam Navigation Company (P&O), started taking passengers from Britain to Spain and Portugal in 1822, and then as far as Malaysia and Hong Kong in 1844.

These developments in the UK and the USA were mirrored elsewhere. For example, Thomas Cook had his equivalents in other industrializing countries, such as the Stangen brothers in Breslau, Germany (now Wrocław in Poland), who from the 1860s operated holidays initially all over Europe, and soon to global destinations. While the UK may have been the first to develop the railways, France, Germany, Switzerland and Italy were among those to follow suit soon after. Steam ships operated in German waterways and on Lake Geneva in Switzerland from the 1920s.

One idea that infused leisure and travel in Victorian Britain, and had its parallels elsewhere in Europe and North America, was the association of tourism with self-improvement, as a part of 'rational recreation'. It was in this spirit that the sometimes carefree and drunken behaviour of holidaymakers was frowned upon by supporters of temperance and self-improvement through travel. Note that, infamously, Thomas Cook's first ever tour, from Leicester to Loughborough, involved attendance at a temperance meeting (Brendon, 1992).

Yet despite the moral proscription of the times, and the derogatory view held by some who deemed travel to be wasted on the uncultured masses, mass travel was celebrated and enjoyed by increasing numbers. Holidays were becoming part of the cultural life not just of the landed elites or the business class, but of larger swathes of society. This is sometimes referred to as the 'democratization' of leisure – a process key to tourism becoming a truly mass phenomenon. Workers – the masses – progressively obtained greater holiday entitlement and were able to claim a share of the surpluses generated by industrialization, through class struggle and a little paternalism. With the general expansion of national electoral franchises in the economically developed countries to progressively include workers, women and eventually all citizens above a legal voting age (a process begun in the 19th but mainly of the 20th century), governments were starting to become more accountable to and representative of the masses, and hence progressive expansions of leisure time were to become a product of democracy too.

It was Thomas Cook himself who expressed forcefully the democratizing potential of tourism's growth. Tourism was for 'the million' who could 'o'erleap the bounds of their own narrow circle, rub off rust and prejudice by contact with others, and expand their sails and invigorate their bodies by an exploration of some of nature's finest scenes' (cited in Withey, 1997: 145). Piers Brendon's *Thomas Cook: 150 Years of Popular Tourism* (1992) and Lynn Withey's *Grand Tours and Cooks Tours: A History of Leisure Travel* (1997) are excellent books that capture the optimism associated with these times.

The period witnessed the growth of commercial business and entrepreneurs who brought together the new opportunities for travel with the expanding demand for it, and developed the modern tour operations industry. Entrepreneurs – most notably in the UK, Thomas Cook and then his son John Mason Cook – pioneered leisure travel for those who could afford to pay. Thomas Cook is rightly famous as the 'father of the package holiday'. The idea of a 'package' – combining transport, accommodation and other services – was pioneered in the mid-1800s. Cook even invented the travellers' cheque – redundant now, of course, but vital then in facilitating travel abroad (see Brendon, 1992).

Cook's approach was tailored to the growing commercial environment and the increase of incomes to the level where some working-class families enjoyed a little disposable income – principally in the second half of the century. The seaside holiday, along with the music hall, books from the lending library and football, became part of mass British leisure culture. A similar pattern, albeit taking different forms and at different times, was evident in other economically developing societies.

Over the century, a great deal of the infrastructure of tourism was established: railways, the development of sea ports, passenger ships, grand hotels and even the idea of tourist attractions (sites designed specifically around the needs of the tourist). In more wealthy societies, demand was growing beyond the rich to those of more humble means. The 19th century witnessed the birth and rapid development of mass society, and mass tourism was part of that.

## Turning point 2: Mass tourism from 1900 to 1945

For the majority of travellers in this period, visits to foreign countries were with kitbag and gun as a soldier in the army. Yet despite the massive disruption of two world wars, coupled with economic depression, tourism grew, premised principally on the basis of supply-side developments: advancing technology (notably the motor car) and infrastructure. On the demand side, economic privation for the masses meant that tourism's growth over this period was far more limited than it was to become during the post-WWII package holiday revolution (Raitz, 2001).

The period witnessed the further development of inter-city and intra-city transport systems, grand hotels, cameras, guidebooks and tourist bureaus; both the infrastructure of modern travel and the demand for information for tourists continued to grow. They have continued to develop apace ever since.

The advent of the motor car and its affordability for a wider section of society contributed to the growth of the motoring holiday, although it was after WWII that this became a truly mass phenomenon. Driving was associated with freedom and pleasure; it opened up much wider possibilities for visits of all kinds to beauty spots and attractions in the vicinity and further afield. Henry Ford claimed to have been motivated to invent his Model T Ford in order to get out of the small US town where he was brought up (Watts, 2006). The Model T was standardized and manufactured on a production line to make it affordable. 'Fordism' was later coined as term for mass-produced and standardized products that take advantage of economies of scale and bring new products to the masses, a principle that was, as we shall see, to be applied to mass tourism too.

In France the growth of car ownership spawned the *Michelin Guides*, advising on good food en route. In the United States the road trip was born and would later, post-1945, become a staple for middle-class families. Across the developed world the motor car became an icon of freedom, opening up new opportunities for day trips and vacations (Zuelow, 2015). Its immense popularity has led to a situation today where the cultural associations of the motor car are often less with freedom than with congestion and traffic jams!

Air travel remained marginal to tourism in the interwar years. Yet, ironically, it was the two world wars that involved states in concentrating resources on aviation technology, technology that would later be exploited for more benign pursuits.

## Turning point 3: Post-1945 – international tourism as a mass phenomenon

It is in the period since the end of WWII that mass international tourism has become an established feature of the culture of many developed societies. There was a certain mood of anti-elitism after the war. Having fought and died in their millions in two world wars, the masses were less likely to accept 'their place' or the poverty that many had experienced in the interwar years. Reflecting this, votes for Left-wing parties after the war were motivated by the promise of a fairer settlement for the masses, settlements that would come to include greater holiday entitlements and a fairer share of society's wealth.

A number of features are key. The global economy experienced consistent growth through the 1950s and 1960s – the period sometimes referred to as the 'post-war boom'. The Marshall Plan, the USA's initiative to assist financially with post-war reconstruction in Europe, actually had a 'tourism' component (Zuelow, 2015). Growing levels of disposable income meant that greater numbers of people had the financial means to become international tourists for the first time, especially in the more economically advanced countries. For example, in West Germany real wages tripled between 1953 and 1973, and continued economic prosperity in West Germany (and subsequently Germany) has enabled Germans to become the most prolific of international holidaymakers (Judt, 2010).

New possibilities for middle- and working-class people resulted from increases in paid holiday entitlement and a shorter working week, a result of the ability of trade unions to win pay increases and also of governments amenable to the interests of the masses who had voted for them. For example, by 1960 most employees in Europe had two weeks of paid holiday per year (with three weeks in Sweden, Norway, Denmark and France). Tourism was to grow over the next 40 years to become a leading global industry. Rising incomes, growing technology and greater holiday entitlement have enabled people to satisfy a growing desire to travel.

Tourism in general has tended to exhibit income elasticity. As societies have become wealthier and more people are able to meet their basic needs, incremental increases in income have been directed more and more towards leisure pursuits, among which tourism is prominent. Coach companies, charter airlines, holiday resorts, theme parks, campsites and hotels have been among the beneficiaries. In many ways international tourism became a part of the leisure culture of affluent societies, alongside cars, TVs, washing machines, eating out, and later computers and mobile phones (Judt, 2010). For many in wealthier societies, a holiday abroad plus short breaks and days out have become less of a luxury and more of a cultural expectation.

Technology played a key role on the supply side. During the century's two world wars, especially WWII, the technologies connected to air travel developed rapidly. Prior to the world wars air travel was a brave and daring act, parodied in the 1965 British period comedy *Those Magnificent Men in Their Flying Machines*. Yet during the wars air warfare was seen as potentially crucial to gaining an advantage (and it indeed became so in WWII), and so was prioritized. The advances in air travel gave rise to the use of aircraft for tourism in the interwar years, although this was the prerogative of a tiny number of people.

After WWII, the aircraft, landing strips, pilots and air technology would come to serve the rather more benign purpose of holidaymaking. Over time, the airways and technology moved towards civilian use. State airlines were developed. Key for the post-WWII international package tourism model was the advent of the charter flight, which fitted the 'back-to-back charter' model of mass package tourism. Russian émigré Vladimir Raitz ran the very first package holiday flight, from an ex-military airfield in Bournemouth, UK, to Calvi in Corsica in 1950, after a long period waiting to obtain permission to do so from the authorities, and using a converted military Dakota aircraft (Raitz, 2001). This was to become a staple of a rapidly growing package tourism industry.

Jet technology and its adoption for a new breed of jet airliners in the 1960s played an important role in enabling quick, cheap international travel in larger capacity aircraft (Lyth, 2009). For Lyth, these developments in air travel constitute, along with mass car ownership, 'the two driving forces of modern tourism', both 'essentially supply led' (Lyth, 2009: 25).

The growth of airports and aviation in general has been striking. While in the immediate aftermath of WWII air travel was controlled strictly by the military, the charter airlines emerged in the 1950s as a key part of the modern package tourism system in Europe. Airline liberalization – in the United States in the 1970s, in Europe in the 1980s – contributed to the growth of air travel. In the 1990s the budget airline model took off in Europe, and once again provided a boost to mass tourism. Coupled with web-based bookings, the budget scheduled services have made more people more mobile and enabled travel to be self-organized rather than being restricted to the more traditional package holidays.

### Mass tourism and Fordism

An important feature of modern mass society has been the growth of what is often referred to as 'Fordism' in the production of the growing number of goods and services – mass production for a mass society.

Derived from Henry Ford's early 20th century car plants (it was said of his standardized, production line Model T cars that 'You can have any color as long as it's black'), Fordism refers to modern, efficient, standardized production that benefits from economies of scale and scope (Watts, 2006).

Mass tourism has in part grown due to the development of a holiday Fordism based on efficient, mass, integrated production. The mass-market foreign package holiday, which today seems standard and to some even old-fashioned, has involved constant innovation, building on the legacy of Thomas Cook and depending on economic growth and technology. Combined with the rapid development of hotel accommodation and infrastructure, it grew rapidly from the 1960s and became for many in Europe the means by which they travelled aboard for the first time, and indeed regularly thereafter. Its standardization and rational model kept prices low and contributed to a democratizing of international leisure travel – international tourism, if not for the masses then at least for many people in the more developed economies.

The application of rational, economic principles is important in considering mass tourism. It extends to all aspects of the industry. For example, modern mass tourism often involves long-term, extensive economic planning. A good example of this in national tourism planning is the Mexican government's Cancun Project, instigated in 1969 and opened in 1974 (Clancy, 2001). The project involves a zoning of the area to separate out tourism zones, industrial zones and also conservation zones. Cancun has proved an economic asset for Mexico.

AccorHotels' pioneering of cheap, purpose-built roadside accommodation through their Novotel brand, which started in 1967, is another case in point. Taking advantage of cost savings by locating on major arterial routes and initially developed to cater for the growing band of car-driving travelling sales and business people, the brand became increasingly attractive to French families visiting people and places in their own country (Bonin, 2009). The economies arising from standardization and the cost advantage of locating out of town, coupled with the growing car-owning market, led to a strong proliferation of similar brands. New innovations such as this in supply, as well as growing demand, have characterized the mass tourism industry in recent decades.

It is notable that there has been a cultural reaction to the standardization and rationality associated with Fordism expressed in tourism products themselves. In recent decades tourism has moved towards niche tourism: tourism types that seek to appeal – sometimes self-consciously – to specific, niche, ethical and individual preferences rather than a standardized mass market (Novelli, 2004).

Yet in most respects the newer niches are simply variations on mass tourism. They involve the growth of tourism and they rely upon the same modern infrastructure and systems discussed earlier that mass travel has relied upon. Indeed, while tourism's own critics argue that tourism only yields unauthentic 'pseudo events' and staged encounters, or confine the tourist to the 'tourist bubble', the industry has been quick to develop eco-, community- and adventure-oriented niches that take up the spirit of these criticisms and sell them back to the critics as niche holidays.

The Fordist holiday – the 'holiday from the assembly line' – sounds pejorative, but in fact is highly liberating. Lower costs have enabled far more people to join the ranks of holidaymakers. In this respect, the democratization of tourism, from Thomas Cook through to today, has been premised upon the application of rational economic principles to the industry.

*The resort*

With the growth of mass tourism, we also have the growth of the resort – towns and specially developed spaces that focus on tourism. The Mediterranean is strongly associated with the growth of the holiday resort. Some resorts, such as some of those along the French Riviera, have a long tradition of tourism, but expanded rapidly in the 1950s and 1960s with the rapid growth of international tourism. Others, such as Carnon, some 200 km west along the coast from the Riviera, were very much a product of the post-WWII tourist boom itself – in this case developed in the 1960s out of nothing from a pencil of land, to cater for the growth of working-class tourism from the cities.

Resorts – new, modernist, with high-rise developments, crowded yet convivial – are popular, especially on parts of the Spanish coast and islands. Resorts like Benidorm and Torremolinos have become synonymous with mass tourism, and for some critics also with its damaging consequences. Their growth was premised on encouragement from the Spanish government (in part as Franco sought to stabilize the economy under his Dictatorship from the 1950s), the general post-1945 increase in technological and infrastructural development (especially in aviation), and increasing wealth and holiday entitlement in tourism-generating markets such as the UK and Germany.

In 1957 the iconic – and much maligned – mass tourism resort of Benidorm got its first holiday hotel. Notably, Torremolinos and Benidorm are often criticized for the surfeit of concrete, modern buildings, their impact on the local culture and also the behaviour of the tourists (often, it has to be said, British tourists). Giles Tremlett, author of *Ghosts of Spain: Travels through a Country's Hidden Past*, refers to Benidorm as the birthplace of package tourism (Tremlett, 2012), and journalist Julie Burchill rightly berates those who scorn the conviviality, sun and sea that are affordable for the masses as 'travel snobs' (Burchill, 2015).

In the Malaga region, taking in Torremolinos on the Costa del Sol, the economy has shifted profoundly from agriculture and fishing to tourism. In the 1950s tourism accounted for 12% of employment and 5% of GNP, figures that rose to 57% and 24%, respectively, just 50 years later. Some critics cite the Costa del Sol as an example of the cultural and environmental dangers of mass tourism. Yet there is no doubt that the mass tourism industry in this region and others has generated wealth and employment which has played a major role in lifting some of the poorest parts of Europe into relative prosperity (Valenzuela, 1998). In such cases, mass international tourism has facilitated a significant transfer of wealth from the metropolitan core to the poorer, peripheral areas. This is a process evident in many other parts of the world too.

Another telling example of resort development is Rimini on Italy's Adriatic east coast. Rimini underwent rapid development in the 1950s and 1960s. However, the large growth of tourism in other parts of Europe – Spain, Greece and Portugal – led to a decade of stagnation from 1968. The following decade, from 1978, then saw innovations and changes: Rimini associated itself with the disco scene and entertainments and developed a theme park. Then from the 1990s leisure tourism was complemented by a strategy to develop conference tourism (Battilani, 2009). Rimini adapted, with some success, to a changing mass tourism market.

Resorts such as Benidorm, Torremolinos and Rimini are archetypical mass tourism reports. They have played a role in the social, economic and environmental transformation of their respective regions. There is no doubt a balance to be drawn between maximizing the positive benefits and minimizing those impacts of tourism deemed less favourable by residents. However, in all three cases mass tourism has been integral to livelihoods and economic growth, with the attendant benefits that brings.

The growth of mass international tourism has had less favourable implications for domestic tourism in some instances, most strikingly in the UK, where some of the many resort towns that had served the working and middle classes since Victorian times have declined over the last 50 years. They were deserted by aspirational holidaymakers when overseas holidays to places where the weather was favourable for sunbathing and swimming in the sea became increasingly possible and affordable. Some have become economically depressed, although attempts to regenerate them often involve cultural and heritage tourism – my own local east Kent coast, and its resorts of Margate and Folkestone, are excellent cases in point. Domestic mass tourism remains incredibly important in such cases.

*Cultural and political dimensions of mass tourism*

The growth of mass tourism runs alongside the development of national identities, and political attempts to nurture leisure for ideological ends. Authors have written of 'cultures of mass tourism' (Pons *et al.*, 2009) and 'travel cultures' (Koshar, 2000), and many others write about tourism as a sociocultural, and even political, as well as an economic phenomenon.

Mass tourism sites are often connected to or iconic of nations' cultural traditions. British resorts such as Blackpool and Margate, and the

Yellowstone National Park in the United States, are examples of attractions that are connected to the national psyche in different ways. For example, the United States' National Parks were designated to preserve a natural heritage deemed intrinsic to America, and have become very popular tourist attractions. Yellowstone, founded in 1872, was the first national park: a 'public park or pleasuring ground for the benefit and enjoyment of the people' (National Archives Teaching with Documents, 2001: 56–57). By 1916 there were 14 national parks and 21 national monuments officially designated by the Department of the Interior. They constitute a cultural resource linked to the American identity, as well as a tourist resource.

The 'place marketing' of iconic cities, such as Berlin, New York or Barcelona, constitutes a projection of the nations' cultural riches as well as seeking to entice visitors and promote the tourism trade. Even something as banal as the saucy, suggestive postcard of traditional UK resorts says something about the national leisure culture. Mass tourism has, as it has developed, become both a part of culture and a part of the projection of one's culture to others.

There are examples too where travel for the masses has been adopted in support of or linked explicitly to the idea of progress itself. *The Great Exhibition* held in the Crystal Palace, Hyde Park, London in 1851, New York's *Exhibition of the Industry of All Nations* of 1853 (held in the city's own version of the Crystal Palace), various *Expositions Universelles* held in Paris, and numerous others, were attended by tourists of all social classes in large numbers, and were all statements about the technological and cultural advancement of the respective nations and indeed of the world (Zuelow, 2015). These World Fairs are examples of what would now be called 'mega-events', and are indicative of tourism's role in politics and political identities.

A quite different example of the link between mass tourism and wider politics is the former's relationship to colonialism. For example, many of the wildlife parks and designated conservation areas in sovereign African states visited by tourists today were established by colonial administrations as hunting reserves and reactionary symbols of Africa as a wild and untamed land and people. It is also worth noting that colonial links have fed into the shape of tourism flows today. For example, Caribbean islands attract relatively high numbers of visitors from those developed nations that were in the past their former colonizers – UK and Jamaica, Holland and the Antilles, France and Martinique – due to expat links, language links and cultural legacies.

Some even accuse mass tourism today of colonialism (Cohen, 1972; Krippendorf, 1987) or imperialism (Nash, 1996) of a cultural variety, although others see this as specious (Butcher, 2003, 2015).

The rise of authoritarian regimes from the 1930s provides stark examples of how mass tourism – tourism for the masses – has been harnessed in line with the social and cultural aims of governments. In Mussolini's Italy the state adopted *Dopolavoro* (after work), as a theme for trying to organize social life which would respect and live up to the model of virtue favoured by the regime. This included tourist breaks for the workers whose loyalty Mussolini sought (Zuelow, 2015).

Nazi Germany enacted policies on a similar basis, organizing cheap vacations for the workers. The massive brutalist *Prora* complex on the island of Rügen, Germany, was built by the Nazi government for workers as part of the *Kraft durch Freude* (Strength Through Joy) programme, and stands to this day as a monument to the ambition of Fascism to elide private life and leisure with the nation and its goals. Thankfully, today it is used in more benign circumstances.

Stalinist societies, too, developed tourism in part as an attempt to shape the populace (Zuelow, 2015). Proletarian tourism encouraged a utilitarian conception of health and wellbeing, linked to the development of a new man and also with an eye to increased labour productivity.

The very notion of the masses is more than simply empirical. It has been understood in a variety of ways and has carried varied (often negative) cultural assumptions. Williams (1989 [1958]: 154) argued that 'there are no such things as masses, only ways of seeing people as masses'. The idea of 'the mass' has its own history, which works alongside the history of mass tourism itself to shape perceptions of tourism in the public imagination.

In the 1930s sociologist C. Wright Mills differentiated between the idea of a 'public' and a 'mass', the former consisting of individuals actually involved in public discourse and as such a part of their society, the latter more passive recipients of ideas from authorities (Wright Mills, 2000 [1956]). The same point has been developed in relation to culture, consumption and the media: critics have argued that mass consumption has bred a passive culture which treats people in a 'generic and dehumanising way' and also, pertinent for tourism, which has 'give[n] rise to national culture that washes over [the] traditional differences ...' (Macionis, 2009: 498).

Subsequently, the critical theory of the Frankfurt School of social thought, while diverse, emphasized the impact of mass culture, media and advertising upon social and political consciousness. For example, Marcuse claimed that this culture, formed around consumption, created *One-Dimensional Man* (Marcuse, 2002 [1964]) – individuals lacking an authentic sense of themselves and their class interests. Their ideas cross-fertilize with mainstream critiques of 'consumerism' as diminishing authenticity and depth in human relationships. In the case of tourism, writers from the Left (MacCannell, 1999) and from Conservatism (Boorstin, 1962) both, in different ways, challenged a perceived lack of authenticity in mass post-WWII travel.

The critique of mass tourism has moved on. While the Frankfurt School critique of mass culture produced by the 'culture industry' was intimately connected to class and political outlook – the argument was that the culture industry dumbed down class interests and struggle – more recent critiques have focused on the environmental and distinctly cultural impacts of tourism. The end of the Cold War brought into relief the exhaustion of the grand political narratives of Left and Right. A consequence of this was the rise of the politics of consumption.

In particular, mass consumption became problematized as a cultural phenomenon. Some mass-society theorists 'fear that the transformation of people of various backgrounds into a generic mass may end up dehumanizing everyone' (Macionis, 2009: 498). Accusations against tourism of cultural levelling along these lines are commonplace, although certainly contested.

One feature of this has been the assumption of 'good' tourism and, by implication or sometimes explicitly, 'bad' tourism – indeed, there are two books titled *The Good Tourist* (Popescu, 2008; Wood & House, 1992). This distinction is made in different ways in university textbooks, in popular culture and in marketing for the newer, 'good' niches that play up their ethical credentials (Novelli, 2004).

A counter-accusation against mass tourism's cultural critics is that they are treating culture as an immutable phenomenon, while in practice adaptation, borrowing and development are intrinsic to human culture. Also, where critics suggest that mass tourism is detrimental to culture, they may also be implying a 'development freeze' – the idea that a society should stay as it is rather than seeking economic advancement (Butcher, 2007).

## Future Turning Points

### Future turning point 1: China and India as global powers

Recent decades have witnessed the rise of China and India as global economic powers and as societies with growing wealth and a growing class of people with the disposable income that tourism – international tourism especially – relies on. Domestically, too, China is developing to cater for increasing domestic and overseas tourism, including beach and ski resorts and city-based and rural tourism.

The cruise ship industry is witnessing a substantial new market among the Indian middle class. The growth of air travel is rapid in both. China's rate of infrastructure growth has been phenomenal in recent decades and the budget airline model which had a profound impact on intra-European tourism from the mid-1990s has more recently grown rapidly in India.

While China and India, due to their size, are notable, other Asian economies have been enjoying levels of growth well in excess of those in Europe and North America. Countries such as Vietnam, for example, have been experiencing large increases in inbound (and also outbound) tourism as a result of modernization and growth, with attendant advances in incomes and the infrastructure that serves the tourism industry.

### Future turning point 2: African potential

As well as being dramatic, the growth of mass tourism has reflected the combined yet uneven patterns of development across the world. The experience of the more developed economies shares many common features. In the continent of Africa, lower levels of economic development historically have meant that the trends – in terms of infrastructure, technology and disposable income – that shaped Europe have been much less in evidence. Of course, instability and conflict have played an important role, and many would argue that these were shaped by a history of colonial domination that continued to hold back these societies even after formal freedom had been won.

However, it is notable that international tourism from the continent of Africa, and indeed to it in many cases, is growing at a faster rate than in more saturated and developed markets, albeit from a much lower base. Looking to the future, mass international tourism of the type many enjoy in the developed world is likely to grow as poorer countries become richer. This is true both in terms of generating tourists and in terms of receiving them.

Of course, the preconditions for this in Africa, just like anywhere else, are stability and safety. Conflicts and terrorism have very badly impacted tourism in countries such as Egypt and Tunisia. Libya, where the regime had big plans to develop coastal tourism 25 years ago, is unlikely to be visited by tourists for some time due to the chaos and attendant dangers that characterize this failed state.

### Future turning point 3: Overtourism

Of late, 'overtourism' has been adopted by many commentators as a term to describe the cultural and environmental issues that have arisen as mass tourism has rubbed up against the limits of cities, coastal resorts and rural attractions to cater for and cope with it. Small 'antitourism' protests have been publicised prominently as some demand controls on tourism's growth. A focus for this has been the growth of Airbnb and similar peer-to-peer platforms, which facilitate greater numbers of visitors and impact on rents and prices of accommodation for resident populations. There are also fears that the growth of such platforms may contribute to undermining the character and culture of places, as popular cities become overly tourist focused.

Yet the legacy of modern mass tourism has been a positive one. The opportunities for global travel, to relax, to have fun, to learn or to visit friends and relatives, have been an inspiring achievement of modern societies.

Amid the discussions of overtourism, it is worth bearing in mind that people's desire for greater leisure mobility has always accompanied greater affluence. Also, tourism looks set to grow and become even more of a mass, global phenomenon. The large majority of the world's population are not yet internationally leisure mobile, but we can safely guess that many of them would aspire to be. Perhaps the East African coast (more a case of undertourism than 'overtourism') – at least those areas that are deemed safe and stable – will witness more all-round economic growth and establish a presence on the tourist map. Perhaps the Caribbean island of Haiti, devastated by natural disaster and political instability, will emerge as a popular holiday island if it can establish a dynamic of economic growth and greater political stability.

It is important, in looking forward, to celebrate the achievements of the past and champion the democratization of tourism. On an optimistic note, a creative, innovative and ambitious approach to this expanding industry and aspect of our culture may yield new possibilities and ways in which the privilege of being a tourist can be extended to greater masses of the population.

## References

Battilani, P. (2009) Rimini: A mass tourism resort which based its success on an original mix of Italian style and foreign models. In L. Segreto, C. Manera and M. Pohl (eds) *Europe at the Seaside: The Economic History of Mass Tourism in the Mediterranean* (pp. 104–124). Oxford: Bergahn.

Bonin, H. (2009) The French group Accor and tourism (since 1967): Business tourism without a mass tourism strategy. In L. Segreto, C. Manera and M. Pohl (eds) *Europe at the Seaside: The Economic History of Mass Tourism in the Mediterranean* (pp. 144–173). Oxford: Bergahn.

Boorstin, D. (1962) *The Image: A Guide to Pseudo-Events in America*. New York: Vintage.

Brendon, P. (1992) *Thomas Cook: 150 Years of Popular Tourism*. London: Secker & Warburg.

Burchill, J. (2015) Lie back and think of Benidorm. *The Times*, 24 May. See https://www.thetimes.co.uk/article/lie-back-and-think-of-benidorm-d52wbp0df2l.

Butcher, J. (2003) *The Moralisation of Tourism: Sun, Sand and Saving the World?* London: Routledge.

Butcher, J. (2007) *Ecotourism, NGOs and development: A Critical Analysis*. London: Routledge.

Butcher, J. (2015) *Volunteer Tourism: The Lifestyle Politics of International Development*. London: Routledge.

Clancy, M. (2001) *Exporting Paradise: Tourism and Development in Mexico*. New York: Pergamon.

Cohen, E. (1972) Towards a sociology of international tourism. *Social Research* 39, 179–201.

Feifer, M. (1985) *Going Places: The Ways of the Tourist from Imperial Rome to the Present Day*. London: Macmillan.

Judt, T. (2010) *Postwar: A History of Europe Since 1945*. New York: Vintage.

Koshar, R. (2000) *German Travel Cultures*. Oxford: Bergahn.

Krippendorf, J. (1987) *The Holidaymakers: Understanding the Impact of Leisure and Travel*. London: Butterworth-Heinemann.

Lyth, P. (2009) Flying visits: The growth of British air package tours 1945–75. In L. Segreto, C. Manera and M. Pohl (eds) *Europe at the Seaside: The Economic History of Mass Tourism in the Mediterranean* (pp. 11–25). Oxford: Bergahn.

MacCannell, D. (1999) *The Tourist: A New Theory of the Leisure Class*. Berkley and Los Angeles, CA: University of California Press.

Macionis, J.J. (2009) *Society* (10th edn). Upper Saddle River, NJ: Prentice Hall.

Marcuse, H. (2002 [1964]) *One-Dimensional Man: Studies in the Ideology of Advanced Industrial Society*. Routledge Classics. London: Routledge.

Nash, D. (1996) *Anthropology of Tourism*. Oxford: Pergamon.

National Archives Teaching with Documents (2001) *Westward Expansion: 1842–1912*. Santa Barbara, CA: The National Archives and Records Administration and ABC-CLIO, pp. 56–57.

Novelli, M. (2004) *Niche Tourism: Contemporary Issues, Trends and Cases*. London: Routledge.

Pons, P.O., Crang, M. and Travlou, P. (2009) Introduction: Taking Mediterranean tourists seriously. In P.O. Pons, M. Crang and P. Travlou (eds) *Cultures of Mass Tourism: Doing the Mediterranean in the Age of Banal Mobilities*. London: Routledge.

Popescu, L. (2008) *The Good Tourist*. London: Arcadia.

Raitz, V. (2001) *Flight to the Sun: The Story of the Package Holiday Revolution*. London: Continuum.

Tremlett, G. (2012) *Ghosts of Spain: Travels Through a Country's Hidden Past*. London: Faber & Faber.

Valenzuela, M. (1998) Spain: From the phenomenon of mass tourism to the search for a more diversified model. In A.M. Williams and G. Shaw (eds) *Tourism and Economic Development: European Experiences, European Perspectives* (pp. 43–73). London: Wiley.

Watts, S. (2006) *People's Tycoon: Henry Ford and the American Century*. New York: Vintage.

Williams, R. (1989 [1958]) *Resources of Hope: Culture, Democracy, Socialism*. London: Verso.

Withey, L. (1997) *Grand Tours and Cook's Tours: A History of Leisure Travel, 1750–1915*. London: William Morrow.

Wood, K. and House, S. (1992) *The Good Tourist*. London: Mandarin.

Wright Mills, C. (2000 [1956]) *The Power Elite*. Oxford: Oxford University Press.

Zuelow, E. (2015) *A History of Modern Tourism*. London: Palgrave.

# Part 2

# The Development of Destinations

# 4 The Historical Future of Tourism: The Case of Malta's Policies

Marie-Louise Mangion

### Introduction

Malta, located in the central Mediterranean at the southern periphery of the European Union (EU), is a small independent island nation and the EU's smallest Member State. The country's geographical features and its history have played a key role in determining and shaping the nature of tourism activity in Malta. As a destination it is attractive primarily due to an agreeable climate, a rich history and it being a Mediterranean island – which does, however, amplify the challenge of accessibility. It has an extremely limited land area of just 316 km², having hardly any natural resources except for the sun, sea, stone – and the 'Grand Harbour', whose deep waters and capacious shoreline has, over centuries, made it attractive to naval and commercial shipping taking advantage of its geostrategic positioning. The island is densely populated with over 1455 people per km², compounded by an additional 143 tourists per km².[1] This distinctive combination of features, with complexities condensed into a small-territory context, presents a singularly fascinating case study within which to examine evolutionary patterns of tourism and critical development points in order to obtain insights into what may shape the future of a destination such as Malta.

Accordingly, this chapter identifies key turning points in Malta's tourism activity which, it is argued, occurred primarily as a consequence of the country's tourism policies and the private sector's response. Its originality and value lies in delineating, precisely, the singularity of Malta as a case study within tourism research on the one hand, but also, revealingly, its import for broader extrapolated reflection in the field, on the other. On the basis of consideration of Malta's tourism policy history, the chapter outlines what turning point transformations have taken place in the sector over a 60-year period, what governed the threshold below which a decline in tourism activity would not be acceptable to policy makers, and what triggered change and associated policy responses. This study acquires further trenchancy as it keeps consistently in view the already-noted challenges of accessibility, the regular vulnerability to Mediterranean geopolitical and cross-cultural sensitivities, relatively smaller budgets with which to contend powerfully in a cut-throat market, and competing, and often conflicting, demands for the island's limited land space, reflective of the destination's constrained resources. The distinct relevance to Futurism-based approaches arises primarily through the detection of common threads drawn from the identified historic turning points which, when projected onto the future, highlight critical developments that may occur and that policy makers need to be aware of. As Samet (2012: 509) contends: 'The greatest value from futures research and policy analysis arises from the identification of critical intervention or leverage points in the macro systems, at which investment capital or the release of resource constraints can achieve a disproportionate and beneficial change to the system'.

## The Historical Turning Points of Malta's Tourism

Malta's tourism story is captivating in its complexity of cycles of survival and adaptation, culminating incontrovertibly in overall tourism growth, as depicted in Figure 4.1 – a trend that has been more discernible still in the consolidation of tourism arrivals at record levels in recent years. The destination can boast many worthwhile, diverse and enriching experiences within a very small land territory. It competes in an ever-expanding, highly competitive, demanding and dynamic market, which over the years

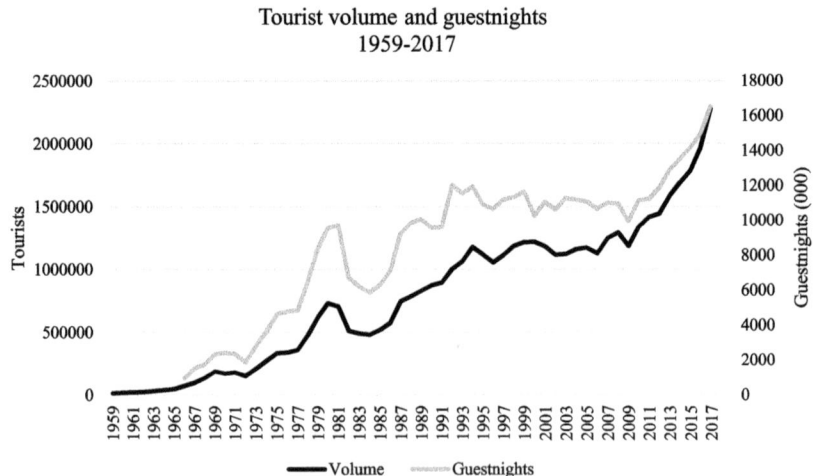

**Figure 4.1** Malta's tourism volumes and guest nights, 1959–2017
Source: National Statistics Office.

threatened tourism activity to the destination. Yet particular turning points, primarily driven by tourism policy, regularly returned the little archipelago onto the tourist map, at times, though, at the cost of irreversible impacts. This in itself highlights the importance of reflecting on the future of this destination, with possible lessons for its own policy practice and for many other island destinations.

### Turning point 1: Conscious and active initiation of tourism as an economic activity due to the political-economic scenario and potential economic downturn

The opportunity for Malta to tap into the international tourism market arose from the need to survive and adapt to a change in the overriding political-economic scenarios and technological advances that undermined the island's strategic importance. The focus on tourism as a potential export activity for Malta was established in 1959 as documented in the *Development Plan 1959–1964* (Malta Government, 1959). The strategy announced there, of diversifying into this labour-intensive economic area of activity (as well as into shipbuilding and manufacturing), was triggered by the run-down of British and NATO military bases on the island and the anticipated move towards national independence. At the time, one-third of the active persons were employed with the British presence and it was evident that the evolving political-economic context would not otherwise sustain the local population.

Malta's history, its geographical position, its harbours, its colonial legacy and a population proficient in the English language acted as strategic assets for tourism. Policymakers recognized that Malta had to penetrate the international tourism market beyond attracting those arrivals who visited friends and relatives working at the British military base. Investment in the publicly owned attractions as well as a reconfigured institutional setup were required. A programme of capital spending on improving beach access was initiated. The Malta Government Tourist Board (MGTB) was set up through Emergency Ordinance XIII of 1958 with the aim of promoting Malta, utilizing a mere Lm0.5 million (£0.8 million) over a five-year period (Pollacco, 2003).

In 1959, official statistics indicate that 25 hotels offering 1200 beds were available; 12,500 tourists visited the islands spending Lm765,000 (£1.18 million). The *Second Development Plan 1964–1969* (Malta Government, 1964) once again stressed the importance of tourism. Grants and interest-free loans to incentivize investment in tourism were offered, encouraging private enterprise and home-grown capital to participate in this economic activity. Some of the individuals who responded to this incentive and were then entrepreneurial are today among the major hotel owners on the island. Consequently, a 20% average annual increase in bedstock (101 hotels by 1969 offering over 7500 beds) and a 30% average annual increase in tourist volumes (reaching over 186,000 tourists) and earnings (over Lm10.8 million) were registered (COS, annual). This performance was also due to the British Government's travel currency restrictions permitting British citizens an annual allowance of £50 for travel outside the sterling area. This restriction did not apply to Malta (Davis, 1973), and the UK market, given the historical and colonial connections, accounted for 75% of tourist arrivals to Malta. Such developments did not come without a price, as prime coastal locations, particularly towards the north and northwest of the island, were taken up.

### Turning point 2: Tokenism

The plan for tourism as outlined in the *Third National Development Plan 1969–1974* (Malta Government, 1969) was to register further growth, diversify into new geographical markets beyond the UK and spread tourism activity throughout the year. A change in administration halted this vision due to different political ideologies and a more interventionist approach. Tourism, viewed as a highly vulnerable economic activity, was afforded a lower priority in favour of industrial policy, resulting in a downturn in tourism performance in the early 1970s and a slowdown in private investment in serviced accommodation.

In contrast, a major turning point during this period of time was the Resolution of Malta's House of Representatives on 21 March 1973 to set up the national airline, AirMalta (www.airmalta.com), which still today plays a pivotal role in Malta's tourism development. The first scheduled flights were operated on 1 April 1974 to London, Birmingham, Manchester, Rome, Frankfurt, Paris and Tripoli (Air Malta, 2018), all of which, except for the latter given the current political situation, are still key airports for the airline.

The decline in tourism performance was overturned by the political scenario in other Mediterranean destinations, particularly Spain and Cyprus, and by international travel trends which engulfed the country through package travel offering sun-and-sea holidays. Although increased State intervention characterized the new administration, little effort was afforded to regulate or control the fast and haphazard development of self-catering accommodation in localities such as Buġibba and St Paul's Bay on the northwestern coast (Mangion, 2011). That impact can still be seen and felt today. This period changed the character of tourism in Malta: it became a predominantly summer, sea and sun destination, with 60% of available tourist accommodation being in self-catering by the end of the decade.

### Turning point 3: Policy crises

Lack of planning, the unmanaged influx of tourism activity and the lack of investment in public infrastructure resulted in massive problems not only for tourists but also for the resident population. Water shortages, for example, were frequent. This, coupled with the international recession, forced primarily British tour operators to reduce their Malta programme, resulting in an overall average yearly decline of 10% in arrivals and 13% in tourism earnings, with 1982 registering the worst decline of 27% in total volumes and earnings. These declines could have been worse were it not for the compensatory increases from the German, Italian and Libyan source markets, which made up for the average annual decline of 18% from the British market over a five-year period. Inevitably, accommodation establishments suffered large declines in occupancy rates. Consequently, a protectionist policy was adopted, halting additions to the stock of serviced accommodation. Numerous self-catering apartments went out of business or were sold to Maltese residents for domestic use or upgraded to holiday complexes. This turning point is marked by the impact of past decisions and a lack of effectively concerted economic strategy and action, but also by pressures on the perceived democratic credentials of the island at a time of political unsettlement and the consequences of a command economy. These broader circumstances as well as the intractability of the slump to corrective measures acted as a turning point for the evolution of tourism in Malta. In the jargon of evolution, this led to a mutation (Charlesworth & Charlesworth, 2003), leveraging a permanent alteration of the attention subsequent governments afforded to tourism policy.

## Turning point 4: Path dependence – proactive policy support for tourism

Such mutation (reflected in a five-year decline in performance as well as in projections indicating even harsher declines as British tour operators continued downsizing the Malta programme, and an unfavourable exchange rate market for potential British travellers) pushed the Maltese government to act to entice the UK market. An election was also looming, due in 1987. Introduced in 1986, a Forward Buying Rate (FBR) scheme targeting British tour operators was developed to provide a subsidy on the exchange rate. This scheme, although originally intended as a short-term measure, was retained for 10 years until 1995. British tour operators reacted negatively to the Maltese government's announcement that the FBR scheme was to be removed. Consequently, a similar yet differently designed subsidization policy, known as the Tour Operator Support Scheme (TOSS), was adopted from 1996 to 2000 to stabilize demand from the UK market (Mangion, 2011). The British market responded to the subsidies through higher output levels; demand elasticities were stabilized, with the lowest own-price elasticities registered during the TOSS period. In the long run, though, Malta had a more price-elastic demand (Mangion *et al.*, 2012). This policy of subsidization acted as a foundation underlying the tourism structure of that time period. In anticipation of EU membership, however, this layer had to be transformed as it was not in line with EU competition law.

In parallel, following the 1987 election, the Maltese government sought to upgrade the island's infrastructures and instigate private investment for tourism. This leveraged 'path dependency', whereby:

> Economic development takes place through investment waves of successive layers of economic infrastructure rather like geological strata in the natural world, and this generates structural changes in the economy with increasing per capita assets and transitions in the composition of the workforce. (Samet, 2012: 506)

Waves of public investment in major infrastructure facilities, including the airport, water supply and energy generation, as well as a conscious attempt at holistic tourism planning, were evident. Efforts were made to address the weaknesses in the Maltese tourism industry, namely through market diversification, deseasonalization and tourist/product upgrading (WTO, 1989). To strengthen contact with operators and to increase visibility in these markets, Malta opened tourist offices in Frankfurt, Paris, Amsterdam and Milan, complementing that which had previously been opened in London. To attract a higher spending tourist, only five- and four-star accommodation developments were approved. This policy meant that Malta could be marketed to new niche markets such as the conference and incentive travel market, but it also meant changes to the landscape of primarily coastal areas. An additional layer of investment was made through the setting up of the Institute for Tourism Studies, which aimed to train potential entrants to the tourism labour market to a professional level, providing the right level of service. Consequently, structural changes started being registered, making a difference in the diversity of jobs on offer in the tourism industry.

### Turning point 5: Institutional adaptation and policy leadership

The drive to tap into a series of niche markets indicated that although international marketing was essential, it was not sufficient, requiring institutional adaptation. The marketing organization, then the National Tourism Organization – Malta (NTOM), was transformed into the Malta Tourism Authority (MTA), which acquired a wider remit incorporating international marketing but was also charged with responsibility for tourism product development, industry human resource development and regulating the industry. This adapted structure triggered the need for strategic planning for the MTA, a process not previously adopted by its predecessor, the NTOM. A further adaptation, enshrined in the legislation establishing the MTA, was the inclusion of private sector representatives on the MTA Board, in recognition of the role the private sector plays in tourism. This fostered a closer relationship between the public and private sectors and allowed the private sector representatives direct influence and participation in national decisions relating to tourism, whether for better or for worse.

As Malta joined the EU in 2004, the public institutions involved in tourism, particularly the ministry responsible for tourism and the MTA, took on more of a policy leadership role both at a national level and within the EU. These institutions had to adapt to playing a critical role related to strategic policymaking within EU structures. The Maltese tourism authorities sought to influence and participate in the EU-level debates and decisions which were not just about tourism, but also about aviation, consumer rights, competitiveness, environment, education, employment and regional policy – policy fields which affect tourism.

At a national level, the tourism authorities took on a more leading role at four levels:

(a) by identifying or instigating public and private investments to improve the overall tourism offer utilizing appropriate EU funding programmes;
(b) by adopting a policy of minimizing the impact of tourism activity, particularly of large-scale tourism projects;
(c) by adopting a market-oriented revitalization policy, aggressively tapping international market trends; and
(d) by seeking collaboration with tour operators and airlines.

This policy leadership set the foundation for tourism in Malta to reach new levels, acting as a turning point for the next years.

### Turning point 6: Air connectivity policy – independent travel and low-cost airlines

As Yeoman *et al.* (2009) state, global trends matter. Malta's case indicates that, without government intervention, global trends may not necessarily touch every destination or they may take time to reach the shores. The growth registered by low-cost airlines and by independent travel in Europe was not experienced by tourism in Malta during the first years of the millennium. Tourism's performance in Malta was registering minimal or negative growth, partly also due to the impact the 9/11 terrorist attacks had on international travel. Package travel, upon which Malta's tourism depended, was in decline Europe-wide due to alternative travel options available to consumers, instigating further reductions for Malta's tourism. Independent travel was now routinely supported by online booking systems which allowed flexibility and self-customized arrangements. Malta's travel and tourism service provision throughout the value chain, on the other hand, was designed primarily for organized travel and groups. This meant that the private sector needed to adapt quickly if it was to capture a share of that independent travel market. In parallel, low-cost airlines were registering double-digit growth rates, but Malta was not a destination for any of the leading low-cost airlines (Mangion, 2017). Recognizing the gap between these changing market trends and Malta's tourism offer, and with an election looming in two years, the Maltese government adopted a more market-oriented revitalization policy.

Following a series of debates, studies and negotiations, a decision was taken not simply to attract low-cost airlines but to adopt a wide-ranging strategy to develop an air accessibility and connectivity policy seeking the development of new and under-served routes which would contribute to addressing the seasonality issue. This would significantly supplement the already existing wide network of airports Malta was connected with. Consequently, the first new routes (including Luton and Pisa) following this policy, commenced in the last quarter of 2006. This approach to route development was retained throughout the next 10 years such that even during 2017, '8 new airline routes from 7 countries commenced operations thus adding to the already extensive network linking Malta with its source markets' (MTA, 2018). This aggressive approach to attracting airlines to operate to Malta required continuous collaboration with the distribution channels, both with airlines and with tour operators; indeed, notwithstanding the local and international trends already referred to, tour operators still generate 35% of Malta's incoming tourism (NSO, 2018).

This connectivity policy facilitated the development of other economic sectors which have also registered growth. Malta has experienced an influx of foreign workers from the EU and third-country nationals – around 37,000 or 18% of the workforce – and it is estimated that a further 20,000 are required to sustain the current economic growth (JobsPlus, 2018). This in itself fuels further tourism as people come to Malta to visit their friends and relatives – calling to mind the main motivation for visiting the islands 60 years ago. This additional influx of human activity accentuates density and pressure on infrastructure and resources.

These turning points have resulted in changes in the tourist profile visiting Malta, although some elements, such as the agreeable climate acting as the main attractor, are constant. Although the UK remains Malta's main source market, generating 25% of incoming tourists, the typical arrival is no longer the British senior citizen opting for a 14-day package holiday; rather, it is a much younger European citizen who independently books a trip to Malta, as official tourism statistics indicate, seeking to explore and experience the country's history and activities, and quite possibly scouting the country for work opportunities in the context of good economic performance on the island and constricted employment opportunities at home. Unsurprisingly, increased tourism activity and burgeoning numbers of economic migrants are leaving an impact on the destination, particularly on certain localities such as Sliema and St Julian's, both of which are among the island's more cosmopolitan areas and attractive to younger travellers, and on Valletta, the capital, which at the time of writing is European Capital for Culture and ever more prone to seeing its traditional economic activities overtaken by a fast-evolving hospitality scene (notably through the explosion of the boutique-hotel offer) and accompanying shifts in café and restaurant cultures. More recently, the debate on saturation, on carrying capacity and on residents' attitude to tourism is once again becoming more strained, as are the threats to the natural environment at a time of increased development. Indicative of limitations to carrying capacity is the interest of entrepreneurs to disperse to other localities which are not as associated with tourism. These qualitative shifts, whether viewed for their positive and/or negative facets, appear to take place as a result of internal or external forces, pushing tourism activity to a different state of play.

## Evolutionary Paradigm: The Case of Malta

The evolutionary paradigm does appear to apply to the case of Malta's tourism. Using Mika Mannermaa's (1991) postulates of the evolutionary paradigm, it transpires that:

(1) Malta's tourism is complex, dynamic, influenced by its internal and external environments and composed of various players, individuals (tourists, residents) and organizations (airlines, tour operators, tourism service providers, public authorities) which interact to organize the activity of tourism, as shown by the six turning points. Such organization, although always dependent on government subsidization of one form or another throughout the years, has shifted from one which is primarily driven and designed by private firms to one which is primarily in the hands of the individual consumer who yet still requires the services of those private firms. This has resulted in further shifts in the role of public organizations, facilitating the provision of services by firms to meet and attract the independent tourist.
(2) The turning points indicate that transformations will be difficult to reverse, or are reversible only at high economic and political costs. Each turning point seems to place leverage on the next stage, whereby tourism's outcomes depend on the path of previous outcomes – history matters and has a persistent influence.
(3) Malta's tourism story illustrates 'the evolution of societal systems', with increased complexity and dynamicity as well as the interaction and building of a closer relationship among the components of the system. (Mannermaa, 1991: 358–362)

The variations and divergences experienced throughout the past 60 years of Malta's tourism are suggestive of evolution. But they are also indicative of powerfully organic connections between policy and outcomes within the tourism sector, doubtless facilitated and amplified by the context's very specific scale factors and the immediacy effects that can obtain across policy decisions, implementation of strategy and strategization, with discernibly consequent outcomes. Malta as a case study is therefore telling in important and revealing ways that all come down to a simple yet very significant point: policy *can* nudge evolution, and *provably* so. It would, admittedly, take a longer study to render that provability incontestable, but it is hoped that enough has been articulated here to give an idea of the contours of the significance of the Maltese example for broader – and deeper – analysis and reflection. Each turning point detailed above has indicated considerations governing the threshold below which a decline[2] in tourism would not have been acceptable, triggering government intervention.

## Conclusion: Malta's Evolutionary Threads and Future

As Smart (2015: 117) argues, 'parts of our future are unpredictable and creative, whereas other parts are predictable and conservative ... both processes are always operating at the same time'. How does this apply to the Maltese context studied, and to the futures it faces?

Drawing together the threads of the six identified turning points, three dominant trends can be identified, namely:

(a) the deeper appreciation of the broader context and the influence of external factors, generally of either a political-economic nature or a technological development, on the market;
(b) the move towards concerted active involvement and interaction of public institutions (often through the provision of subsidies and other financial support) and private tourism service providers – the 'transacting entities', as termed by Samet (2012: 505);
(c) the effects of policy leadership (or lack of it), particularly in the fields of spatial planning, provision of public infrastructure and air connectivity.

Given the historical evolution of Malta's tourism, these three dominant trends are expected to extend to the future – but with variations. Malta's specific characteristics (being an archipelago with limited land area and densely populated) are expected to be even more determining in the future. Two particular turning points can be envisaged:

### Future turning point 1: Society's reaction and involvement

Malta's tourism future is expected to continue on the evolutionary path whereby public policy and private sector response play the primary shaping roles. However, as tourism activity continues to grow rapidly, to spread inland and to acquire variations in tourist motivations and profile, one expects a third player, namely civil society (or parts of it), in its broadening diversity and its own ongoing changes, to increasingly influence the nature of tourism development in Malta. The high population density exacerbated by tourism inflows and migrant workers, the resulting harsher competition for very limited resources and a public infrastructure which was not designed for this level of activity (which is not solely generated through tourism but through other economic activities) could trigger protest – and in fact there are distinct signs of this already – from various social players who feel that the fabric of familiar existence is, more and more, textured too differently from what they are familiar with and what they can adapt to. In particular, the sense of economic principles trumping other considerations is one countenanced by players doing well out of that, and denounced by others alert to longer term downsides. Society's representatives, whether self-elected or elected, are likely to become more vociferous still, calling for more direct involvement, beyond consultation, in shaping the nature of tourism activity. The private sector and policy makers would oblige if, and when, tourists' and society's acceptance of tourism activity reached a threshold which backfired on the destination's performance over a number of years. Past patterns and Malta's political systems suggest that public discourse and rhetoric, media attention and governmental reaction all tend to be more direct *and* incisive if the perception grows that such a threshold has been reached in the run-up to a looming election.

This future path does assume the current paradigm of seemingly infinite growth. Residents' expectations may continue to embrace this current paradigm, but Raworth's (2017) *Doughnut Economics* may shift that current paradigm. Society may eventually request 'turning today's divisive and degenerative economies into ones that are distributive and regenerative by design' (Raworth, 2017: 287). If it does, then this future path is inevitable.

### Future turning point 2: Technology, digital supremacy and competition from the digitalization of virtual visiting experiences

Technological developments, including as far back as 1959, have forced Malta's tourism into changes. More recently, technology has affected the distribution channels' market share, in favour of airlines but to the detriment of tour operators. Given that air transport is essential for Malta's connectivity, one could envisage a future path based on the route development policy being extended to long-haul direct flights to Malta tapping new or underserved markets. Already, there have been signal attempts to explore and open routes from China, among other destinations. In addition, digital and online technologies, together with the reach and impact of social media, have allowed the potential tourist to be the prime decision maker with direct access to the required services and combining the experiences s/he desires. This might be taken further.

Digital technology may trigger a future state of Malta's tourism – offering the experience of virtually visiting a site through, for example, 3D visualization. The availability of such technology online could push Malta's tourism to another state, particularly given that culture is a major reason for people to choose Malta and that such technologies are currently (although not exclusively) applied in such contexts. Thus, for instance, a tourist intent on visiting prime locations like the Neolithic temples at Ħaġar Qim and Mnajdra may in future do so not *in situ*, but seated at home. If such trends catch on – and it is not unthinkable that they might – the consequences for the tourism offer in Malta and beyond need hardly be pointed out.

Yet the battles of competition among destinations may in future be won not by those who possess digital supremacy but by those that adapt faster to such technologies, which anyway lure the consumer to travel to experience those additional elements present on site and which no simulatory technology can transmit. As Psaila (2012) remarks:

> Perhaps no technology can ever replace the experience of being on site and seeing with your own eyes, touching with your own hands and, if in a preserved context, smell and hear what the ancients would have smelt and heard. However, 3D Visualization aims to invite the viewer to feel a part of the scene as much as possible through the available and ever-growing technologies.

Will tourism become virtual – opting for more frequent 'visits' but being very selective about the actual trips undertaken to physically explore that destination? Or will the digital age spur more people to travel to reconnect with people, nature and the world? Such a future would require, once again, the involvement of the private sector in collaboration with public entities, coupled with policy leadership to prod and support such transitions. The implications for the current distribution channels and supporting governments are vast, as market dynamics would change.

In conclusion, the evolution of tourism in Malta offers a particular lesson for its future. Acute economic and political scenarios or slumps (noted not only through lower performance and activity but also through the composition of such activity) act as turning points to leverage government intervention and policy action shifting to the next stage of development. Within 60 years, tourism activity has evolved from coinciding with people just visiting their friends or relatives working at the British military base in Malta to being an indubitably successful destination: one which tourism professionals, researchers and academics in different fields can discuss in relation to overriding questions concerning controlled or managed growth. This transformation has taken place as a result of a series of critical junctures often triggered by slumps and election cycles, linking one turning point to the next – a chain, a linking of perceived and anticipated pattern, policy intervention and discernible consequences. Policy responses were initially related to: creating or improving the tourism offer, addressing the supply side; followed by a focus on attracting demand through subsidization; moving on to a more holistic approach of instigating demand through ensuring accessibility and improving the tourism experience through further private and public investments and collaborations. Malta's next challenges lie in striking the right balance between saturation and tourist satisfaction, in adapting fast and effectively to digital developments, in understanding and acting upon consumer trends (Yeoman, 2013), in being attractive to the distribution channels and to the diversity of tourists' interests, in ensuring environmental sustainability (Uzzell *et al.*, 2002), social cohesion and residents' satisfaction. It is all enough to suggest that a rethink about growth, or at least about the rate of growth and its contrasting and evolving desirabilities, may one day be prompted – within contexts relating both to tourism in Malta and beyond. And Futurism, as this study has sought to bear in mind, is precisely one of the contexts in which such reflection is most apt.

## Notes

(1) Calculations for population and tourist density are based on National Statistics Office Malta data.
(2) The decline is not necessarily in the form of tourist numbers but it could be in terms of the quality of experience on offer or the impacts of tourism activity on the social fabric or in terms of a number of years of decline with a looming election.

## References

Air Malta (2018) Air Malta commemorates 45 years since its setup. See https://www.airmalta.com/information/about/news-overview/news-detail/031-2018 (accessed 15 June 2018).
Central Office of Statistics (Annual) *Statistics for Tourist Arrivals, Guestnights, Total Tourist Expenditure 1959–2002*. Valletta: COS.
Charlesworth, B. and Charlesworth, D. (2003) *Evolution: A Very Short Introduction*. Oxford: Oxford University Press.
Davis, D. (1973) *Sector Report: Tourism in Malta*. Valletta: Unpublished.
JobsPlus (2018) Foreign nationals employment trends. See https://jobsplus.gov.mt/resources/publication-statistics-mt-mt-en-gb/labour-market-information/foreigners-data#title1.1 (accessed 15 June 2018).
Malta Government (1959) *Development Plan 1959–1964*. Valletta: Government of Malta.
Malta Government (1964) *Second Development Plan 1964–1969*. Valletta: Government of Malta.
Malta Government (1969) *Third National Development Plan 1969–1974*. Valletta: Government of Malta.
MTA (2018) Over two million reasons to celebrate. Valletta: Malta Tourism Authority. *Media Release*, 1 February. See https://corporate.visitmalta.com/en/file.aspx?f=31641 (accessed 15 June 2018).
Mangion, M.L. (2011) Evidence-based policy-making: Achieving destination competitiveness in Malta. PhD thesis, University of Nottingham.
Mangion, M.L. (2017) Tourism in Malta: Policies and performance 1958–2015. In M.T. Vassallo and C. Tabone (eds) *Public Life in Malta II: Papers on Governance, Politics*

*and Public Affairs in the EU's Smallest Member State* (pp. 175–211). Msida: Department of Public Policy, FEMA, University of Malta.

Mangion, M.L., Cooper, C., Cortés-Jimenez, I. and Durbarry, R. (2012) Measuring the effect of subsidization on tourism demand and destination competitiveness through the AIDS model: An evidence-based approach to tourism policymaking. *Tourism Economics* 18 (6), 1251–1272.

Mannermaa, M. (1991) In search of an evolutionary paradigm for futures research. *Futures* 23 (4), 349–372.

NSO (2018) *Inbound Tourism*. Valletta: National Statistics Office Malta. See https://nso.gov.mt/en/News_Releases/View_by_Unit/Unit_C3/Tourism_Statistics/Pages/Inbound-Tourism.aspx (accessed 15 June 2018).

Pollacco, J. (2003) *In the National Interest: Towards a Sustainable Tourism Industry in Malta*. Valletta: Fondazzjoni Tumas Fenech Għall-Edukazzjoni Fil-Ġurnaliżmu.

Psaila, S. (2012) 3D reconstructions and simulations on Ħaġar Qim: Visualising and testing the doorway and facade through archaeoengineering and 3D visualisation techniques [Poster]. See https://www.academia.edu/3104387/3D_Reconstructions_and_Simulations_on_Hagar_Qim_Visualising_and_Testing_the_Doorway_and_Facade_through_ArchaeoEngineering_and_3D_Visualisation_Techniques (accessed 15 June 2018).

Raworth, K. (2017) *Doughnut Economics: Seven Ways to Think Like a 21st-Century Economist*. London: Random House Business Books.

Samet, R.H. (2012) Complexity science and theory development for the futures field. *Futures* 44, 504–513.

Smart, J.M. (2015) Humanity rising: Why evolutionary developmentalism will inherit the future. *World Future Review* 7 (2–3), 116–130.

Uzzell, D., Pol, E. and Badenes, D. (2002) Place identification, social cohesion and environmental sustainability. *Environment and Behaviour* 34 (1), 26–53.

WTO (1989) *Tourism Development Plan for the Maltese Islands 1987–2010 for the Government of Malta*. Madrid: World Tourism Organisation.

Yeoman, I. (2013) A futurist's thoughts on consumer trends shaping future festivals and events. *International Journal of Event and Festival Management* 4 (3), 249–260.

Yeoman, I., Greenwood, C. and McMahon-Beattie, U. (2009) The future of Scotland's international tourism markets. *Futures* 41, 387–395.

# 5 Jules Verne as a Key to Understanding Irish Tourism

Klaus Pfatschbacher

## Introduction

The history of Irish tourism is intertwined with the notorious Grand Tour which encouraged young aristocrats of the 18th century, in particular, to widen their cultural horizons by travelling to the European continent. Italy was ranked in the top position among their destinations: the famous sights of Antiquity in Rome, the Renaissance wonders of Florence and the Baroque churches of Naples put a spell on the noblemen. They returned with an entirely different mindset and commenced collecting treasures of times gone by; they carried home precious items from their continental journey and they even started touring relatively nearby areas on the British Isles and collecting antiques illustrating the local history of the regions. In this context, Ireland in particular played an outstanding role as it was rather unknown in touristic respects. Travellers were dependent on slow-moving coaches to discover the remote areas of the country, especially those culturally interesting western regions where the Gaelic heritage literally unfolded before their eyes (McCarthy, 2016).

Irish tourism thus came into existence thanks to researchers who saw their destination first and foremost as a study object. Only decades later did it start to exert its charms of unspoiled nature on the guests. Consequently we can observe, as we do in economics, a trickle-down effect: Irish culture initially only drew the top of a demographic pyramid to the Green Island; just a small and affluent minority who possessed the intellectual and financial means was drawn to the country. This triggered a downward movement that gradually aroused general interest and attracted, over the years, the middle classes and people relying on smaller incomes.

This is best illustrated by Alexis de Tocqueville's journey to Ireland in 1835. He was France's eminent sociologist of the 19th century and undertook an enriching tour to talk to political representatives, gain societal insights and gauge the degree of poverty in the country. This typical attitude of a researcher is completely in line with the elite visiting the country in the context of the Grand Tour. Here is what Tocqueville, who achieved a massive impact on his contemporaries, noted about his object of analysis:

> Pretty country. Land very fertile. Beautiful road. Toll gates far apart. From time to time some very beautiful parks and rather pretty Catholic churches. Most of the dwellings of the country very poor looking. A very large number of them wretched to the last degree. Walls of mud, roofs of thatch, one room. No chimney, smoke goes out the door. The pig lies in the middle of the house. It is Sunday. Yet the population looks very wretched. Many wear clothes with holes or much patched. Most of them are bare-headed and barefoot. (de Tocqueville, 1835, translated by Larkin, 1990: 39)

> From Kilkenny to Mitchelstown the country has the same appearance as before. Hills despoiled of woods; cut up into a vast number of small fields. From time to time some great moors. Few villages, no belfries. One encounters churches without parishioners and one does not see those that have them. The habitations scattered along the road. The same kind of house, perhaps more wretched still than those in County Kilkenny. Houses of mud, roofs of thatch, often falling down. No chimney, or chimney so imperfect that nearly all the smoke comes out the door. No windows. A little dung hill. (de Tocqueville, 1835, translated by Larkin, 1990: 87)

Tocqueville's eagerness to study the country coincided with scientific attempts to establish dictionaries, thus providing better opportunities for the traveller to gain more profound insights into the island. One of these publications was Samuel Lewis's (1837) *A Topographical Dictionary of Ireland*. The title itself reveals the truly felt desire to categorize and measure the place, so that other values and activities, among them notably mobility and transport, might be enabled. The first tourists benefitted from such a scenario. The latter was thoroughly altered towards the end of the 19th century. Visiting Ireland no longer occurred against a backdrop of poverty and the superior attitude of the affluent traveller; on the contrary, the Irish service industry in those days counted on a much more self-confident position which was encouraged by the Celtic Revival movement. Its members cherished the cultural productions of the Celtic past, seeking artefacts of outstanding skills to solidify Irish identity. William Butler Yeats ranks among the most renowned representatives of this Celtic School. When analyzing his poem 'The Lake Isle of Innisfree', the reader becomes aware of the key notions establishing the basis for tourism at the end of the 19th century (see Yeats, 1888).

An interpretation best stems from the ending of the poem which depicts the background to tourism at the end of the 19th century: people already live in large cities which have developed thanks to technological progress and are marked by concrete patches and a dense array of roads. If people want to experience unity with nature they have to, as in our age, leave the urban area and go west. Yeats suggests spending time by a picturesque and peaceful lake where you can even build your own cabin. Such a self-catering holiday was completed with attentive observations of nature, diverse acoustic impressions and, overall, a journey to oneself: the traveller can really detect his/her true self. These essential characteristics of a tour through Ireland could easily be felt by the growing number of visitors who benefitted from well-functioning steamship services to Britain and from the steadily advancing railway network connecting Dublin with Galway, for example:

> In the great cities we see so little of the world, we drift into our minority. In the little towns and villages there are no minorities; people are not numerous enough. You must see the world there, perforce. Every man is himself a class; every hour carries its new challenge. When you pass the inn at the end of the village you leave your favourite whimsy behind you. (Yeats, 1893)

Yeats confirms the soul-searching opportunities country life can provide. A tourist not only experiences the treasures and soothing effects of the countryside, but s/he also grasps his/her own identity and manages to find the right position in society. As village life unites diverse social classes in one place, the traveller can leave his fragmented city life behind and approach the true sense of a fulfilled life in a rural setting. Tourism thus acquires a very Rousseau-like meaning: it is based on the contrast between populated areas and the fascination of the sparsely inhabited regions, particularly of the west. People tend to travel to get in line with authentic societal values, to mingle with the locals and to identify with their way of life. This very modern, almost sustainable type of tourism differs enormously from the initial concept of scholarly and aristocratic research which shaped the Grand Tour spirit of earlier years. A veritable milestone has been achieved.

The years of the romantic Celtic Revival were followed by a succession of crises which hit Ireland severely: the country gained full independence from the UK but had to sustain numerous victims. The island was divided and a period of instability commenced, which lasted until the end of the 20th century and saw terror attacks occur, particularly in the 1960s and 1970s. Subsequently, the image of the country suffered not only in political but also in touristic respects. It was seen as a place of social turmoil and unrest where a traveller could potentially get involved in aggression and attacks. Yet, the Irish did not give in to this tendency and managed to turn the situation around. This modification was based on a common effort of the population and a shared mindset which is difficult to grasp, as it constitutes a psychological phenomenon. Literature, however, can provide an insight into a development that started in the second part of the 20th century, in particular. *Angela's Ashes*, for instance, is regarded as one novel which best illustrates the changes:

> My father and mother should have stayed in New York where they met and married and where I was born. Instead, they returned to Ireland when I was four (...). When I look back on my childhood, I wonder how I survived at all. It was, of course, a miserable childhood: the happy childhood is hardly worth your while. Worse than the ordinary miserable childhood is the miserable Irish childhood, and worse yet is the miserable Irish Catholic childhood. People everywhere brag and whimper about the woes of their early years, but nothing can compare with the Irish version: the poverty; the shiftless loquacious alcoholic father; the pious defeated mother moaning by the fire; pompous priests; bullying schoolmasters; the English and the terrible things they did to us for eight hundred long years. Above all – we were wet. Out in the Atlantic Ocean great sheets of rain gathered to drift slowly up the River Shannon and settle forever in Limerick. (McCourt: 1997: 1)

> I want to get pictures of Limerick stuck in my head in case I never come back. I sit in St. Joseph's Church and the Redemptorist church and tell myself take a good look because I might never see this again. I walk down Henry Street to say good-bye to St. Francis (...). Now there are days I don't want to get to America. I'd like to go to O'Riordan's Travel Agency and get back my fifty-five pounds. (McCourt, 1997: 415)

The first positive impression arises despite a significant number of negative terms, in the second paragraph. Although the narrator complains about the hardship of his childhood, he creates a contrary effect by using superlatives that prove the statement to be full of irony: the miserable Irish Catholic childhood does not turn out to be that dramatic; at least, it allowed the narrator to become a renowned writer.

The same applies to the passage where the author criticizes the misery he had to go through as a child. He enumerates diverse aspects, ranging from the pernicious influence of the English to parental neglect. However, the criticism loses its acuity when the reader encounters the climax of the passage: the most serious plight was the humid climate. We can thus go so far as to maintain that the whole complaint just serves to illustrate Irish humour: it encourages people to master life perfectly and to remain optimistic.

Self-confidence and courage become manifest in the second part of the quotation above. The narrator is on the verge of leaving his native country in order to emigrate to the United States. As he is considering staying there for good, he turns nostalgic. However, this nostalgia is based on a very cherishing attitude to Ireland; it implies values such as patriotism, pride and loyalty. Understandably, the speaker in the passage refers to some places and items he will never forget. He even considers cancelling his trip. Such a thought can only cross the mind of a citizen who is attached to his region and appreciates it thoroughly. The reader who establishes a link to tourism consequently views Ireland as a very attractive destination which must be seen. Inspired by the main character, the reader attempts to identify with the interest of the narrator and shares the desire to remain in the westernmost part of the British Isles.

Such a prestigious position in literature at the end of the 20th century corresponds with liberal measures taken by the Irish government in economic terms. The authorities adopted the open sky policies in 1986, providing international airlines with easy access to the national airports. Old trade and travel barriers were lifted and arrival numbers began to soar dramatically. Ireland secured the first position of tourism growth among all the OECD countries. It acted as a kind of role model of how successful innovation in the service sector works: according to the Irish example it is based on a courageous liberalization of the market so that even other areas of the economy can look up to the adopted measures; the latter consist of abandoning monopolies and restrictions which had prevented numerous travellers from indulging in the Irish heritage. Moreover, reforms also tackled the marketing segment: Northern Ireland and the Republic made a common effort to promote the country in various respects, thus paving the way for further integration. This reached its climax in 1998, when the Good Friday Agreement settled the year-long conflicts between the opposing parties and contributed largely to a new image of Northern Ireland in particular on the international stage. Thanks to the peace treaty, travellers were reassured that a journey to the Emerald Isle would leave them safe and sound. Subsequently, a massive boom that shaped the years of the Celtic Tiger set in and were not even interrupted by the financial crisis, as it lead to decreasing prices and lured more guests to the country.

## Innovation through Literature

When it comes to innovating Irish tourism we have to depart from what can be found. This has to be structured, evaluated and taken to higher levels of modernization. Another option, maybe a more relevant one, is to trust storytelling. This means grasping the gist of Irish touristic identity by asking holidaymakers in Ireland about their experiences, gaining feedback and finally listening to the tale they have to present on the visit to the Green Island. This unveils all the layers of what a country can offer to its guests and provides tracks allowing experts to improve the current system by focusing on flaws and further developing already existing patterns. The problem with this approach is finding a sufficiently high number of interviewees willing to describe their own experiences. To avoid such a problem, experts could turn to literature, as texts – often famous novels, short stories or plays – regularly show the way to the essence of national touristic offers: Pierre Loti and France go together, Robert Burns and Scotland, or finally Peter Rosegger and Austria.

Ireland prides itself on a very special link between tourism and literature. Jules Verne (1893) describes Irish idiosyncrasies outstandingly well in his novel *Foundling Mick*, and thus puts astonishing information at the disposal of the Irish agency of tourism. He sums up all the details of a stay in Europe's most western nation, offers points of comparison (how the descriptions relate to the current status) and encourages experts to pay even more attention to certain dimensions in order to reach better results of matching Verne's points with innovative touristic plans for the future. Subsequently, various aspects of progress are distilled through literary analysis.

## Turning point 1: Luxury accommodation is recommended

In Verne's story the protagonist experiences numerous types of accommodation so that the reader gets a very palpable impression of staying on the Green Island. First the hero spends his time wandering around, sleeping under the stars or in dreadful circumstances (e.g. derelict sheds). Then he slowly discovers middle-class flats before entering the world of aristocratic manor houses which are perfectly located in picturesque places in the south. Finally the protagonist returns to farm life, living a decent existence in a solidly equipped building. Even city life and its corresponding bourgeois accommodation is not spared in Verne's novel.

In these descriptions the emphasis is placed on the luxurious dimension. The hero spends a considerable amount of time with an English gentleman; although the latter acts like a villain, the audience discovers an outstanding world of exotic evasions (see Chapter 17). Despite the stress on luxury, staying on a farm is also recommended. This accommodation option attracts Verne's particular interest; he can perfectly identify with it and evaluates it even more positively than the luxury segment (Verne, 2011, position 195566).

What should tourism experts learn from such storytelling? Accommodation is intertwined with identity. A customer will truly feel at home if his room/flat or rental house is completely in line with what really constitutes the Irish way of life. Hotels or guesthouses must therefore either be anchored in country life, offering nevertheless a decent standard of amenities (reflecting Verne's taste for rural areas), or illustrate the extraordinary side of the nation, catering to the high-end segment (for example, giving visitors the opportunity to act like former landlords at ancient estates that have carefully been renovated).

Other types of accommodation might prove problematic as their uniqueness might not be recognized. They would perhaps be mixed up with international offers such as Airbnb products which flourish all over the world. Investments should therefore affect areas promising enough potential. Low-budget products or standardized urban flats can definitely not be taken into account.

## Turning point 2: Round trips more and more relevant

Verne's text constitutes a picaresque novel. The hero goes on a lot of excursions leading him through various parts of the country and the theme of travelling makes up the essential structural thread of the book. An orphan grows up with a wandering artist living at the edge of poverty. They barely have enough to survive, playing music in the streets and begging. But one day the orphan – called 'Foundling' in the novel – takes up lodgings with a rich lady who considers him as her foster child. This makes the protagonist discover Limerick, a region he had not known as he had grown up on the west coast. Later the journey continues to the southeast, the area around Cork. Finally, the main character discovers the urban centre of Dublin where he becomes a successful businessman.

Here is how the novel starts, a potential starting point even for tourists of our era:

> The town of Westport, in the province of Connaught, is situated on Clew Bay, which resembles Morbihan on the coast of Brittany in the number of islets – 365 – which dot the surface of its waters. This bay is one of the most beautiful along the entire seaboard of Island; its capes, promontories, and points are ranged like so many sharks' teeth which bite the incoming rollers. (Verne, 2011, position 194395)

What do we learn from such recommendations? Visitors are invited to obtain a variety of impressions during their stay. In contrast to other nations it is not enough to just discover the capital to grasp the essence of the country; on the contrary, a round trip seems to be typically Irish. Farms, hills, mountain ranges and above all coastlines combine to provide unforgettable stretches of landscape. A tourist should plunge into such a myriad of impressions, looking for constant changes of scenery.

If we look at current efforts being made to promote Irish tourism, we can observe that investments are going in the right direction. The Wild Atlantic Way, for instance, is being advertised worldwide (https://www.wildatlanticway.com/home); however, the emphasis often focuses too much on urban centres where infrastructure is well developed. In the countryside, by contrast, public transport is sorely absent, forcing visitors to constantly rely on cars. An appeal for more governmental commitment thus appears to be unavoidable.

What is missing in Verne's description is the north; he limits his portrait of the island largely to the territory of the Republic of Ireland. This symbolizes an attachment to typically touristic areas, which form the heart of the nation and should not be underestimated. Innovative concepts are welcome, of course, but traditional tourist areas such as Killarney, the Ring of Kerry or the Cliffs of Moher need to be protected and even be more developed. They form a solid foundation stone that the service area can depart from. Verne thus distinguishes between a core of touristic offers and a field of potential projects that have to be exploited carefully without damaging the splendour of the basis.

## Turning point 3: History guarantees a rising numbers of visitors

Ireland has become famous for its history. Worldwide, members of the Irish diaspora and also artists and students have told the story of the Irish fight for independence numerous times. It is probably one of the best-known accounts of dramatic events marking history in Western Europe. This struggle consisted of opposing Britain. The Irish sought to get rid of British oppression, particularly in the 19th century. By blocking parliamentary debates, organizing mass rallies and complicating everyday relations with their neighbours, leaders such as Daniel O'Connell and Parnell attempted to establish permanent autonomy or even independence. Such a commitment included a variety of measures which finally led to crimes and a tragic war. Such disastrous but also complex developments can nowadays serve as a field of interest for tourism. Verne stresses the importance of the topic constantly; consequently, the struggle for independence emerges as a central theme of Irish identity which must be considered for tourist development.

The centennial of the Easter Rising of 1916 offers a perfect example of how history can be turned into a marketing tool to eventually attract large numbers of visitors. The General Post Office, Dublin, was completely renovated, transforming the historic place into an ultramodern museum landscape that illustrated the story of the revolution appealingly. After its opening, the masses flocked to this very personal account of the historical event (many attractions were based on eye-witness reports) so that the guests could easily identify with the protagonists a century earlier. The location of this exhibition contributed a lot to the success of the celebrations. Situated in the heart of Dublin, the GPO could not help but become a tourist magnet.

Similar scenarios are imaginable for other tenets of the independence movement. Exhibitions in Irish manor houses – former possessions of British landlords – could highlight the gap separating the foreign invaders from local farmers. Furthermore, guided tours through abandoned rural areas might shed light on the issue of poverty which marked long periods of the Irish past. Finally, hiking tours could familiarize visitors with what nature yields – especially the potato crop and its links to the famine of 1847. This theme could be completed with excursions to the notorious ports from where thousands of citizens had to leave their country to emigrate to the United States.

These descriptions entail a certain target audience, probably a class of culturally interested people aiming at learning or revising something essential during their holiday. This group of visitors almost certainly belongs to a very affluent category who will not refrain from spending significant sums of money throughout their stay. An emphasis on the staging of historical events consequently demands considerable investment in accommodation. Well-off tourists require more than acceptable standards, so even in the countryside infrastructure projects should be carried out, potentially also in the context of EU subsidies. In addition, the historical events have to be highlighted so that a sufficiently large numbers of guests will be attracted. If Ireland respects these suggestions, history will serve as a tourist magnet.

### Turning point 4: Britons are Ireland's essential target group

A special relationship links the Irish to the UK, reminding us again of an extraordinary impact of history: for centuries British landlords found themselves in a position of power, exploiting Irish farmers by pushing them to hand over large parts of their harvest and income. The property owners, by contrast, led a life of luxury, spoilt by this easy access to money, victuals and other items. In his novel, Verne sheds light on such a situation when the protagonist is confronted by a merciless landowner who oppresses a farmer's family that provides shelter and education to the hero (a foundling). The latter even meets the aristocratic landowner, another opportunity for the narrator to emphasize the terrible flaws of the 'bad guy'.

As a result of this complicated and even damaging relationship, the reader can deduce a very special connection. However, despite the horrors of the British reign, Verne suggests a closeness between the Irish and the British. For centuries they have struggled and worked together. For tourist experts this implies that the hosts have to display outstanding care. Britons, like Germans in Austria, have to be catered for with passion. They should be given the impression of playing an important role in the tourist business, not only in terms of earnings but also regarding a warm welcome which they deserve.

How can businesses lure Britons to the Green Island? At first sight we might say that too many similarities exist between the two countries to interest English visitors. After more careful analysis we can of course see the differences, promising enriching discoveries: the Celtic heritage, Early Christian times including outstanding artefacts and festivals (e.g. St Patrick's Day), or Dublin as a capital city of literature (Joyce's 'Bloomsday'). These areas all point in one direction: it is all about offering sophistication. The high-end cultural products mentioned above entail elaborate guided tours, detailed explanations and simply accessibility. In this manner the Irish might particularly appeal to UK visitors, who might for example feel flattered if they receive enlightening information on the *Book of Kells*.

Such a procedure naturally necessitates numerous marketing efforts which should also aim at further improvements to transportation. Flights and ferry connections, already at a high level, have to emerge as an efficient gateway to a holiday in Ireland.

### Turning point 5: Authenticity boosts Irish tourism

Verne's text is essentially marked by two basic terms: poverty and simplicity. In other words, the story, which assembles all the relevant notions regarding Ireland at the end of the 19th century, is based on a plot which highlights poverty and a simple lifestyle as basically Irish. Of course, the economic situation has changed and progressed dramatically. However, as key concepts for marketing purposes they still serve an important purpose – allowing the customer to actually seize the core of intercultural differences and the clichéd perceptions of the nation. By being exposed to a staging of simplicity and poverty in touristic contexts, the visitor might have the impression of better understanding the country.

How does Verne actually illustrate these two key concepts? The hero grows up in the streets of western Ireland, begging and relying on the mercy of other people. He cannot afford to buy any new clothes; even his survival seems to be at risk. His accommodation usually ranges from sleeping rough to living in a simple farmhouse. What does this mean for touristic enterprises? Apart from the choice of the place to stay for the night, tourists have various other ways to approach the gist of a typically simple Irish lifestyle: tasting traditional dishes of the poor (prepared with a lot of potatoes); experiencing the cultural pleasures of country folk (e.g. Irish dancing); or simply socializing in the way so many Irish people have done in the past, by spending one or two hours at the pub or going for a walk through the countryside.

These suggestions appear to be extremely simple; however, when it comes to defining strategies for touristic marketing, looking at the storytelling of the past pays off. It prevents experts from engaging in elaborate or far-fetched categories that appear to be efficient at first sight, but are too artificial after several checks. Storytelling conveys the natural strengths of a country and thus should not be neglected.

### Future Turning Points: National Identity and Individualism

Verne deserves a special status in the history of literature. He not only developed far-reaching genres, but also delivered insights into modern tourism, which allow us to categorize his influence in several fields. Utmost importance can be accorded to his inspirational dimension which could trigger significant research. The first question raised by his novel *Foundling Mick* relates to national identity. How far does tourism represent national idiosyncrasies and how should experts stage national features? How far can they go without running the risk of succumbing to clichés? What can be sold as authenticity? These are just some questions which can be derived for the future from the novel analyzed above.

Further approaches refer to the analysis of accommodation. Verne's text raises this issue as a key question in tourism and thus researchers should be tempted to launch best practice reports covering the impacts of different accommodation offers. Which target groups do luxury resorts appeal to, for instance? How far can very expensive rooms compensate for the decrease in booking figures as the more modest clients tend to avoid the region? Can one destination simply base its services on one single branch or does variety seem to be unavoidable?

Another focus could be placed on the significance of history. Scientists should target the question of the exploitation potential of history. A limited historical influence becomes manifest in many projects, but is it possible to develop concepts to boost tourism on a large scale? The Irish commemoration of the centennial in 2016, which attracted thousands of visitors to the GPO, Dublin, for instance, could serve as a guideline to imitate. Possible further ideas could revolve around The Titanic, the Great Famine, Daniel O'Connell and Parnell, industrial heritage, seafaring, the Early Christian Period or whiskey production. These historical themes demand important investments which turn interest in the past into an entertaining and intellectual experience at the same time.

Reading Verne's novel, tourism professionals could also question the impact of landscape on travellers. In this regard they could examine how far picturesque and breathtaking countryside could compensate for deficits in infrastructure, lodgings or food. Can the beauty of a region allow for a certain negligence in other realms of interest? An interesting approach would be provided by the analysis of Irish coastal areas, for example parts of the Burren, which impress tourists massively, but do not offer enough incentives for a longer stay.

In addition to the features mentioned above, Verne's text also draws our attention to the aspect of round trips. The novel's hero travels back and forth throughout the Emerald Isle and sets an outstanding example of the curious traveller who does not shy away from making strenuous efforts to understand even the remotest parts of the country. Does such a category of traveller still exist? Is editing dense and insightful guidebooks about the different counties with their attractions, which should be discovered during one longer stay (probably in a rental car), still worthwhile? Has society not evolved dramatically and decided on short trips, forcing professionals to almost completely drop traditional concepts of tourism? Is it not sufficient to cater solely for the requirements of a prolonged weekend?

Apart from these issues, researchers finally have to deal with another aspect arising from *Foundling Mick*: individualization. The whole novel is geared towards an outstanding character whose perspective and interests dominate the choice of destinations and their perception. The reader partly discovers the country as if s/he is reading a long feedback published on social media with a focus on individual wishes and preferences. Experts consequently have to wonder how much they should concede to the impact of sites such as TripAdvisor or Instagram. Should they constantly bear these personal reactions in mind and for example underline their response in the advertising leaflets, on their menus and websites? Or can we still recognize approaches which do not put professionals under pressure and allow them to set priorities according to their own needs? Does individualized tourism offer more opportunities or does it hamper tourism? The future of Irish tourism might hinge on this dimension. Verne already stressed it more than 100 years ago.

## References

Larkin, E. (1990) *Alexis de Tocqueville's Journey in Ireland, July–August, 1835*. Washington, DC: Catholic University of America Press. See https://www.jstor.org/stable/j.ctt284x3v (accessed 8 February 2018).

Lewis, S. (1837) *A Topographical Dictionary of Ireland in 1837*. London: Lewis & Co.

McCarthy, F. (2016) *The History of Tourism in Ireland*. See https://www.academia.edu/10142055/The_History_of_Tourism_in_Ireland (accessed 8 February 2018).

McCourt, F. (1997) *Angela's Ashes*. London: Flamingo.

Verne, J. (2011) *Delphi Complete Works of Jules Verne*. Hastings: Delphi Classics. See www.amazon.co.uk/kindle/ebooks (accessed 30 June 2016).

Yeats, W.B. (1888) The lake isle of Innisfree. See https://www.poetryfoundation.org/poems/43281/the-lake-isle-of-innisfree (accessed 8 February 2018).

Yeats, W.B. (1893) *The Celtic Twilight: Faerie and Folklore*. Mineola, NY: Dover Publications. See https://www.goodreads.com/book/show/194204.The_Celtic_Twilight (accessed 8 February 2018).

# 6 The Growth, Decline and Resurgence of the City-State

Brian Hay

## Introduction

This chapter aims to explore the growth, decline and resurgence of the city-state through an examination of three themes: reasons for the growth of city-states; historical turning points in the development of city-states and their impact on tourism; and the future growth of city-states and possible different visions of tourism in city-states. It also sets out to demonstrate that the growth of tourism has been closely linked with the development of the city-state as the optimal form of government.

In the 18th, 19th and 20th centuries there were profound changes in the structure of Western society and these changes have continued into the 21st century, such as:

- from an agrarian-based society to an industrial and urban-based society, and then to a post-industrial society;
- from an almost universal acceptance of religiously grounded faiths, to the growth of territorial-states, based both on a secular society and on a single religion;
- from a mainly rural-based to a largely urban-based society, resulting in different forms of community and governmental arrangements;
- from a society based on the production of goods, to one based on the provision of information and services, and – in the future – to a society where experiences will be the main driver of social and economic development;
- from a society based on the acceptance of the power of a territorial-state and a managed economy, to a society where the people increasingly demand more autonomy from the central territorial-state's government.

Today, however, perhaps the greatest change facing society is the unrelenting growth of the world's population, along with its aging demographics and increasing urbanization. In 2015 the world's population was

7.5 billion; it is forecast to be 9.7 billion by 2050, with some 66% (54% in 2015) of the 9.7 billion living in cities. Also by 2050, the number of 'older persons in the world will exceed the number of young for the first time in history' (UN, 2015: 28). This huge growth in the population will have a profound impact on how and where people live. City governments are likely to demand more control over their expanding populations and in the future it is probable that they, rather than the territorial-states, will be the new political power brokers. This growth in the population of cities is expected to develop in two directions: the continued expansion of mega cities, when 'by 2030, 41 urban agglomerations are projected to house at least 10 million inhabitants each', and the growth of smaller cities where 'close to half of urban dwellers [will] reside in relatively small settlements of less than 500,000 inhabitants' (UN, 2014: 2). Porter (2017), however, argues that smaller cities may well be over-dependent on redundant industries, such as steel making or car manufacturing, and that the territorial-states will try to protect such industries by imposing protective trade barriers and restricting the number of immigrants.

There are also likely to be demands by the cities for more political power to manage their own populations and their increasing temporary workforce, immigrants and tourists. This growth in the world's population has been mirrored by the growth in international tourism, which increased from 531 million trips in 1995 to 1322 million trips in 2017, and is forecast to grow to 1800 million trips by 2030 (UNWTO, 2018).

## Tourism and the Development of the City-State

Tourism as we know it today is a relatively modern concept, but since the beginning of civilization people have travelled. They have done so in order to wage war, to search for food, to undertake religious pilgrimages and to trade with other people. Although the growth of city-states and their commercial trading has historically been associated with developments in the countries around the Mediterranean, this is a narrow view. Over 2000 years ago Omani traders sailed to India and China, 1500 years ago Polynesians sailed across the Pacific, and 1000 years ago the Vikings sailed to North America, while at the same time the Chinese developed the Silk Road (Burkart & Medlik, 1976; Richardson, 1999).

To understand the development of the city-state and its implications for tourism, it is necessary to recognize that city-states are much more than a collection of buildings, configured in such a way as to enable the efficient trading of goods and services. Ancient Greece is regarded as the first political model of the city-state, and Aristotle (4th century BC) suggested that 'men come together in the city to live; [and] they remain there in order to live the good life' (Aristotle, 1959); this statement could equally apply to the city-states of the 21st century. This 'good life' was reflected in the emergence of tourism between the Greek city-states; for example, religious tourism grew through people's visits to the Oracle in Delphi, health tourism developed in Cos and sports tourism developed in Olympia. After the decline of the Greek city-states, the Romans developed holiday and spa resorts outside their cities, making use of the roads that had been built for military use. By the time of the Middle Ages, Arab tourists were travelling with their harems to Asia, and religious tourism was thriving within all the world's major religions: Christians travelled to Rome and Jerusalem, Muslims to Mecca, and Hindus to the Ganges. At the height of the development of the city-states in Europe in the 17th and 18th centuries, the Grand Tour was well established as a form of cultural tourism between city-states, to be enjoyed by wealthy young Europeans.

However, travel even between city-states in Europe was slow, expensive and sometimes dangerous and, as noted by Richardson (1999: 6), 'Napoleon's armies moved no faster than did Caesar's'. The invention of the steam engine in the 18th century and the subsequent Industrial Revolution resulted in an increased demand for people to be able to trade goods between city-states more easily, so rapid improvements were made in the infrastructure, resulting in new roads, seaports and canals; however, the greatest change was the development of the railways. Railways enabled people for the first time to travel relatively safely and cheaply between city-states. When travel was linked with other activities, by Thomas Cook in 1841 with an organized trip between two cities for a fixed price, the package tour was born. Later, at the beginning of the 20th century, technological improvements in travel also had a profound impact, when the invention of the car and the aeroplane led to the creation of the mass tourism market.

## Historical Turning Points in the Development of City-States and the Implications for Tourism

### Turning point 1: The growth of city-states was driven by both their political and their economic power

City-states did not spontaneously develop in isolation, but grew from existing cities. However, as in life, not all cities are equal, and with the growing centralization of power in Europe from the 10th century onwards (Scott, 2012), the importance of one city over another led in time to the creation of recognized principal cities. The creation of these principal cities was driven not only by the increase in population, but also by the requirement to maintain written copies of legal decisions and the need to collect taxes (Tout, 1934). Since writing skills were the domain of the Church, cities with cathedrals began to dominate those without such skills.

In the late Middle Ages (13th–15th centuries), as a reaction to the growing dominance of principal regional cities, the less powerful cities began to develop loose confederations; however, these failed because each city wanted the benefits of confederation and free trade, but they were unwilling to lose their powers of taxation over goods moving through their city (Toynbee, 1934). As suggested by Braudel (1982), city-states in the late Middle Ages grew in power not only because they had successful economies, but also because of the creation of central banks.

Although there is no agreed set of reasons why such city-states developed in Europe, Scott (2012) argued that four factors drove this development. First, public jurisdiction of power was gradually transferred from the centralized control exercised by the Church (Roman Catholic and Protestant) to the control of local bishops in the dioceses and this transfer was also accompanied by the latter's growing temporal authority. Secondly, the development of legal land ownership outside the control of the nobles and the Church, and the encouragement by the city-states to the farmers and peasants living outside the city-states to sell their surplus to the city-states, in exchange for being granted citizens' rights by the city-state. Thirdly, city-states expanded their external trading links through the development of coastal and river cities and long-distance trading by sea. Fourthly, a legal structure was developed to support both their trading arrangements and their judicial authority over the citizens of the city and its hinterland.

*Implications for tourism*
- The influence of the Church in the early city-states encouraged the development of fixed and recognized holy days; that is, they established the first universally accepted days free from work – the first holidays.
- Because of the need for trade between city-states, the earliest forms of tourism were business tourism and political tourism.
- The development of a central banking system and an official, tradable currency (notes of exchange, central banks and promissory notes) enabled people to travel between city-states without the need to carry precious commodities to trade goods and services.
- As the city-states increased their power over their citizens, they raised taxes to provide public services, such as places for celebration and public entertainment.

### Turning point 2: The decline of the city-state was driven by the expansion of territorial-states and their growing commercial interests

All good things come to an end and the demise of independent city-states in Europe in the 19th century was driven by a combination of four issues. First, as a result of the numerous wars in Europe, city-states realized that their survival could be achieved only 'by combining numerous small territories to form a number of new states' (Sutcliffe, 1981: 6). Over the course of the 18th and 19th centuries, the territorial-states, with their stronger, managed economies and, in particular, their military dominance (Lane, 1979) proved to be much stronger than the city-states.

Secondly, commercial trading interests in Europe saw the benefit for businesses in the creation of a territorial-state, as demonstrated by the unification of the UK in the early 18th century. This allowed for the free movement of goods across the borders of England, Scotland, Wales and Ireland without tax or trade barriers. This was difficult to achieve in other parts of Europe because there were so many small city-states, where goods were subject to taxes every time they crossed a boundary.

Thirdly, in the late 19th century in Europe, the demise of city-states was hastened by the growth of national unification movements in Germany, Italy, Austria, Switzerland and the Netherlands (Scott, 2012). Although there was no universal model, in Germany and France in particular, power was split between the central government and the regional/state governments, with the latter developing from already established city-states. In the United States, Canada and Australia, a different model was developed, that of a federal system of government which favoured the development of large states/provinces/territories over the city-state model.

Fourthly, the first efficient, mass public transport system was developed, namely the railways. This enabled the cheap transport of goods and people between city-states (Burkart & Medlik, 1976), and so strengthened the business, cultural and language links between cities, a consequence of which was the development of both a standardized universal time and currency, within the boundaries of each territorial-state. It also provided, for the first time, an affordable means for people to travel for purposes other than for trade or business.

*Implications for tourism*
- The emergence of territorial-states resulted in the free movement of people within a larger area, without the need to cross boundaries.
- The territorial-state also provided a focus for a common social identification of all its people within a much larger geographical area, which encouraged communication.
- The development of railways provided an inexpensive means for people to travel between cities within a territorial-state and between a city and its hinterlands.
- The provision of travel documents (passports) became a necessity for travel between these larger territorial-states, as the governments began to control the flow of tourists and immigrants.

### Turning point 3: The emergence of world/global trading

In the 20th century the consolidation of the territorial-state was further driven in part by the global impact of the two world wars, after which territorial-states believed that regulatory policies were necessary in order to develop, manage, protect and control their weakened internal markets.

In the late 20th century, however, cities began to realize that their territorial trading policies, often accompanied by subsidies for declining manufacturing industries, were not always appropriate for their development. Cities began to look outwards for their markets and, with the development of globalization, they realized that their markets, like those of the 19th century city-states of Europe, lay externally, not within their own boundaries. These new, outward-looking cities, such as Singapore, Shanghai, Buenos Aires, Hamburg and Dubai (Sassen, 2003), often known as world-cities/global-cities, were much more dependent on trade, investment, information services and tourism with other world-cities/global-cities, rather than acting as service centres for their local or regional hinterlands (Friedmann, 1986). External trading, along with the development of the technological infrastructure (Fainstein, 1994), expansion of long-haul air transport links (Keeling, 1995) and the growth in budget airlines offering many more direct city-to-city routes, was central to the development of such cities.

This emergence of world-cities/global cities, as suggested by Halperin (2017: 104), arose because of their strong 'socio-economic, political and cultural relationships' with each other rather than with their hinterlands. Castells (1994: 30) submits that they 'could be the necessary complement to the expansion of a global economy'. The re-emergence, therefore, of both service and manufacturing trading between these new world-cities/global-cities, with their duty-free zones, sovereign funds, private military contractors and cabinet forms of governments, could represent an early indication of a return to the city-state model of government.

*Implications for tourism*
- Technology (e-passports and biometric security) will drive the development of tourism in these world-cities/global-cities.
- City-to-city tourism drives the marketing of holiday tourism in world-cities/global-cities while, conversely, the territorial-state as a tourism destination is increasingly seen as an obsolete marketing concept.
- Reflecting the importance of trade between such cities, international conferences/meetings/exhibitions form the basis of much of their high-yield tourism product.
- The airports of world-cities/global-cities and other parts of their tourism infrastructure are seen as the cutting edge of contemporary design, acting as a magnet to encourage growth in tourism.

## Future Turning Points in the Creation of a City-State

## Future turning point 1: Efficient governance is assured through the free election of an illiberal democracy

The core concept essential to establishing a successful city-state is efficient and effective governance and, as suggested by Jones et al. (1995) and Perry et al. (1997), this may well overrule the traditional ideals of a freely elected, democratic territorial-state, with frequent but peaceful changes in government. It may well lead instead to the creation of a freely elected but illiberal democracy. As suggested by Levitsky and Ziblatt (2017), this will be welcomed by the both the government and the people, because good governance is essential for the successful management of a city-state.

To achieve this requires, first, the continuing and peaceful re-election of a single, dominant political party (Singapore, Hong Kong), where re-election is regarded as a form of feedback and endorsement of the current status quo (Perry et al., 1997), and the primary role of government is the management of social affairs in the national interest, rather than in the interest of the individual. Secondly, the individual citizen has limited recourse through the legal system to challenge the decisions of the city-state. The legal system provides a code of conduct for both the city-state and its citizens, rather than a means to protect the citizens from the city-state. Thirdly, as a result of the development of a managed civil society, the political structures are not burdened with the trappings of a formal democratic process, because any political opposition is absorbed into the controlling political system. That is to say, good governance in the city-state is dependent on the government's development of a managerial solution to any political opposition.

*Implications for tourism*
- In the future, city-states will adopt a managerial and restrictive approach to tourism, where the government acts to ensure that the development of tourism will not be at the expense of the quality of life of its residents or its external economic and trading arrangements.
- There will be specific laws and regulations to govern interaction between tourists and residents, and tourists will have to agree to these before gaining entry to the city-state. Opposition to them by either tourists or residents will not be tolerated by the government.
- The creation of new tourism-permitted zones within city-states will form part of a policy to develop the infrastructure of the city-state and protect its residents from an excess of tourism.
- The city-state will ensure that the preservation of its assets, both built and cultural, will not just be for the tourists' enjoyment, but also for the enhancement of the lifestyles of its citizens.

## Future turning point 2: Technology will be used to control and manage the behaviour of residents and tourists

The city-state of the future will not operate in isolation from its hinterland, nor will it become some idealized back-to-nature model operating with no relation to reality. It will, however, become a place where technology will deliver services to mitigate any labour shortages. In the early 21st century, the emerging global-cities/world-cities already have technology embedded into everyday activities, such as the delivery of services, transport systems and ID/payment cards. Such technology will need skilled innovators and maintenance staff, which suggests that a technically elite workforce, along with an army of temporary, low-skilled immigrants, will live alongside the residents of the city-state, but probably in their own residential zones.

The use of technology to provide personal services will involve a loss to some degree of freedom and privacy and, to paraphrase Spock in *Star Trek*, in the city-states of the future 'the needs of the many outweigh the needs of the few'. Much of this new technology will operate unseen, such as: a health system that monitors every individual's activity; an information system that provides a work/play plan each morning which all citizens are expected to follow, because they will be closely monitored; and a daily nutritional and consumption plan.

People in the new, model city-states will be aware that technology controls their social activities, with the aim of encouraging the development of both their bodies and their minds. Through data mining, the monitoring of their social activities will ensure that such activities not only benefit them as individuals, but will also add to the cumulative quality of life in the city-state. In the future, therefore, the city-state will both direct and manage the social activities of its citizens.

*Implications for tourism*
- The city-state will seek to ensure that tourism benefits its citizens by using technology to guide the tourists' activities. For example, before entering the city-state, tourists will be required to outline their proposed recreational activities, as well as their food preferences, physical limitations and cultural interests. A personalized, daily programme of activities will be provided to each tourist, based on these preferences and also taking into account the carrying capacity of the city-state's facilities. Tourists will be required by law to follow the programme given to them.
- Such programmes will ensure that the tourists' activities will have only a limited detrimental impact on the lives of the residents. Tourists will be monitored to ensure that they follow their official daily plan.
- In order to maintain the quality of life of the citizens, tourists will have no choice but to use the transport provided by the city-state, which will take them to the places listed in their daily approved programme.

## Future turning point 3: Development models for city-states of the future

Given the constraints on existing land, the continuing growth of the world's population and the mechanisms by which city-states will be able to control their citizens, we can expect schematic shifts in the policies of city-states to control their future development. There could be a radical shift in the power structure from the current city-state model, with the private sector taking control away from the public sector. This could lead, for example, to: new, private enterprise city-states based on islands floating in international waters, free from state control (Hoyle, 2017); airport city-states where people meet and live for the sole purpose of exchanging ideas (Avakian, 2018); undersea city-states; or even city-states circling the Earth (Gamble, 2014). Other models of city-states could be based on the creation of like-minded utopian communities, such as university city-states, religious city-states, sexual-orientation city-states, gender-neutral city-states or even ethnically based city-states. A less radical vision could include the development of autonomous city-states, created by territorial-states in order to ease the pressures of their growing populations (Boyes & Hipwell, 2017).

It could be argued that there are already such self-contained, semi-private, functional city-states, such as theme parks, with their own entertainment, recreation, shops, accommodation and policing, which by any definition are, in reality, entertainment city-states. City-states in the future will not even have to be permanent; today temporary cities already exist, based on religious events such as visits to Mecca, refugee camps in Bangladesh, and festivals such as the Burning Man festival in Nevada.

### Implications for tourism

The development of such diverse city-states would require a tightly controlled mechanism in order to manage their inflow of tourists.

- The current model of free and open movement of people and their almost universal visa-free travel across the world will in the future be regarded as naïve. A paid-permit system could be developed to control access by tourists to what will be gated communities, so that only approved tourists whose social, economic and political profile matches that of the city-state would be admitted.
- Controlled tourism zones will be created in which tourists will be able to enjoy their experiences, without negatively impacting on the culture or the lives of the people in the city-states.
- Through regulated access to city-states, tourists could also be monitored to assess their suitability to be citizens of the city-state in the future. That is, the city-state could use tourism to control the future selection of its residents, and so control its population growth.

- The development of privately managed city-states, whether on land or sea or circling the Earth will result in them being able to offer unique tourism experiences outside the limited experience of those still living in territorial-states.
- The establishment of travelling or temporary 'pop-up' city-states based on mega tourism events and functions may not be too far in the future.

## Visions of Tourism in the Future City-State

The distinction between tourism and non-tourism businesses will become increasingly irrelevant, as tourism in the city-state may not be regarded as a product that needs to be provided for tourists, but rather as a service which should be managed for the benefit of its citizens. That is, the city-state's attractions, restaurants, recreation facilities, hotels, transport, etc., will be seen first and foremost as facilities for its citizens, and only secondly for tourists. This shift in thinking will be seen as good business management, because it fits into the concept of a city-state of the future, one which is managed for the benefit of its citizens.

This business management approach could also lead to a restriction in the number of businesses providing tourism services, with city-states developing a licensing system to control both inbound and outbound tourism. In addition, the approach could be used as a means to control prices of goods and services for both tourists and residents. The idea of territorial-states providing equal access to all services and facilities to tourists and citizens alike will probably be rejected by city-states, on the grounds of unfairness to its citizens. As part of a business management approach, tourism will be regarded as an external cost which brings little direct benefit to the city-state's citizens, and tourists will be expected to pay the full economic cost of their use of public facilities. Conversely, there may be disincentives for the citizens of city-states to travel elsewhere, because this could be seen as being detrimental to the quality of life of those citizens who do not travel, because such trips may not be not be perceived as contributing to the social cohesion of the city-state.

In most territorial-states, destination management/marketing organizations (DMOs) exist either to market tourism or to work with tourism businesses to manage and develop the product. The function of future city-state DMOs will be to ensure ongoing improvements to the quality of life of its citizens, rather than developing services for tourists.

### *Implications for tourism*
- The word 'tourist' may become redundant in the city-state of the future, and instead they may be called 'non-citizens'. These 'non-citizens' will be regarded as a drain on resources and, therefore, may be heavily taxed, if they want to enjoy the same quality of life as its citizens.

- Businesses which provide services to 'non-citizens' may be subject to increased taxation; conversely, businesses that provide services to citizens may be more lightly taxed.
- Control of the right to run a tourism business/service will be through a licensing system similar to the trade guilds in the city-states of Europe in the Middle Ages.
- The services managed by DMOs will be fully chargeable to the 'non-citizens'; indeed, the DMOs may even make a profit, with such profits being used for the benefit of the citizens of the city-state.
- The DMOs' role will change from that of managing or marketing the destination into one of monitoring the destination, and their function will be to control the activities of 'non-citizens' visiting their city-state, so as to minimize their impact on its citizens.
- As for 'non-citizens' who contravene the rules of the city-state, they will likely be segregated from the citizens in special, but comfortable, 'tourist jails', which will be more like hotels, rather than prisons as we know them.
- Specific universal city-state-only e-passports will be developed, separate and different from the traditional territorial-state passport.

## Conclusions for Futures Thinking

The power of future city-states can already be seen, not in the few city-states that exist today (Vatican City, Macau, Andorra) but in the new, emerging world/global cities which already operate at arm's length from the territorial-states (London, Dubai). Such cities could form the kernel of the development of new city-states, but this will not happen in the near future. As with the growth of the early European city-states, they will gradually emerge over the next few decades as territorial-states, and world organizations such as the United Nations will come to recognize the global trading power of these city-states. While some of this power will be through traditional trading activities, their real dominance will be in the exchange of services, ideas and new technology. The governments of city-states in the past recognized that not all growth could be supported by their working populations and they needed temporary workers to support the lifestyles of their citizens. In the world-cities of the 21st century, this workforce of temporary workers will, on the whole, not be seen by the citizens of the city-state, and their access to public spaces will be controlled.

Cities have always been and will always be in a state of transition; they are not static groupings of buildings but dynamic places that nurture human development. The transition of the city-states of the past into today's territorial-states is part of this continuous cyclic process of development (Figure 6.1). The current emergence of world-cities/global cities is just another step in this transition. The next stage, as with the settlements

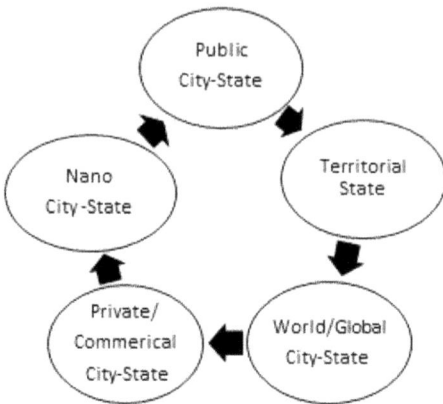

**Figure 6.1** The continuous cyclic city-state model

in the old American West, will be a demand for the creation of self-reliant communities, free from the government's control – privately managed city-states where money and/or technical skills will be the entry permit to a lifestyle environment. These privately managed city-states may gradually evolve into pseudo territorial-states, or nano-states, and they will show that there are 'better ways of living together that will change the governance structures of old continental nations' (Hoyle, 2017: 21). The first signs of this development can be seen in ideas such as the proposal to develop 'Neom', a US$500bn 'i-city' in Saudi Arabia (Boyes & Hipwell, 2017). These nano-states will be managed not as democracies, but rather as managed communities, where decisions are taken by an appointed government for the collective benefit of all its citizens, not individual citizens or small groups of citizens. The first indication of the recognition of this trend will be their admission as associate members of the United Nations, similar to other quasi-autonomous states such as Palestine. When – or if – they fail, they will revert to being public city-states, and so the cycle will begin all over again!

Finally, it would be wrong to conclude that, as postulated in this chapter, the city-state will not welcome tourists unreservedly; they will be open to tourism, but only on their own terms and under their own control. They will not welcome tourists as strangers who need to be supported and guided, but rather as like-minded individuals who will conform to the visions and ideals of the city-states' citizens.

## References

Aristotle (1959) *Politics* (Book III, Chapter VI) (with English translation by H. Rackham). London: William Heinemann. See https://ryanfb.github.io/loebolus-data/L264.pdf (accessed 20 March 2019).

Avakian, T. (2018) *This Airport will have Its Own City with Driverless Cars and Paprachute Rides.* 13 March. See https://www.travelandleisure.com/travel-news/oslo-airport-city-norway (assessed 4 September 2019)
Boyes, R. and Hipwell, D. (2017) City in the sand that could save a kingdom. *The Times,* 28 October, pp. 32–33.
Braudel, F. (1982) *The Wheels of Commerce, Civilization and Capitalism, 15th–18th Century,* Vol. III. New York: Harper & Row.
Burkart, A.J. and Medlik, S. (1976) *Tourism: Past, Present and Future.* London: Heinemann.
Castells, M. (1994) *Technopoles of the World: The Making of 21st Century Industrial Complexes.* London, New York: Routledge.
Fainstein, S. (1994) *The City Builders: Property, Politics and Planning in London and New York.* Oxford: Blackwell.
Friedmann, J. (1986) The world city hypothesis. *Development and Change* 17 (1), 69–83.
Gamble, J. (2014) How to build a city in space. *The Guardian,* 16 May. See https://www.theguardian.com/cities/2014/may/16/how-build-city-in-space-nasa-elon-musk-spacex/ (accessed 10 January 2018).
Halperin, A. (2017) The imperial city-state and the national state form: Reflections on the history of the contemporary order. *Thesis Eleven* 139 (1), 97–112.
Hoyle, B. (2017) Floating cities aim to colonise the sea. *The Sunday Times,* 18 November.
Jones, D.M., Jayasuriya, K., Bell, D.A. and Jones, D. (1995) Towards a model of illiberal democracy. In D. Bell, D. Brown, K. Jayasuriya and D. Jones (eds) *Towards an Illiberal Democracy in Pacific Asia* (pp. 163–167). Basingstoke: Macmillan.
Keeling, D. (1995) Transport and the world city paradigm. In P. Know and P. Taylor (eds) *World Cities in a World System* (pp. 115–131). Cambridge: Cambridge University Press.
Lane, F.C. (1979) *Profits from Power: Readings in Protection Rent and Violence-controlling Enterprises.* New York: State University of New York.
Levitsky, S. and Ziblatt, D. (2017) *How Democracies Die: What History Tells Us About Our Future.* London: Penguin.
Perry, M., King, L. and Yeoh, B. (1997) *Singapore: A Development City State.* Chichester: John Wiley.
Porter, E. (2017) Here's why big cities are thriving, and smaller ones are being left behind. *Las Vegas Sun,* 12 October, p. 5.
Richardson, J.I. (1999) *A History of Australian Travel and Tourism.* Elsternwick, Australia: Hospitality Press.
Sassen, S. (2003) Globalization or denationalization. *Review of International Political Economy* 10 (1), 1–22.
Scott, T. (2012) *The City-State in Europe, 1000–1600.* Oxford: Oxford University Press.
Sutcliffe, A. (1981) *Towards the Planned City.* Oxford: Blackwell.
Tout, T.F. (1934) *The Collected Papers of Thomas F. Tout: With a Memoir and Bibliography.* Manchester: Manchester University Press.
Toynbee, A.J. (1934) *A Study of History.* London: Oxford University Press.
UN (2014) *World Urbanization Prospects: The 2014 Revision, Highlights.* New York: United Nations, Department of Economic and Social Affairs, Population Division.
UN (2015) *World Population Prospects: The 2015 Revision, Key Findings and Advance Tables.* Working Paper No. ESA/P/WP.241. New York: United Nations, Department of Economic and Social Affairs, Population Division.
UNWTO (2018) *World Tourism Barometer* 16 (1). Madrid: UNWTO.

# 7 Geohistorical Analysis of Coastal Tourism in China (1841–2017)

Benjamin Taunay

### Introduction

Since the government of the People's Republic of China (PRC) introduced fixed-date paid holidays in 1999 (known as 'golden weeks' or *huangjinzhou*), there have been persistent comparisons with Japan (which established this policy in 1966) or France (where 'paid holidays' began in 1936). Articles on the subject consistently point to this delay, synonymous with China lagging behind in adopting sea-bathing, effectively comparing coastal tourism activities in China to those witnessed elsewhere in the world (Taunay, 2015). Rather than involving ourselves in this controversy, this chapter wishes to open up a new perspective, hypothesizing that sea-bathing activities existed in China well before the granting of paid holidays at the turn of the 20th century.

A large number of documents refer to sea-bathing undertaken by Chinese nationals from the beginning of the 20th century, often in the form of postcards as well as several videos or eye-witness reports written in travel memoirs. The former foreign concessions established in China following the two Opium Wars (1839–1842 and 1856–1860) were the sites where such activities were most frequently documented. In these enclaves, where the weakening power of the Qing dynasty (1644–1911) did not apply, many wealthy Chinese came to visit the leaders of the current colonial power in place. This sometimes allowed them to adopt the activity of sea-bathing, imported by 'foreigners' (*waiguo ren*) from Europe or the United States. Later in the 20th century, these foreign concessions also saw sea-bathing undertaken by Chinese people from different social classes. Although numerically limited, this bathing did take place and we need to look at the evolution of these activities at these specific sites to understand how a relationship with the coast in general, and sea-bathing in certain local areas of the country in particular, was formed.

Here we propose to trace the history of Chinese beach activities within China by outlining the sites where they take place. In this sense it is a geographical, localized, approach to the historical phenomena, a geohistory of tourist activities on the Chinese coast. Our questions are therefore: What are the currents – political and economic – which underlie the history of Chinese sea-bathing within China? At which specific sites do these activities take place? Why here and not elsewhere? The location of the phenomena is therefore essential in understanding their logic, as well as for situating the cases studied in a wider context, on the global scale. We will pay particular attention to the relationship between the places described within China and those outside the country during the same period, in order to highlight how the processes of activities and body usage spread from their traditional bases (European and American), sites which became familiar with coastal tourism activities far earlier than in China. This spread has witnessed various trajectories and periods of permanence and inertia, as the historian Fernand Braudel (1966) has shown. More generally, this text is part of a perspective on global history, which is a recent trend in this discipline, wishing in fact to distance ourselves from the latent nationalism of publications in the Chinese language which prefer to emphasize the 'Chinese characteristics' (*zhongguo tese*) of current activities and seek to link them to others which are older in China than elsewhere in the world. We will instead highlight the 'Circulation of human, material and technological capital, which could be defined as "strange", "exotic" or "foreign"... With geographical delimitation and space' (Perez Garcia & De Sousa, 2018: 9–13). We will insist on the weight of institutions, controlling the area of possibilities: 'The importance of informal institutions and social networks that regulated the political power and control of knowledge' (Perez Garcia & De Sousa, 2018: 13).

Several turning points are identifiable in the subsequent development of this geohistory. We propose (1) to trace the origins of sea-bathing activities in China, arguing that they coincided with the arrival of Western powers in China in the second half of the 19th century (1841–1911). We will then focus on the hesitant spread of bathing activities during the 20th century, returning (2) to the period preceding the establishment of the PRC (1911–1949), and then (3) to the first two decades of the Communist regime (1949–1966). After the end of the Cultural Revolution (1966–1976), we suggest that there was a final historical phase (4), that of the opening up of sea-bathing in areas other than those of the foreign concessions and allowing mass access to the country's different beaches (1979–2007). After a conclusive synthesis, we suggest two possible routes for future beach activities in China, insisting (5) on political control over body uses and (6) on the individualization in progress in sites reserved for certain populations but to the detriment of others.

## Historical Turning Points

### Turning point 1: Arrival of Western powers

In Europe, beaches represent the contemporary culmination of a historical construction of views and social activities, the beginnings of which date back to seaside visits by the Dutch from the 17th century onwards (Knafou, 2000). The English then played a fundamental role by developing a number of resorts on the south coast of England beginning in the 1750s, preceding a spatial distribution across the rest of the country in the first half of the 19th century (Towner, 1996). In his major work, *Le Territoire du Vide : l'Occident et le Désir du Rivage, 1750–1840*, Alain Corbin (1988) thereby reconstructed a cultural history of this territory, long abandoned as it represented an area of shipwrecks for a population who had not yet mastered swimming. It was only in the 19th century that the beach became a space, first used for therapeutic practices and then – gradually – for fun, where individuals might indulge in multiple activities (Charpentier, 2013). During the first half of the 19th century, and with bourgeois access to tourism, an initial densification of activities appeared, leading to 'strategies of distinction and distance' between different co-present populations, and as such 'the development of individual practices related to the self' (Corbin, 1988: 103).

Later, in the second half of the 19th century, the curative role of the seaside, which had prevailed until then, lost out in favour of a more leisure-related purpose. The invention by the English in the second half of the 19th century of sports such as swimming introduced another relationship to the aquatic element where sport and play were involved. Despite the first swimming club being founded in London in 1837, during the first edition of the modern Olympic Games, in 1896, 13 of the 14 participating countries took part in the sea-swimming competitions.

From then onwards, swimmers gradually began to teach swimming to children, and thus became lifeguards. In terms of non-aquatic activities at the end of the 19th century, beach games (ball-games, tennis, kite-flying, croquet, beach-casting, sand-castle competitions) became more popular than water-based activities (Urbain, 1994). This beach culture is therefore strictly Western in nature, where the influence of the English is fundamental as the situation in Australia proves, being similar in many respects to European activities of the time (Booth, 2001).

This link between bathing and the sea, invented in Europe and spread throughout its colonies, did not at first find its way to China. However, at a time when Europeans were discovering sea-bathing, many Chinese were taking advantage of hot baths (*wenchuan*), which were found in the interior of the country and were similar to the *Onsen* type of mineral water baths (the name given to the same activity in Japan). Although China has 14,500 km of coastline, when the Germans for example seized the Qingdao peninsula (in 1897), Ersnt Hesse-Wartegg (an Austrian traveller and writer, 1851–1918), described the population of Shandong in writing by explaining that the Chinese do not have a culture of sea-swimming (Andreys & Taunay, forthcoming). During the reign of the Emperors (from the Qin dynasty to the fall of the Manchu, 211 BC to AD 1911), China had in fact not developed any relationship between pleasure and the seaside: no aesthetic idea of the sea or the beach emerged. To understand how tourism activities developed on the Chinese coast, one must understand that this only became visible in China following the arrival of colonials. The introduction occurred through the establishment of foreign concessions, especially on the coast (Table 7.1), after the Opium Wars and the succession of 'unequal treaties' (*bu pingdeng tiaoyue*), qualified as such by Chinese leaders for being signed under pressure applied by European and Japanese gunboats.

Table 7.1 Foreign coastal concessions and ports open to foreign powers between the Nanking Treaty and the end of WWII (1842–1946)

| Provinces/ municipalities | Cities | Port/ concession | Dates | Controlling country |
| --- | --- | --- | --- | --- |
| Liaoning | Yinkou | P | 1858–1945 | UK, USA, France, Russia |
|  | Dalian | C | 1898–1945 | Japan |
| Hebei | Tianjin | P/C | 1860–1945 | UK, USA, Germany, Austro-Hungarian Empire, France, Italy, Belgium, Russia, Japan |
| Shandong | Yantai | P | 1858–1945 | UK, USA, Germany, France |
|  | Qingdao | C | 1897–1922 | Germany, Japan |
|  | Weihai | C | 1898–1940 | UK |
| Jiangsu | Zhenjiang | P | 1861–1940 | UK |
| Shanghai | Shanghai | P/C | 1843–1946 | Great Britain – USA – France |
| Zhejiang | Ningbo | P | 1841/1842 | UK |
|  | Wenzhou | P | 1876–1937 | UK |
| Fujian | Fuzhou | P | 1842–1945 | UK then Japan |
|  | Xiamen | P/C | 1842–1912/ 1852–1930 | UK |
|  | Gulangyu (Xiamen) | C | 1903–1945 | International |
| Guangdong | Shantou | P | 1858–1945 | UK then Japan |
|  | Zhanjiang | C | 1898–1946 | France |
|  | Haikou | P | 1858–1945 | UK, France, then Japan |
| Guangxi | Beihai | P | 1876–1940 | UK, USA, Germany, Austro-Hungary, France, Italy, Portugal, Belgium |

Notes: P, port; C, concession.
Source: China Knowledge (http://www.chinaknowledge.de/index.html) and different sources collected by the author.

The role of the UK was crucial in this introduction of sea-bathing in China, as it was the country with the largest number of 'foreign concessions' (*zhimin di*) dating from the signing of the original treaties (Table 7.1). The UK was in particular developing the protectorate of Hong Kong which was, however, a special case since it remained independent from China for 150 years (1841–1997). However, the other concessions, even if they benefited from the principle of extraterritoriality (if foreign nationals committed an offence or crime, they were judged by the colonial government), remained in touch with the Chinese people. The other exception was Macao, formerly a Portuguese territory, which remained out of Chinese control for a century (1897–1999).

## Turning point 2: The hesitant spread of bathing activities (first half of 20th century)

In the various archives that we consulted, we could find no trace of sea-bathing undertaken by Chinese nationals before the implantation of foreign concessions. On the tropical island of *Hainan*, we find (in the *Haikou* concession) the same testimonies as those provided by the first German settlers quoted above for *Qingdao*. This latter city is an exemplary case for bathing which took place in the second half of the 19th century and the early 20th century in China. Although not initially devoted to sea-bathing (*Qingdao* is a city which grew in line with the rise in German trade in the region, and is not a seaside resort), beach activities developed in the early years of the colony (Andreys & Taunay, forthcoming). These activities, as well as the equipment used for them, are very similar to what existed along the coastlines of the North Sea and the Baltic during the same period: the city was in fact served directly from German ports (Hamburg in particular), allowing settlers' families to join them at the beaches during the summer holidays. The German authorities also supported the development of the seaside resorts because they attracted summer visitors of all nationalities. The authorities hoped to boost knowledge of the benefits of the city for both businesses and trading companies. The government approved the construction of beach huts (of which there were 140 in 1904), and a bathing commission was created in 1903, showing that bathing was spreading from the beginning of the 20th century onwards (Andreys & Taunay, forthcoming).

This copying of the forms and uses of the European holiday resort underlines the synchronicity of the phenomena at that time (a large hotel based on the style of European seaside resorts opened in *Qingdao* in 1904) and shows how colonization reinforced the spread of tourism along the Chinese coastline. We should, however, point out that in *Qingdao*, as in the other foreign concessions located in China, with the exception of the elite party members, ordinary Chinese were forbidden from visiting. In *Qingdao* in particular a strict distinction existed between the German city and the Chinese city located further north. The Chinese population was therefore not legally allowed access to the coast of *Qingdao* until 1911 when the Qing dynasty collapsed. Between 10,000 and 12,000 Chinese (from the most privileged sectors of society) then settled in the city. The Chinese in exile considered *Qingdao* to be both a safe and a comfortable place of residence, while the Germans saw in it a way to ensure their political and economic supremacy in the region by allying themselves with influential mandarins. Also, from 1912 onwards, many Chinese began buying land that was for sale in order to build themselves houses. There, as in other foreign concessions, it then became possible for the most fortunate citizens to rub shoulders with foreign populations, sometimes going as far as copying their bathing activities.

However, the outbreak of the WWI (1914) quickly led to a reordering of the colonies. Japan subsequently seized a large number of territories (with the northeast of the country being fully colonized), including the city of *Qingdao*. During this colonization, which lasted until 1922, the Chinese were no longer allowed on the beaches, and a strict rule forbade them from entering the former German city, then reserved for the Japanese only. These new masters were often to be found bathing, as evidenced by numerous vintage postcards. The various epithets attributed to *Qingdao* up to that point, such as the 'Brighton of China' and the 'Ostend of the Far East', continued to boost the city's reputation as 'The best bathing place of the Orient'. However, the Japanese were defeated eight years after their arrival in *Qingdao* and China's nationalist government took possession of part of the Chinese coast (except for the northeast of the country) following the Treaty of Versailles (1922), for 15 years (until 1937). As evidenced by the various plans of the time (in 1928 in particular), the nationalists appropriated the beaches and infrastructure created by the Germans (the beach huts appear on city maps from the period). Becoming Chinese once again, the city developed into a destination chosen by those rich Chinese wishing to spend an exotic holiday at the seaside from 1929 onwards. There was an evolution in the scale of seaside infrastructure, as well as the opening of new beaches. All this leads us to believe that this period was conducive to Sino–foreign meetings and the appropriation of beach activities.

It should be noted, however, that here as elsewhere, in *Beidaihe* for example (in the Gulf of *Bohai*), Chinese holidaymakers very rarely went swimming. Although there were links between the two cities – both of which were connected to Shanghai from where foreigners (Europeans and Americans) living there came to spend their summer holidays – contacts between the Chinese and foreigners did not lead to a mass adoption of sea-bathing activities for the Chinese population. Only the most well-off Chinese, who gradually settled there, were able to swim in *Qingdao* and other foreign concessions in China. The ensuing political context did not really allow leisure activities to develop. The fall of the Qing dynasty

(1911), the attempted restoration of the empire in 1915–1916 (by President Yuan Shikai) and the wars against Japan paralyzed and limited bathing to the foreign concessions only.

In 1937 the second war with Japan began, which led in 1940 to the colonization of almost all the foreign concessions in the country. Despite Japan's defeat in 1944, it took until the end of the civil war between Communist adversaries (led by Mao Zedong) and nationalist leaders (controlled by Chiang Kai-shek) for the PRC to emerge (on 1 October 1949), thereby opening a new historical chapter for sea-bathing.

## Turning point 3: Beaches reserved for the Chinese Communist Party (1949–1966)

When Communist rule took effect at the beginning of the 1950s, foreigners were particularly unpopular. Not content with supporting Chiang Kai-shek and recognizing the island of Taiwan as representing China (the Republic of China – ROC) at the United Nations (UN), the Americans outlawed the presence of Chinese in *Qingdao* – until 1945 on the 'American Beach' – by means of a sign announcing 'For officers only. Civilians, Chinese and dogs forbidden' (just like in the Bund in Shanghai). The Chinese Communist Party, which governed the PRC, then turned its back on the West and looked to countries in the Soviet bloc. At the same time, the foreign concessions, considered as being an affront to China by the West, were re-appropriated: from 1950 onwards, the 'American Beach' in *Qingdao* was renamed Beach No. 2 and was reserved for the exclusive use of senior executives in the Chinese Communist Party.

If the tourism system changed in the interwar period in Europe and the United States – with a decline in the power of the aristocracy, seasonal changes on the coasts (with the summer becoming the most popular period; Granger, 2009) and the first policies favouring mass tourism – these changes no longer affected China, which was cut off from the Western world and its changes in terms of seaside activities. The link that had existed until then (increasingly tenuous given the wars mentioned above) was broken with the banning of Western activities because they were synonymous with the humiliation of China over the previous century. The sites of 'humiliation' did not, however, disappear. On the contrary, there was a resilience among these former colonial areas, where the Communist Party perhaps asserted its power more than elsewhere. The end of the Korean War (1950–1953) which resulted in a non-aggression pact (between North Korea supported by the PRC and the US-backed South Korea) coincided with the creation of China's first state-owned tour operator, CITS (China International Travel Service). Under the control of the Ministry of Foreign Affairs, tourism thus became a diplomatic tool and one strictly controlled by the central state. At that time, China brought in a large number of foreign technicians to assist the government in implementing its First Five-Year Plan (1953–1957). In order to look after its official visitors and to train staff in the tourism sector, Beijing sent them via CITS to various destinations around the country. The tourist visitors came from Eastern Europe and the Soviet Union (representing 95% of the 5439 tourists in 1960). They were not, however, sent to the country's beaches. On the contrary, those who were on all-expenses-paid trips were directed to the sites which the government considered important in terms of their historical cultural significance. Notable among them was the city of *Guilin*, known for its landscapes (Taunay, 2009), and described as beautiful with its combination of 'mountains and waters' (*shanshui*; Sofield & Li, 1998) since ancient times.

The country's beaches were reserved for two categories of people: soldiers of the People's Liberation Army (*zhongguo renmin jiefang jun*), accompanied by senior Chinese Communist Party leaders. The beaches at *Beidaihe* were at that time the most famous because they hosted visits by key party dignitaries each year in the run-up to their congress (held in October). Several images show the various visits Mao Zedong made to the beaches and he also travelled there to swim in the summer and to write poems for the rest of the year. Soldiers were accommodated in dedicated buildings, well away from party cadres. They came to relax in sanatoriums like those found in mountain resorts, sites which were also set up by European settlers (Spencer & Thomas, 1948) and whose systematic reoccupation once again confirmed the country's rejuvenated nationalism. This situation was, however, ephemeral: first, because the Great Leap Forward (*da yue jin*, 1958–1960), and its millions of deaths linked to the famines in 1960 and 1961, put the brakes on the nascent tourism industry at both international and domestic levels; and secondly, because disagreements between China and the Soviet Union in the early 1960s caused the number of tourists to drop by 70% between 1960 and 1963. At the time, China turned temporarily to the West and in 1965 the number of tourists increased to 4519 visitors, 85% of whom were from Western countries (Uysal, 1986).

Above all, the Cultural Revolution (*wuchan jieji wenhua dageming*, 1966–1976), launched by Mao Zedong in an effort to regain the power he had lost due to the strategic mistakes made during the Great Leap Forward, completely blocked the process of tourism development. For 10 years the country's President sent those he considered as opponents, the 'bourgeois intellectuals', to the countryside. But they were not alone. CTIS employees who had previously been students were therefore targets of the purges, separated from their families and sent to the outlying rural areas of the country to be 're-educated' by the peasants (Uysal, 1986). Considered a bourgeois activity, tourism during this period was a forbidden activity. As with other tourist sites, all the beaches in the country were empty. Even as neighbouring countries such as Japan established their own 'golden weeks' in 1966 (*goduren wiku* in the indigenous language), the PRC was distancing itself from the changes in seaside tourism taking place elsewhere in the world.

### Turning point 4: A progressive and rapid opening (1979–2007)

The beaches of the PRC only became popular once again in 1979, three years after Mao Zedong's death. The Chinese Communist Party's new First Secretary, Deng Xiaoping, launched a policy of modernization conceived a few years earlier (1975) by former Prime Minister Zhou Enlai (1949–1976). Tourism was one of the tools used in this complete modernization of the country and that was why, from 1979 onwards, deserving workers were sent on holiday to the best-known beaches, by which we mean those located in the former foreign concessions. Those who were fortunate enough to have an all-expenses-paid visit to the beach for a few days were selected because they had served the 'homeland', and in that respect were 'exemplary' citizens and therefore deserved a rest on the coast. Interviewed in 1999 as part of a Franco-German documentary ('the Chinese on the beach', Hudelot & Vassort, 1999), Ms Sun said she went to *Beidaihe* every year between 1979 and 1986. Created by the UK, the main colonial power in the 19th century, the foreign concession that became a Chinese seaside resort was during the 1980s the most popular coastal destination in the country. As a site for excursions for the inhabitants of the capital (Beijing being located just over 150 km west of the resort), *Beidaihe* hosted several million visitors each year in the mid-1980s (Boulet, 1986), of which 60% would come for one day only (Xu, 1999).

Another policy also allowed coastal tourism to restart. In 1984, in the context of the Sixth Five-Year Plan (1981–1985), the Chinese state launched a reform to create open cities, allowing 14 municipalities to accept direct foreign investment. These cities were mainly those of the former colonial powers, established during the second half of the 19th century. This was the case with *Beihai*, a small town in the south of the *Zhuang* autonomous region in *Guangxi*, on the *Beibu* Gulf coast. In addition to building a new port to the west of the city, the municipality was instructed to develop a tourist district primarily on the north coast of the peninsula where the foreign concessions were located. The old and only remaining state-run hotel, opened in 1958, was situated there too. As of 1985, a new hotel was built and coastal development works began to build a dedicated footpath. President *Jiang Zemin* visited the city on 21 November 1990 to see how the work was progressing. However, here as elsewhere in the country at the time, numerous obstacles were hindering the development of tourism based on sea-bathing. First, up until 1989 tourist stays were mainly funded by employers (up to 50% of all trips; Wen, 1997), preventing the free movement of travellers as they were restricted to a pre-arranged programme. Next, activities on the coast were primarily based on the concept of discovery, with sea-bathing not being well-known at the time. As a result, the new zone built to the south of the city, based on an 'international' model, remained empty for a long time and some of the buildings have never been completed (Liu, 2010; Taunay, 2008).

Until the mid-1980s, all the coastal sites were evidence of the beaches developed during the period of foreign concessions. In 1987, however, *Hainan* became the first coastal tourism area that was not a Western creation. Although part of the city of *Haikou* had been a foreign concession, tourism planning was initially focused on the south of the island in *Sanya* Prefecture. Historically part of *Guangdong* Province, *Hainan* Island became a fully independent province in 1986, on the eve of the Seventh Five-Year Plan. The central government's stated objective, in overseeing the work with the newly created provincial government, was to build an 'international' destination, meaning that it would meet Western standards of comfort and offer comparable activities (the island was renamed the 'Hawaii of the East'), while at the same time being accessible for international tourists without any visa requirements. Thanks to a series of national policies aimed at developing a tourism- and leisure-orientated economy (including the creation of 'golden weeks' in 1999, 33 years after Japan), the beaches of *Sanya* municipality are today the busiest in China: more than 40 million Chinese tourists visited them in 2016, but in the same year there were barely 1 million foreign tourists. That means the Chinese political framework has produced, in a short time frame (30 years up to 2017), the most frequently visited coastal region in China. The Chinese economy, through careful planning, has therefore produced a place and an associated imaginary which, although built on Western international logic (the model of the Blue Lagoon; Vacher, 2016), nonetheless attracts an almost exclusively Chinese clientele. The shorelines around *Sanya* are now influencing the island's other municipalities, as has been the case in *Wanning* (Guibert & Taunay, 2013) and as has happened in the west of the island since the high-speed train service was introduced (2017). The era of tourism for all is looming on many of the coastlines, and local governments have clearly understood the challenge. The *Hainan* model has frequently been emulated and tourism on the coast, not just the idea of sea-bathing in the most famous places, is therefore becoming a territorial marketing issue.

## Conclusion – Future Turning Points

The history of coastal activities and bathing in China can only be understood on an international scale. The use of the beach, a European invention, arrived with the settlers when they created their foreign concessions in China from the second half of the 19th century onwards. This imposition of extraterritorial places consequently limited the spread of recreational activities to some of the country's coastlines for nearly 100 years. Acting on the nationalist movement in response to the 'humiliation' caused by 'foreign powers', Communist leaders subsequently reinvested in the areas but reserved the sites for those who deserved them according to their own criteria. After a period when they categorically rejected any Western influence (the Cultural Revolution), the Chinese government nevertheless references the Western international model of the turquoise lagoon when defining the country's main 'international' coastal tourism destination, the island of *Hainan*. Whatever the most zealous Chinese writers say about current nationalist politics ('one belt, one road'; *yi dai yi lu*), this game of transfers between global sites is what today makes up the DNA of coastal activities and sites in China. However, it must be said that current activities on Chinese beaches differ from what is to be found elsewhere in the world. The beach in China is primarily a place reserved more for bathing than for tanning oneself, with more than half of all the tourists in the water regardless of the time of day (Taunay & Vacher, 2018). Nowadays, and since the arrival of Xi Jinping (2012), the strengthening of state control over tourism activities continues apace. In addition to the Tourism Act of 2013, a 64-page guide to good manners was published the same year.

### Future turning point 1: Controlling bodies and increasing standard setting

Encouraging Chinese tourists to behave well when travelling both internationally and within China, this setting of standards is probably one of the key steps for future beach use in the coming years. In *Sanya*, topless tanning is no longer allowed on the beach, and neither is full nudity even though it was still possible a short time ago. China's beaches are becoming more and more regulated with opening hours and lifeguard patrols making sure people respect morality rather than checking on the safety of bathers. In this way the current trend is for body control and therefore, to a certain extent, the refusal of everything that is 'Western' – a concept that is no longer regarded as a model but rather as something outdated and depraved. In the future this idea could be reinforced, influencing beach practices. Nationalism is indeed being encouraged at the highest level since the 19th Congress of the Chinese Communist Party. In his speech of more than three hours (October 2017), the Chinese President wished that economic development would lead China to take its place among the great nations of the world (Xi, 2017). China and the Chinese should assume the status of the country's new power, refuting the Western hegemony that reigned until the beginning of the 21st century. This should lead tourists to assume their ways of doing rather than conform to the old, Western, world tourist order.

### Future turning point 2: Banalization

Finally, access to beaches is becoming increasingly restricted, with gates at the entrances (and often an entry fee to be paid), as well as checks and inspections. Although the wish to create financial benefits for the local economy is evident, it is becoming clear that the process of individualization and personalization of activities is taking shape. In such a populous society, a new model allowing the better-off to get away from the masses may well be in the making. In the future it is therefore possible that new frontiers appear on the Chinese beaches. These will separate the populations according to their financial capacities, the richest setting aside the middle and popular classes. The banalization of beach practices could therefore be accompanied by a restriction of access to certain beaches or parts of them.

## References

Andreys, C. and Taunay, B. (forthcoming) Origins of beach practices in Qingdao. Research article.

Booth, D. (2001) *Australian Beach Cultures: The History of Sun, Sand and Surf.* Sport in the Global Society. London: Routledge.

Boulet, M. (1986) *Dans la peau d'un Chinois.* Paris: Bernard Barrault Editions.

Braudel, F. (1966) *La Méditerranée et Le Monde Méditerranéen à l'époque de Philippe II.* Paris: Armand Colin.

Charpentier, E. (2013) *Le Peuple du Rivage: Le Littoral Nord de la Bretagne au XVIIIeme Siècle.* Rennes: Presses Universitaires de Rennes.

Corbin, A. (1988) *Le Territoire du Vide: L'Occident et Le Désir du Rivage, 1750–1840.* Paris: Flammarion.

Granger, C. (2009) *Les Corps d'été: Naissance d'une Variation Saisonnière XXe Siècle.* Paris: Autrement.

Guibert, C. and Taunay, B. (2013) From political pressure to cultural constraints: The prime dissemination of surfing in Hainan. *Journal of China Tourism Research* 9 (3), 365–380.

Hudelot, C. and Vassort, F. (1999) Les Chinois à la plage. Documentary film. See http://www.film-documentaire.fr/4DACTION/w_fiche_film/6754_1

Knafou, R. (2000) Scènes de plage dans la peinture Hollandaise du XVIIe siècle: L'entrée de la plage dans l'espace des citadins. *Mappemonde* 58. See https://www.mgm.fr/PUB/Mappemonde/M200/Knafou.pdf (accessed 20 March 2019).

Liu, J. (2010) A comparative study on coastal resort development model: A case study of Yalong Bay in Sanya and Beihai Silver Beach. *Human Geography* 25 (4).

Perez Garcia, M. and De Sousa, L. (2018) *Global History and New Polycentric Approaches: Europe, Asia and the Americas in a World Network System.* Singapore: Springer Nature.

Sofield, T. and Li, S. (1998) Tourism development and cultural policies in China. *Annals of Tourism Research* 25 (2), 362–392.

Spencer, J.E. and Thomas, W.L. (1948) The hill stations and summer resorts of the Orient. *Geographical Review* 38 (4), 637–651.

Taunay, B. (2008) Le développement du tourisme national chinois: L'exemple de la région autonome du Guangxi. *Mappemonde* 89. See https://mappemonde-archive.mgm.fr/num17/lieux/lieux08103.html (accessed 20 March 2019).

Taunay, B. (2009) Le tourisme intérieur Chinois: Approche géographique à partir de provinces du sud-ouest de la Chine. PhD thesis, Université de La Rochelle. See https://tel.archives-ouvertes.fr/file/index/docid/538154/filename/ThA_se_TAUNAY_complA_te.pdf (accessed 20 March 2019).

Taunay, B. (2015) Comment faire avec les espaces des bains de mer en Chine contemporaine? In P. Duhamel, N. Talandier and B. Toulier (eds) *Le Balnéaire: De la Manche au Monde. Actes du Bolloque de Cerisy* (pp. 207–222). Rennes: Presses Universitaires de Rennes.

Taunay, B. and Vacher, L. (2018) Pratiques et organisation spatiale de la plage Chinoise: L'exemple de Dadonghai à Sanya (île de Hainan, Chine). *Mappemonde* 123. See http://mappemonde.mgm.fr/123as1/ (accessed 20 March 2019).

Towner, J. (1996) *An Historical Geography of Recreation and Tourism in the Western World (1540–1940)*. Chichester: John Wiley.

Urbain, J.-D. (1994) *Sur La Plage: Moeurs et Coutumes Balnéaires (XIX-XXe Siècles)*. Paris: Payot.

Uysal, M. (1986) Development of international tourism in PR China. *Tourism Management* 7 (2), 113–119.

Vacher, L. (2016) The marvel of tropical waters: The invention of an imaginary at the pace of technological advances. In M. Gravari-Barbas and N. Graburn (eds) *Tourism Imaginaries at the Disciplinary Crossroads: Place, Practice, Media* (pp. 197–209). New York: Routledge.

Wen, Z. (1997) China's domestic tourism: Impetus, development and trends. *Tourism Management* 18 (8), 565–571.

Xi, J. (2017) Secure a decisive victory in building a moderately prosperous society in all respects and strive for the great success of socialism with Chinese characteristics for a new era. Speech delivered at the 19th National Congress of the Communist Party of China. See: http://www.xinhuanet.com/english/download/Xi_Jinping's_report_at_19th_CPC_National_Congress.pdf (accessed 24 September 2019).

Xu, G. (1999) *Tourism and Local Economic Development in China: Case Studies of Guilin, Suzhou and Beidaihe*. Richmond: Curzon Press.

# Part 3
# Mobility

# 8 The Future Past of Aircraft Technology and its Impact on Stopover Destinations

Rafael Castro, Gui Lohmann, Bojana Spasojevic, Carla Fraga and Thiago Allis

## Introduction

Ever since its earliest inception, aircraft technology has amazed humans, with the flying successes of aviation pioneers such as the American Wright brothers and the Brazilian Santos-Dumont capturing people's imagination. With advances in aircraft technology and increases in speed and safety, the numbers of air transport passengers have soared year on year. This growth has been underpinned by significant technological advances in aircraft technology, including the Boeing 747, Concorde and, more recently, the Airbus A380 and the Boeing 787 Dreamliner. As identified by Spasojevic *et al.* (2018), aircraft technology is a key theme highlighted in the academic literature as underpinning tourism development. Long-haul flights across the world have changed as aircraft have increased their range, becoming technically able to fly directly between any two points on the globe.

This chapter aims to present the impact of aircraft technology since the 1930s, in particular examining the changes in travel patterns from multi-stop trips to the implementation of direct flights worldwide. Notably relevant to this chapter is how stopover destinations have evolved because of the increase in stopover traffic, which can make a significant contribution to local tourism economies. Using the 'Kangaroo Route' between Australia and Europe as a case study, we examine the historical turning points that have shaped aviation and the tourism business as a result of the development of aircraft technology, even leading to the threat of eliminating stopover traffic as non-stop flights become an increasing global reality.

## Aircraft Development, Distances and their Impact on the Kangaroo Route Stopover Destinations

When tourists decide to travel, their choices regarding destinations, air routes, airlines and other providers most commonly reflect an attempt to minimize stress, including those related to perceived distance and travel times. Cooper and Hall (2008) list eight types of distance that can influence travel behaviour: Euclidean (physical) distance, time distance, economic distance, gravity distance, network distance, cognitive distance, social distance and cultural distance. Larsen (2017) identified three categories of representation of distance: distance as resources, distance as accessibility and distance as knowledge:

- Distance as *resources* refers to the amount of resources needed for a given distance to be transcended by the tourist in terms of money and time. So, the more expensive a trip is and the more time demanded by the trip, the higher the distance perceived by the potential tourist. Flights with stopovers tend to be cheaper compared to direct flights but, on the other hand, they require more time.
- *Accessibility* is an important deciding factor for tourists in choosing a particular destination. Thus, the accessibility of a destination is related to the distance perceived by the tourist, as the places that are easier to reach are perceived as closer than those that are more difficult to get to, even when the physical distance is shorter for those more inaccessible places. A direct flight route, for example, can give the tourist the feeling of closer proximity to the destination, unlike flights with stopovers.
- *Knowledge* (and information) about places can influence how they are perceived as being closer than they are. Moreover, it is argued that the more familiar the tourist is with the transport mode and the route itself, the more comfortable the travel will be; hence, unfamiliar stops along the route can increase the level of stress during the trip. A non-frequent traveller or someone who has never made stopovers at major airport hubs such as Singapore Changi or Dubai, for example, may feel insecure and willing to avoid such stopovers.

Over time, concepts of distance and remoteness have played an important role in the development of Australia's history. Taylor (1919) sought to demonstrate the isolation of the country by drawing a map with a 4000-mile radius circle centred on the city of Canberra, the capital city. The result indicated that this circle would pass through Antarctica, near the South Pole, but would not include any other large land territory. In this sense, the development of new aircraft and air routes to/from Australia (such as the Kangaroo Route) have been paramount in connecting the country to the rest of the world.

The Kangaroo Route is a term that has been trademarked by Qantas and refers to air routes flown by this particular airline between Australia (especially to/from Melbourne and Sydney) and the UK (London) via the eastern hemisphere. Over time, the development of new aircraft technology has had a substantial impact on decisions regarding the stopover destinations along the route, creating a series of historical turning points as

Table 8.1 Qantas' Kangaroo Route over time

| Year | Aircraft | Number of stops | Stopover destination(s) | Average flying time (hours)/journey time | Seats |
|---|---|---|---|---|---|
| 1947 | Lockheed Constellation | 6 | Darwin, Singapore, Calcutta, Karachi, Cairo and Tripoli | 55.07 hours/4 days with overnight stops in Singapore and Cairo | 38 |
| 1954 | Super Constellation | 6–7 | Darwin, Jakarta, Singapore, Colombo, Dubai, Cairo, and Rome, among others | 51.30 hours/4 days | 57 |
| 1959 | Boeing 707-300 (introduction of the jets) | 5 | Darwin, Singapore, Dubai, Cairo and Rome | 26 hours/34 hours | 220 |
| 1971 | Boeing 747-200 | 2 | Singapore and Bahrain | 22 hours/25 hours | 356 |
| 1989 | Boeing 747-400 | 1 | Singapore | 22 hours/23.25 hours | 371 |
| 2009 | Airbus A380 | 1 | Singapore | 22.2 hours/24 hours | 484 |
| 2013 | Airbus A380 | 1 | Dubai – the beginning of the Qantas–Emirates alliance | 21.5 hours/23.1 hours | 484 |
| 2018 | Airbus A380 | 1 | Singapore | 22.2 hours/24 hours | 484 |

Source: Qantas (2018); Rimmer (2005); Rodrigue et al. (2017).

summarized in Table 8.1. The rest of this section presents more details about the six different historical turning points of the Qantas Kangaroo Routes.

## Historical Turning Points

### Turning point 1: Beginning of the Kangaroo Route

In 1947 Qantas started flying the Kangaroo Route with a Lockheed Constellation carrying 23 passengers and 11 crew members from Sydney to London, with stops in Darwin (Australia), Singapore, Calcutta (India), Karachi (Pakistan), Cairo (Egypt) and Tripoli (Lebanon). The flying time was around 55 hours on a four-day journey with overnight stops in Singapore and Cairo.

### Turning point 2: Introduction of the Super Constellation

In 1954 Qantas started flying the Super Constellation on the Kangaroo Route. The average flying time was 51.3 hours on a four-day journey with six to seven stops in cities such as Darwin, Jakarta (Indonesia), Singapore, Colombo (Sri Lanka), Dubai (UAE), Cairo and Rome (Italy), among others throughout the time. The number of seats available increased to 57.

The introduction of the Super Constellation on the route did not represent major advances regarding travel time or the number of stops, but rather an increase in the available seating capacity.

### Turning point 3: Beginning of the jets operation

The introduction of jets to the Kangaroo Route occurred when Qantas flew the Boeing 707-300 for the first time in 1959. This aircraft enabled a decrease of approximately half of the travel time, even though it still required five stops: Darwin, Singapore, Dubai, Cairo and Rome. The number of seats available increased by almost 300% and the company could offer 220 seats on each flight, allowing a more significant number of people to fly between the two continents.

### Turning point 4: The Super Jumbo takes over

The fourth historical turning point in the Kangaroo Route marked the beginning of the operation of the Boeing 747-200 (the Jumbo), in 1971. With this new aircraft, Qantas was able to fly from Sydney to London with only two stops: Singapore and Bahrain. The flight time decreased by four hours, with total travel time falling by 11 hours due to the reduced number of stops along the route. Regarding passenger capacity, the Boeing 747-200 had 356 seats over two decks.

### Turning point 5: The Boeing 747-400 era

In 1989 Qantas introduced the -400 version of the Boeing 747. This aircraft was a longer range aircraft enabling the carrier to finally connect Australia to the UK with only one stop, in Singapore, on a 24-hour journey. The number of seats kept increasing, albeit timidly. The possibility of making this trip with only one stop represented, therefore, a decrease in the distance perceived by the travellers.

### Turning point 6: The Airbus A380

As with the introduction of the jets and later the appearance of the wide-body jets (Boeing 747), the start of the operation of the double-decker Airbus A380 was, in itself, a real turning point for the aviation world. This aircraft transformed the way the industry operated, demanding changes in airport infrastructure on both air and land sides. Runways had to be long enough for take-off and landing procedures. Although the length of many runways and taxiways in major hub airports could accommodate the dimensions of this aircraft, most were not wide enough. The main issue, however, was the lack of terminal space. Until then, most airport terminals did not have enough space to park an A380 or the facilities to process such a large number of passengers embarking and disembarking. In 2009 Qantas started operating a 484-seat A380 on the Kangaroo Route with one stop in Singapore, serving passengers in economy, premium economy, business and first classes.

A few years later, on 31 March 2013, two Airbus A380s flew over Sydney Harbour, celebrating the beginning of the Qantas–Emirates alliance. This agreement involved operational and marketing aspects, the key one being the codeshare agreement to Europe, via a stopover in Dubai. After five years, in 2018, the airlines decided to change the terms of their partnership. In an interview for the Australian news portal ABC, Qantas Chief Executive Officer, Alan Joyce, stated that recent consumer surveys showed that passengers expressed a stronger preference for an Asian stopover, rather than the Middle East, on flights to Europe. More than that, it was mentioned that their partnership had evolved to a point where Qantas no longer needed to fly its aircraft through Dubai, leaving this operation to Emirates (ABC, 2017). On the Kangaroo Route, Qantas would return its stopover location to Singapore. From this point, it seems that the discussion about stopover destinations has become more strategic in terms of operational and commercial contexts, rather than being impacted by aircraft technology as it used to be. This led to the emergence of the Sunrise Project, a topic covered in the next section.

### Turning point 7: The Sunrise Project

On 11 December 2016 Qantas announced that direct flights between Perth, on the west coast of Australia, and London would commence in March 2018. The first QF9 flight departed Perth on 24 March at 18:50, landing in London Heathrow on 25 March at 05:00, after flying for 17 hours non-stop. By using their recently purchased Boeing 787-900, this initiative marked the first time the Kangaroo Route was flown directly between Australia and Europe. These flights operate out of Terminal 3 in Perth, rather than the traditional T1, to facilitate seamless transfers from Qantas domestic flights. The aircraft flying this route originally departs from Melbourne, with a stopover in Perth.

Flying directly between Australia and the UK is part of a broader strategy developed by Qantas. 'Project Sunrise' was announced on 25 August 2017, with the possibility of flying non-stop from the east coast of Australia (including Sydney, Melbourne and Brisbane) to London, Paris, New York or Rio de Janeiro by 2022. To achieve this goal, Qantas has challenged Boeing and Airbus to adapt current aircraft models to stretch their flying range. Days later, on 31 August 2017, Qantas also announced that its flight on the Kangaroo Route from Sydney to London would revert to a stopover in Singapore, instead of Dubai, from 25 March 2018. This decision marks the return of Qantas' Airbus A380 to

Changi Airport, five years after the change to fly through Dubai, as mentioned in the previous section.

## Ultra Long-Haul Air Travel

Making the ultra long-haul air travel decision takes into account airlines, passengers, aircraft manufacturers, airports and destinations. Weber and Williams (2001) state that geography (transport networks and centrality of hubs), regulation (aviation laws and air sovereignty), manufacturers (technological barriers), passengers (stress factors inside the cabin, including barometric pressure, noise, crowding, seating comfort, inactivity, jet lag, etc.) and the airlines' operating attributes (alliances and worldwide network collaborations, frequency of flights) are relevant factors in shaping ultra long-haul air travel.

Regarding the initiatives put forward by the two main aircraft manufacturers, Airbus proposes the use of its A350-900ULR (ULR: ultra-long range). It hopes to make the model fly for 20 hours, carrying a payload that would satisfy Qantas regarding the economic viability of these routes. The model was launched in 2015 to enable Singapore Airlines to fly non-stop to the United States (e.g. New York, 9534 nautical miles (nm) and Los Angeles, 8770 nm), travelling for 19 hours. Airbus had to make aerodynamic improvements, modify the fuel system to increase fuel capacity and increase the maximum take-off weight (Airbus, 2018). In 2017 the world's longest route by distance was operated by Qatar Airways between Doha and Auckland, a service of 7848 nm, operated by a Boeing 777-200LR.

While private and government organizations can facilitate the regulatory environment, the strategic cooperations, the technological advancement and the financial viability payloads of ULR routes, it is not clear yet whether there is a significant market of passengers willing to fly ULR flights. To sum up, if on the one hand one celebrates higher mobility enhanced by technology evolution, on the other hand the 'darker sides of hypermobility' can also be highlighted, 'including the physiological, psychological, emotional and social costs of mobility for individuals and societies' (Cohen & Gössling, 2015: 1661).

On one hand, ULR flights have the primary benefit of reducing the total travel time of journeys, by avoiding unnecessary stopovers and connections; on the other hand, ULR operations (those with 16 or more hours) can have many effects on the human body. Both passengers and crew can experience aggravated symptoms of flying, including: lack of oxygen (hypoxia) and pressure (affecting blood circulation, which can even lead to disabilities); ear and sinus pain caused by the difference in pressure blocking gas in the Eustachian tube; gastric distress produced by food, drinks or illness; dental pain due to the reduction of atmospheric pressure; decompression sickness; fatigue and lack of sleep; thermal stress due to differences of temperature at various destinations and inside the cabin; and nutrition (including alcohol consumption) (Martinussen & Hunter, 2017).

Regarding the crew, Holmes *et al.* (2012) summarize some safety issues related to fatigue during ULR flights. An extra crew is required on these flights, with the recommendation that pilots and crew will have the chance to alternate in-flight sleep, usually in bunk facilities provided with horizontal beds, in a quiet, dark and temperature-controlled environment. In their study, Holmes *et al.* (2012) analyzed the Houston–Doha route and considered a number of attributes contributing to sleep and sleepiness, including: the conditions of the aircraft and the rest environment; the pilots' role as operating or relief crew as indicated in their rosters; sleep and rest guidance (sleeping recommendations prior to flight, diet, exercise, light exposure, etc.); and direction of flight (outbound or inbound). The results of this study show that 'sleep and sleepiness levels were similar between operating and relief crew, but there was a high degree of inter-individual variation between pilots' (Holmes *et al.*, 2012: 31). One piece of evidence was that pilots with a 'siesta culture' were more likely to nap during rest days, layover and recovery, with greater sleep duration, requiring less sleep after their second rest period.

Griffiths and Powell (2012) provide a wider approach to the occupational health and safety analysis of flight attendants, including their greater exposure to injury risk, higher incidence of respiratory disorders, permanent disability, including musculoskeletal, psychiatric disorders, heart disease and cancer (mainly breast and malign melanoma) due to ear, nose and throat (ENT) disorders. Air quality, acute injuries as the result of working unrestrained in the cabin during turbulence or emergency evacuations, and radiation exposure are some of the obvious contributors. Also, excessive exposure to 'emotion work', fatigue, isolation, demanding work and work–home imbalance are frequently cited as contributors to mental health issues, which are only aggravated by ULR flights.

Thus, understanding how stopover destinations have evolved due to the impact of the increase (and later decrease) of stopover traffic is paramount, as this can have a significant contribution to the local tourism economy.

## Checkmate on Stopover Destinations?

From the perspective of transport networks, the relationship between air hubs and tourism destinations occurs within the context that places concomitantly have a diversity of nodal functions. In this sense, stopover destinations, in particular, can play a crucial role in the development of tourist destinations (Lohmann *et al.*, 2013). With passengers passing through hubs and gateway airport destinations, there is an increased opportunity to convert these passengers into tourists for their surrounding destinations.

Historically, on the one hand, competition appears to exist among those airlines that hold the longest air routes in the world, as this achievement carries the reputation of technological evolution of the aircraft itself, creating value for the consumer who is saving time. However, on the other hand, destinations that no longer serve as stopovers end up needing to reinvent themselves if their goal is to continue to attract tourists. Table 8.2 shows the trend of direct ULR flights.

With the growth of aviation since the 1950s, airports have become primary places in the economic and urban agendas. Until recently, the main air hubs were located in Europe, the United States, Australia and some countries of Asia (Japan and China). With new players like Singapore and Dubai emerging on the centre stage of aviation aeropolitics, the redefinition of global economic flows and, accordingly, the distribution of air traffic to/from other parts of the globe has shaped the air global aviation market and has led to the proliferation of new airport hubs. Between 1998 and 2008, the highest annual scheduled capacity growth has been observed in South and Central Asia (9.6%), followed by Eastern and Central Europe (7.9%) and North Africa and the Middle East (6.8%). In the same period, growth in North America was close to zero (Bowen & Cidell, 2011: 875). Accordingly, 'over the past decade, new "mega-airports" (which we can define as new airports costing at least $500 million) have been rare outside of the Middle East and Asia-Pacific' (Bowen & Cidell, 2011: 869).

In Asia, for instance, in the wake of national development policies, Southeast Asian countries such as Malaysia, Thailand, the Philippines and most prominently Singapore have invested heavily in the implementation of airport infrastructures, *pari passu* the internationalization of the region's airlines after the 1970s. In fact, 'liberalisation and airport construction comprise two of the most important means through which Southeast Asian governments have sought to influence, either directly or indirectly, the development of airline networks' (Bowen, 2000: 25). Both Singapore

**Table 8.2** The longest hauls

| Route | Airline | Aircraft | Distance (km) | Flight time[a] |
|---|---|---|---|---|
| Doha–Auckland | Qatar Airways | Boeing 777-200LR | 14,539 | 18h05m |
| Perth–London | Qantas | Boeing Dreamliner 787-900 | 14,498 | 17h20m |
| Dubai–Auckland | Emirates | Airbus A380 | 14,200 | 17h15m |
| Los Angeles–Singapore | United Airlines | Boeing Dreamliner 787-900 | 14,113 | 17h50m |
| Houston–Sydney | United Airlines | Boeing Dreamliner 787-900 | 13,833 | 17h30m |

Note: [a]Westbound, times can vary.
Source: The Australian (2018).

(Changi) and Bangkok airports evolved 'as the transit point on long-range intercontinental and transcontinental routes' (Bowen, 2000: 32).

Atatürk International Airport (IST) in Istanbul (Turkey) is also a noteworthy example, considering its rapid growth in the number of passengers, fuelled by the evolution of Turkish Airlines' figures over the last few years, as well as the development of tourism in Istanbul, taking advantage of its location as a transcontinental country between three continents (Asia, Europe and Africa). Indeed, with over 61 million passengers in 2015, IST has become 'a mega-hub that collect air traffic from intercontinental flights and redistributes its traffic […] as the base transfer point' (Cetin et al., 2016: 40). Coinciding with the intense growth of tourism in the city of Istanbul itself, this airport has become an important hub for connections between several continents, powered by transcontinental routes. Between 2003 and 2016, the number of destinations served from IST increased by 377%, from 60 to 286, with significant expansion on long routes connecting Oceania (Australia), South America (Brazil and Argentina) and Africa (several countries) (DHMI, 2017).

Dubai and Singapore are prime examples of destinations that have gained prominence as tourist destinations through a well-orchestrated combination of government policies which were underpinned by tourism initiatives, easy visa arrangements and airport facilities to welcome visitors (Lohmann et al., 2009; Tang et al., 2017). In such cases, 'Hubs were transformed into destinations by the complementary interaction of attractions, transport and accommodation sectors. Both used shopping "paradises" to pull visitors into staying' (Lohmann et al., 2009: 211).

To a much lesser extent, Chengdu (China), Addis Ababa (Ethiopia), Johannesburg (South Africa), São Paulo (Brazil) and many other airports in emerging economies also represent links of growing importance in the air network which have arisen over the last 30 years. Likewise, the presence of these megastructures – with the commercial and business apparatus attached to them – represents an opportunity for local development, in terms of logistics chains, development of specific sectors (catering, supplying services, etc.) and tourism. What is not exactly clear is the relevance of these airports – as transcontinental air hubs – for the development of tourism in the cities/regions. In fact, '[l]ittle is known about the transit component of tourism systems, including the transit hub, despite its indispensable role in connecting origin regions with destinations' (Tang et al., 2017: 54).

Although some doubts have arisen regarding the 'consistent cost efficiencies' (Wensveen, 2018) of ULR flights, recent developments have shown that these routes could become a trend. From a technological point of view, a new barrier was overcome with the non-stop 17-hour Qantas flight between Perth and London. What will happen to the stopover destinations which, to date and in various ways, have benefited from their air hub condition?

## Concluding Remarks: The Future

Overall, there are some challenges in relation to becoming successful regarding ultra-long range routes. The advantages of arriving faster need to be reconciled with the impacts on the health and discomfort of passengers and crew. As we have seen, the distances, and therefore the times, of travel are perceived and rationalized in different ways (as resources, accessibility and knowledge). Following the aircraft revolution and the development of the jet era (1950–1980), over the last few decades technological improvements have led to gains of scale (regarding fuel consumption, emissions, passenger comfort and profitability, among others).

## Future turning point 1: Future capacity

In this timeline, increasing passenger capacity seems to be a trend in frank consolidation. Compared to the first large-capacity aircraft (DC-8 and B707), the A380 can carry up to seven times more passengers, even if its commercial speed is similar. Conversely, a new speed paradigm is under development: if Virgin Galactic's *SpaceShipTwo* succeeds in reaching up to 3000 mph (Business Insider Nordic, 2017), today's ULR flight duration will drop dramatically. Thus, the speed factor is an issue that is about to be overcome shortly, although operational and commercial issues are yet to be tackled. Indeed, that would reshape the way we fly today, compressing times and shrinking even more relative distances.

ULR flights have increased in frequency, with several routes, operated by different airlines, all over the world. On the one hand, this suggests more direct connectivity between points of interest, in which the elimination – or decrease – of connections seems to be the most significant benefit. In turn, the physiological and psychological side-effects – to ensure sufficient safety for passengers, airlines, insurance companies and the aircraft industry in general – have not yet been adequately clarified.

## Future turning point 2: Airport futures

In spatial terms, the need for larger airports to support the operation of long-haul routes has become an opportunity for local, regional and even national development in several countries. If large and well-known European cities are self-sufficient as tourism destinations, other locations in emerging economies appear to directly benefit from their transit situation, as already shown by the cases of Singapore and the UAE (Dubai and Abu Dhabi). In turn, these countries are also the home of important and thriving airlines which, themselves, can be the point of reference for ULR flights. That is, their condition as stopover destinations can be converted into a new centrality as state policies and marketing strategies (tourism included) redefine the countries' position in the mosaic of global economic forces – including their airlines.

What still needs to be better studied is the willingness of passengers to forego the opportunities associated with stopover destinations. One cannot neglect the fact that, in the current scenario, new tourist profiles make up global tourism and aviation markets, so for holidaymakers, the possibility of adding value to their trips (with experiences and activities in stopover destinations) may continue to be an important element in the commercial strategy of airlines.

The new Qantas flight and other similar ones (Table 8.2) crown the efforts undertaken by airlines and the airline industry to provide more direct and shorter links to distant destinations. It is yet to be understood how a diverse set of global actors (destination stakeholders, local and national policymakers and many other sectors of this complex air industry chain) will behave. After all, an important question still deserves to be addressed: What is the commercial viability of non-stop ULR flights, in rapidly changing air and tourism global and regional markets?

## References

ABC (2018) Qantas extends Emirates deal, switches from Dubai to Singapore as Sydney-London stopover. See http://www.abc.net.au/news/2017-08-31/qantas-extends-emirates-partnership-and-ditches-dubai-london-st/8859790 (accessed 12 March 2018).

Airbus (2018) Airbus launches new Ultra-Long Range version of the A350-900. See http://www.airbus.com/newsroom/press-releases/en/2015/10/airbus-launches-new-ultra-long-range-version-of-the-a350-900.html (accessed 29 January 2018).

Bowen, J. (2000) Airline hubs in Southeast Asia: National economic development and nodal accessibility. *Journal of Transport Geography* 8 (1), 25–41.

Bowen, J.T. and Cidell, J.L. (2011) Mega-airports: The political, economic, and environmental implications of the world's expanding air transportation gateways. In S.D. Brunn (ed.) *Engineering Earth: The Impacts of Megaengineering Projects* (pp. 867–877). Amsterdam: Springer Netherlands.

Business Insider Nordic (2017) Richard Branson says he's 6 months from going to space–but Mars belongs to Musk. See https://www.businessinsider.com/richard-branson-says-hes-going-to-space-but-mars-belongs-to-elon-musk-2017-10?r=US&IR=T (accessed 29 March 2018).

Cetin, G., Akova, O., Gursoy, D. and Kaya, F. (2016) Impact of direct flights on tourist volume: Case of Turkish Airlines. *Journal of Tourismology* 2 (2), 36–50.

Cohen, S.A. and Gössling, S. (2015) A darker side of hypermobility. *Environment and Planning A* 47, 1661–1679.

Cooper, C. and Hall, C.M. (2008) *Contemporary Tourism*. Oxford: Butterworth Heinemann.

DHMI (2017) *General Directorate of State Airports Authority Turkey*. See https://www.icao.int/MID/Documents/2017/Aviation%20Data%20and%20Analysis%20Seminar/PPT10%20-%20Turkish%20DHMI.pdf (accessed 5 April 2018).

Griffiths, R.F. and Powell, D. (2012) The occupational health and safety of flight attendants. *Aviation, Space, and Environmental Medicine* 83 (5), 514–521.

Holmes, A., Al-Bayat, S., Hilditch, C. and Bourgeois-Bougrine, S. (2012) Sleep and sleepiness during an ultra long-range flight operation between the Middle East and United States. *Accident Analysis & Prevention* 45, 27–31.

Larsen, G.R. (2017) Representations of distance: Differences in understanding distance according to travel method. *Journal of Spatial and Organizational Dynamics* 5 (4), 425–442.

Lohmann, G., Albers, S., Koch, B. and Pavlovich, K. (2009) From hub to tourist destination: An explorative study of Singapore and Dubai's aviation-based transformation. *Journal of Air Transport Management* 15 (5), 205–211.

Lohmann, G., Castro, R. and Fraga, C. (2013) *Transportes e Destinos Turísticos: Planejamento e gestão*. Rio de Janeiro: Elsevier.

Martinussen, M. and Hunter, D.R. (2017) *Aviation Psychology and Human Factors*. Boca Raton, FL: CRC Press.

Qantas (2018) *Qantas Airways*. See https://www.qantas.com/br/en.html (accessed 6 March 2018).

Rimmer, P.J. (2005) Australia through the prism of Qantas: Distance makes a comeback. *Otemon Journal of Australian Studies* 31, 135–157.

Rodrigue, J.-P., Comtois, C. and Slack, B. (2017) *The Geography of Transport Systems*. London: Routledge.

Spasojevic, B., Lohmann, G. and Scott, N. (2018) Air transport and tourism: A systematic literature review (2000–2014). *Current Issues in Tourism* 21 (9), 975–997.

Tang, C., Weaver, D. and Lawton, L. (2017) Can stopovers be induced to revisit transit hubs as stayovers? A new perspective on the relationship between air transportation and tourism. *Journal of Air Transport Management* 62, 54–64.

Taylor, G. (1919) Air routes to Australia. *Geographical Review* 4, 256–261.

The Australian (2018) Qantas hopes longest haul will be a dream run. See https://www.theaustralian.com.au/business/aviation/qantas-hopes-15000km-perth-to-london-haul-will-be-a-dream-run/news-story/b06e74e6afddacd7762015e973138e78 (accessed 23 March 2018).

Weber, M. and Williams, G. (2001) Drivers of long-haul air transport route development. *Journal of Transport Geography* 9 (4), 243–254.

Wensveen, J. (2018) *Air Transportation: A Management Perspective*. London: Routledge.

# 9 Forever Young and New: Cruise Tourism

Wendy London and Wallace Farias

## Introduction

Little did Cleopatra know that when she boarded her barge for a cruise down the Nile around 50 BC that she might well have enjoyed the world's first known leisure cruise. A solo, royal passenger on an ornate barge triggered an industry that today caters to more than 26 million passengers cruising on approximately 315 vessels (Cruise Market Watch, 2018) and contributes more than US$126bn to the world's economy (F-CCA, 2018). Cruising continues to be the fastest growing sector within the leisure tourism industry (MacNeill & Wozniak, 2018). That growth can be attributed to two paradigmatic factors. First, leisure cruising is an industry which has embraced, and continues to embrace, technological innovation. Technology continues to change the operational characteristics of passenger ships as well as the nature and quality of passengers' experience at sea. Secondly, the industry continues to demonstrate its remarkable nimbleness by responding to challenging global events. Economic downturns, war and terrorism are just some of the incidents that have tested and ultimately proven the cruise industry's resiliency, i.e. its ability to reinvent itself and grow even stronger (Branchik, 2014). This chapter first identifies the milestones that have defined the evolution of the cruise industry and then speculates as to how the cruise industry might respond to future challenges.

## Historical Turning Points

### Turning point 1: The birth of leisure cruising

It is highly likely that the merchant galleys in the Middle Ages often carried passengers seeking adventure alongside their cargo (Braudel, 1972). However, it was not until 1817 that the Black Ball Line, which was contracted by the government to transport mail and newspapers, also started advertising regular, fixed-date sailings for passengers between Liverpool and New York. The Black Ball Line therefore became the first passenger line to offer regularly scheduled services (Bosneagu *et al.*, 2015). Its majestic, medium-sized packet ships would take at least 23 days from New York to Liverpool and at least 40 days for the return voyage, offering passengers all that the North Atlantic had to offer.

It was not until steam ships were deployed for passenger transport, though, that the first shoots of the modern leisure cruising industry can be detected. The Peninsular and Oriental Steam Navigation Company (P&O) also offered mail and passenger services, but by 1822 steam ships were used for their voyages between England and the Iberian Peninsula (Napier, 1990). P&O and other companies deployed steam ships to carry both mail and passengers across the North Atlantic (Geels, 2002). The most high profile of these companies was the British and North American Royal Mail Steam Packet Company, which became Cunard (Sloan, 1992) – later famous for its regal ocean liners that transported passengers from Dover to New York. The first known effort to advertise cruising with all of the imagery associated with leisure cruising surfaced in 1833, when the *Francesco I*, a Sicilian vessel, advertised a three-month Mediterranean cruise to Taormina, Catania, Syracuse, Malta, Corfu, Patras, Delphi, Zante, Athens, Smyrna and Constantinople (Koteski *et al.*, 2016). Mediterranean cruising became popular.

The first company that appeared to convey a unique selling proposition (USP) in relation to customer comfort and elegance was Cunard. On 4 July 1840 Cunard's first ship, *Britannia*, sailed from Liverpool to Boston with a cow on board to ensure that the passengers had fresh milk (Maxtone-Graham, 1972). This cow would have had no inkling that she had started a trend, a trend which continues to both define and be symbolic of a highly competitive cruise sector that sells images of exemplary service and luxury. She can be considered to be the first of many cows which have proven to be lucrative for the cruise lines as the cruise lines continue to overtake their competitors by offering ever more elaborate and technologically advanced services.

The official start of the modern leisure cruise era, though, is generally thought to have occurred in 1844 when the P&O Steam Navigation Company advertised tours to the Mediterranean from Southampton (Artmonsky, 2012). It also appears that P&O commissioned the first cruise critic when it invited British poet, William Makepeace Thackeray, to travel free on three of those voyages and write about his experiences (Artmonsky, 2012). Soon thereafter, the modern cruise atlas was born (see, for example, Holland America Line, 2018; Princess Cruise Lines, 2017), promoting cruises for pleasure. Services were later extended to India, the Orient, Australia and New Zealand.

The trend started by Cunard's cow continued in the 1850s and 1860s. Significant enhancements were made to passenger comfort (Maxtone-Graham, 1972). No longer did passengers share their vessels with mail or cargo but, instead, they could take advantage of greater deck space.

Passenger vessels began to offer entertainment. Undoubtedly the single most important enhancement during this time was the installation of electric lights. In 1889 the *SS Valetta* became the first ship to use artificial lighting (Koteski *et al.*, 2016). During the 1880s and 1890s, cruising became a popular holiday option, commended for its health benefits, with Mark Twain (in *The Innocents Abroad*, 1869) and the *British Medical Journal* (BMJ, 1899) recognizing the value of spending days at sea while doing nothing. However, this curative environment was tempered by the waves of immigration that were heading to the United States. Passenger ships – luxurious, leisure ocean liners – also carried immigrants in what was known as steerage class. Immigrant passengers provided their own food and slept in the hold wherever they could find room (Branchik, 2014; Maxtone-Graham, 1972).

While electric lighting alone had a profound effect on leisure cruising, the new era of industrialization made bigger, faster and even more luxurious ships possible, leading to robust growth in the industry. In 1880 the *SS Ravenna* became the first ship ever to be built with a totally steel superstructure (Koteski *et al.*, 2016). Advances in marine engineering made it possible to build more extravagant, sumptuous ships which offered such facilities as swimming pools, gymnasia, opulent theatres and expansive dining rooms where passengers no longer had to share tables (Maxtone-Graham, 1972). Industrialization also meant that more people could afford taking a sea-going holiday, both in terms of money and of leisure time.

The year 1896 marked the deployment of three European-owned luxury liners for transportation across the Atlantic, but it was not until the start of the 20th century that a vessel was built exclusively for luxury cruising, the *Prinzessin Victoria Luise*, owned by the Hamburg-America Line (Polat, 2015). Cruise lines soon began to compete for passengers, offering them a choice of speed, luxury or both. German shipping lines were first to market super-liners, designed as ornate, floating hotels.

Speed was Cunard's objective with the *Mauritania* and the *Lusitania*. However, luxury was not forgotten, as passengers were required to dress for dinner, and romance was one of the key images embodied in their advertising campaigns. On the other hand, luxury was the key driver for the *Titanic* and the *Olympic*, owned by the White Star Line. Unfortunately, speed meant that passengers seeking fast transit times had to share dining tables and forego the benefits and enjoyment of large public rooms. Luxury meant that passengers could swim or play a set of tennis.

The popularity of cruising gave rise to competition between the cruise lines, an increase in the number of luxury liners for transport and the start of repositioning cruises. Repositioning is a fundamental, basic operational strategy of the modern cruise industry. Cruise lines such as the Hamburg-America Line sent their transatlantic ships on long Southern cruises during the Northern Hemisphere's winter. Some ships were built specifically to operate as transatlantic liners during the clement weather in the Northern Hemisphere and cruise ships for winter cruising in warmer waters (Karuk & Chala, 2015).

Unfortunately, this bold start to modern cruising suffered two disastrous events. In 1912 the *Titanic* hit an iceberg and sank, and in 1916 the *Britannic* hit a mine and sank. However, despite these events and the winds of war described in the next section, the cruise industry embodied the spirit of Molly Brown and proved to be unsinkable.

### Turning point 2: The winds of war, waves of immigration and the economic downturn

War had a profound effect on the cruise industry. Ships which had carried passengers in great style in the preceding decades were appropriated for troop transport in both world wars (Chalkiti & Sigala, 2006; Maxtone-Graham, 1972). During WWI no new cruise ships were built. Following WWI, German superliners became the currency of reparations, given to both Great Britain and the United States (McCutcheon, 2009). Between the wars, the industry demonstrated its resiliency. The period from 1920 to the 1940s was considered to be the glamour era for transatlantic ships, with rich and famous American tourists replacing immigrants. This growth slowed with the Great Depression when fortunes disappeared and passenger numbers fell. It also resulted in a period of consolidation for the industry. In 1934, the White Star Line merged with Cunard to form Cunard White Star with a fleet of 25 vessels.

At the start of WWII, ships were once again converted into troop transports, putting an end to transatlantic cruising until after the war. British Prime Minister Winston Churchill credited the technologically advanced *Queen Mary* and *Queen Elizabeth* with shortening WWII by a year because of their speed (Chalkiti & Sigala, 2006). The two *Queens* took troops from the United States to their European theatres of war, potentially leaving the ships empty on their return voyages. However, sailing empty was never going to be an acceptable option, so on their westward journeys the ships became lifelines for refugees seeking new lives in the Unites States and Canada. Because there were no American-built ocean liners during this period, the United States Government subsidized the construction of luxury vessels which could be converted to troop carriers if needed (Lane, 2001). In 1947, Cunard launched *Caronia*, reputedly the first purpose-built, tranatlantic cruise ship (Chalkiti & Sigala, 2006).

### Turning point 3: The impact of air travel

The peace dividend arising out of war invariably includes advances in technology. In the case of the cruise industry, these advances had a potentially dampening effect. Regular travel by sea commenced again after

WWII, but in 1958 air transport became the preferred method of travel across the Atlantic (Maxtone-Graham, 1972). On 4 October 1958, a De Havilland Comet 4 owned by the British Overseas Airline Corporation (BOAC) became the first jet aircraft to fly across the Atlantic, from London to New York (with a stopover in Gander, Newfoundland) (Miller, 2011). Three weeks later, Pan American World Airways took delivery of the first Boeing 707 aircraft and flew it in the reverse direction, from New York to Paris (Miller, 2011).

Thus, travellers took to the skies, leaving passenger ships empty and some passenger lines bankrupt. Passenger ships were sold, surrendered to the elements or scrapped (Maxtone-Graham, 1972). More people flew across the Atlantic than crossed it on ocean liners. By 1960, jet travel accounted for 70% of the transatlantic market. By 1970, only 4% of travellers went by sea (Stansfield, 1977). Bruised but undaunted, the cruise lines once again reinvented themselves, and in the 1960s offered vacation trips to the Caribbean (Wilkinson, 2006).

### Turning point 4: Oil shocks and the reinvention of ocean cruising

The next shock came from the oil embargo of 1973–1974. Some passenger lines continued to prosper but others disappeared. Many ocean liners were sold or scrapped (Miller, 1995). The cruise lines were also thwarted by new regulations coming into force, making it difficult to retrofit older ships to comply with them. However, at the start of the 1970s, Royal Caribbean launched its first ship, the *Song of Norway*, which was the first ship specifically designed for warm weather cruising (Valenti, 1998). In 1977, cruising took off through the power of television. *The Love Boat* first aired on 24 September 1977, running until 24 May 1986. The legend of Captain Stubing continues, creating images of high jinks on board and warm tropical settings onshore.

The resurgent popularity of the cruise industry led to substantial investment by the cruise lines. Ship design embraced the new style of cruising, setting the stage for the design of today's cruise ships. For example, passenger cabins were now designed for comfort. Cruising became casual, with on-board entertainment. No longer was the focus on point-to-point destination travel. This casual popularity also marked the end of the opulent and romantic era of cruising. The great ocean liners of the 1930s and 1940s were no longer economic to operate because of their high fuel consumption and deep draught, preventing them from entering shallow ports (Karuk & Chala, 2015). The last transatlantic voyage took place in 1986, except for Cunard, which blended the ethos of transatlantic ocean liners with modern cruise line features and facilities. International celebrities were invited to provide entertainment. A voyage on board Cunard's *Queen Elizabeth 2* ocean liner was 'advertised as a vacation in itself' (Mason, 2017). During this period, many new ships were built, and older ships were repurposed for leisure cruising. Among these was the ocean liner, the SS *Norway* (formerly the SS *France*), the largest passenger ship in the world at that time at 316 m (1035 feet) and carrying approximately 2000 passengers. The *Norway*, the first Sovereign-class cruise ship, found new life as the Caribbean's first mega cruise ship (Chapman, 1993), sporting a multi-storey atrium, glass lifts and the first entire passenger deck to have private balconies (Mason, 2017).

### Turning point 5: Coming of age

The cruise lines came of age in the 1980s, with many new ships being built or refurbished from older ships, new cruise lines being established, and others changing their names through consolidation or mergers. Cruise ships became increasingly self-contained, capturing passenger shopping revenue with new on-board shops and other facilities. Additionally, Royal Caribbean was the first cruise line to quarantine even more passenger revenue by acquiring a private island. Cruise passengers could now enjoy a seamless ship-to-shore experience without wandering into the nearest town and spending their money there. Golden times returned again, until the temporary shock of the Iran-Iraq War, when oil prices increased to US$35 a barrel. And once again, two ships, Cunard's *Queen Elizabeth 2* and P&O's *Canberra*, were appropriated for war duty in 1982, as the British government moved troops to fight in the Falklands War (between Argentina and Britain). The power of television was once again recognized when, in 1984, Carnival Cruise Lines famously used Kathie Lee Gifford to advertise cruises for the very first time on US television.

Terrorism reached the cruise industry on 7 October 1985 when four members of the Palestinian Liberation Front hijacked the *Achille Lauro*, killing a 69-year-old disabled American, Leon Klinghoffer, and throwing him overboard. However, little effect on the cruise industry was felt, and the industry continued to grow. A feeling of safety and security – even during an intensive era of airplane hijackings and land-based terrorist attacks – characterized, and continues to characterize, the cruise industry.

### Turning point 6: The floating resort

The floating resort was born in the 1990s. Until the early 1990s, the SS *Norway* was the largest ship at sea. However, during this period, *Monarch of the Seas* was launched by Royal Caribbean International. Although smaller than the *Norway* at less than 74,000 tons, *Monarch of the Seas* carried 2744 passengers. In 1995, Princess Cruise Line's *Sun Princess* became the largest cruise ship at 77,000 tons. In 1996, Carnival Cruise Lines launched the 101,353 ton *Carnival Destiny*. Although the largest ship to be put into service, she only carried 2642 passengers, fewer than the *Sun Princess*. In that same year, Royal Caribbean announced that it had signed contracts for two 130,000 ton ships, thereby further acknowledging the popularity of the industry.

Thus, ships became small towns. They became even more self-contained than was required by their structure, offering services, features and facilities that were more commonly associated with being on land and not in the middle of vast oceans. Ice-skating rinks, rock-climbing walls, expansive spa facilities, full-service casinos, bowling alleys and shopping malls began to appear (see, for example, Kwortnik, 2008), giving rise to competition within the industry (Paris & Teye, 2011) and between land-based and floating resorts (Kester, 2003). Internet cafes, followed by wireless internet in all areas of the ship including in passenger cabins, meant that passengers no longer had to venture onshore looking for services. This growing 'self-sufficiency' of ships potentially poses a threat for destinations seeking to capture cruise passenger revenue (Braun & Tramell, 2006; London, 2012).

However, it was not only the addition of expansive passenger-oriented facilities and features that signalled a paradigmatic change in cruising in the 1990s. Twentieth-century technology and innovation, the descendants of the previous century's Age of Industrialization, gave rise to many technical enhancements which have set the baseline for the 21st century's ships. Navigational features such as GPS and Voyage Data Recorders and the more fuel efficient Azipod propulsion systems are among the technical advancements implemented by the cruise lines.

Safety also came very much to the forefront. A bundle of industry-wide measures was introduced in the 1990s to improve passenger and crew safety and environmental management in the face of the growth of the industry. Amendments to industry codes such as SOLAS[1] (IMO, 1974), MARPOL[2] (IMO, 1973) and STCW95[3] (IMO, 1978) introduced new safety and pollution control measures, and better training standards. In 2001, guidelines for shore power were also introduced, with cruise lines starting to equip their ships with the necessary facilities to connect (Sulligoi *et al.*, 2015).

### Turning point 7: Terrorism and economic downturn

The first decade of the 21st century represented another significant watershed period for the cruise industry. The tragic events of 11 September 2001 and the Global Financial Crisis (GFC) of 2007–2008 both had a significant impact on the cruise industry. Counter-intuitively, though, both events resulted in a stronger and bigger cruise industry. For example, the initial slowdown in tourism following 9/11 was reversed when 30 new ports were created in the United States (F-CCA, 2009; see also Wilkinson, 2006). This meant that millions of risk-adverse North Americans who were afraid to fly could drive to their cruise ships within one day. In addition, far-flung destinations such as New Zealand and Australia were perceived to be safe, thereby attracting cruise passengers from within those regions who preferred not to fly long distances. During this period, new cruise lines emerged, while others merged.

Six years after 9/11, the GFC once again threatened the viability of the cruise industry. However, the industry's nimbleness, flexibility and ability to withstand economic as well as geopolitical shocks was once again proven (see, for example, Müller & Wieckowski, 2017). Passenger numbers continued to grow, with the cruise industry offering deeply discounted fares which attracted people who had never even contemplated a cruise. The GFC also saw the cruise lines respond by attracting new generations of cruisers, and multigenerational groups cruising together. As a result of the expansion into non-traditional demographic groups, at least one cruise company (Royal Caribbean Cruise Lines) began to deploy increasing features and technology-rich mega cruise ships which could be mistaken for land-based resorts. New passengers continue to be attracted to cruising from the growing middle classes in India, China and Russia (Branchik, 2014), with some ships being retrofitted to cater to Asian tastes.

### Turning point 8: The widening of the Panama Canal

The semi-annual migration of the largest cruise ships from the Northern to the Southern Hemisphere and vice versa could have been severely impeded if not for one extraordinary engineering feat, the widening of the Panama Canal. Repositioning floating cities such as the 228,081 GRT,[4] 6680-passenger *Symphony of the Seas* would have become long, protracted and expensive had the Panama Canal not been widened. Moving post-Panamax ships (i.e. those too big to transit the original canal) from the East Coast of the United States to the increasingly popular Australasian cruise region would have meant long, expensive and somewhat tortuous journeys around Cape Horn. Now, the future's mega ships can transit the Panama Canal, offering attracting and relaxing repositioning cruises between the two highly seasonal cruising regions. Size no longer matters.

### Future Turning Points

### Future turning point 1: Climate change

Cruise itineraries are increasingly being altered mid-cruise to avoid ferocious weather systems, but terrain changes to coastal and island destinations caused by rising sea levels are likely to have a much more profound impact on cruising. Well-constructed concrete and steel piers may or may not survive the inexorable march of the sea (Koetse & Rietveld, 2009), but the wooden jetties which greet cruise passengers on low-lying islands are likely to disappear, as may entire coastal villages. Also, the ease in repositioning ships to the Southern Hemisphere made possible by the widening of the Panama Canal is welcome in the short term, but may result in yet another reinvention of the industry in the future. While the construction of ever-bigger, technologically feature rich mega cruise ships potentially pose a threat to destinations as passengers prefer to stay on board their floating resort, these mobile resorts may pose an alternative means for both the traditional land-based tourists and the cruise passengers to have their 'island' holiday. In other words, the traditional, *land-based* island-hopping tourists may find themselves as passengers on these resort-like *floating* tropical islands when low-lying islands become unviable for tourism. Within this scenario, the ship becomes the island in the middle of the South Seas.

Another scenario can also be considered. Extreme weather will disrupt port operations (Koetse & Rietveld, 2009) and result in significant changes to navigation. For example, port entrances may become increasingly inaccessible by sand bars, high winds may make transit through narrow harbour entrances more and more dangerous, and higher swells may make it difficult or even impossible to keep ships stable enough for passengers using gangways. These events may make the operation of large ships untenable. Moreover, very large mega ships present a particularly exaggerated threat to fragile environments. Therefore, the 'golden age' of mega liners may wane, with an increasing deployment of smaller ships. In fact, this trend appears to have started already, with more and more passengers expressing a preference for smaller cruise ships which offer more opportunities for enrichment and exploration and less emphasis on water slides and glitz (Harpaz, 2018).

### Future turning point 2: Conflict

Continuing global instability poses a threat to cruise tourism. However, if the past is any guideline, the industry will continue to reinvent itself or, perhaps more pragmatically, adopt strategies to cope with the threat. Terrorism, conflict and local tensions can cause itineraries to be changed at short notice, and sometimes on a longer term basis (e.g. the Eastern Mediterranean). The widening of the Panama Canal may prove to be a key ingredient in any strategy given that ships can easily be redeployed to calmer regions. Cruise ships could once again be requisitioned to carry troops. Peacetime dividends could result in the creation of a class or design of ship which eludes today's crystal ball gazers. It is impossible to try to reach any conclusion about what the cruise industry would look like after any protracted, devastating epoch of 21st century conflict; however, given the industry's past track record of reinvention and adaptation, someone, somewhere, will sow the seeds of the next version of leisure cruising.

## Future turning point 3: Technology

Cruise ships are already sporting advanced technology. New propulsion systems including hybrid systems and wind power, on-board recycling plants, new hull paints which lower water resistance and increase speed, and greater use of lighter and recycled materials for ship fit-outs are just some of the advanced technologies that are being implemented to make cruise ships more environmentally friendly (Witthaus, 2018). On board many of these ships, futuristic technology greets passengers, contributing to their cruise escape; in many respects this makes the 'floating island' concept entirely possible. For example, cruise ship design is starting to blend in more with the sea around it. One new ship is dispensing with balconies in favour of cabins with the same design illusion as infinity pools, and movable decks which can take advantage of the time of day and the weather (Fiji Hotel and Tourism Association, 2017). Others offer underwater bars and submersibles (Goldsbury, 2017). Still others are offering virtual excursions, cabins which mimic other environments such as outer space, and dining experiences which take passengers to onshore destinations to match their food through a combination of augmented and virtual reality (Kim, 2017). At least one superyacht cruise ship designer has already produced a concept for a floating island, complete with tiki huts, palm trees and a water-spouting volcano (Souza, 2016).

Is the future of cruising 'non-cruising'? Even today, virtual and augmented reality can offer cruisers the 21st century equivalent of armchair travelling. It makes practical sense, but does it make emotional sense, and does it satisfy the cruise passenger's desire for authenticity, relaxation and the opportunity to purchase mementos of their 'trip' (Guttentag, 2010)? Or is it a gimmick which is reminiscent of the fantasy worlds being created on the ships themselves? Or is it the new cruise ship of the next Millennium?

## Conclusion and Future Growth

From a solo passenger enjoying a leisurely cruise on a barge down the Nile to the more than 26 million passengers who cruise today, cruise tourism has demonstrated consistent, strong growth. The prospect for greater growth is clear. New cruise lines, new ships, new ship designs, new home ports, new destinations and new markets are all factors that signal substantial future growth. The number of new cruise ships alone is a key indicator of the growth in the market. Cruise line capacity at the end of 2018 is forecast to be 537,000, distributed among 314 ships, with 37 more ships to be deployed between 2018 and 2020 (Cruise Market Watch, 2018). These new ships represent an increase in capacity of 99,895 berths, adding US$11.7bn in annual revenue to the industry during the same period (Cruise Market Watch, 2018).

Underpinning much of this growth is the changing demographic of cruise passengers and, to an increasing extent, a response to the need to deploy environmentally sensitive and responsible ships. In terms of the changing demographic, no longer are cruise ships the sole domain of older, well-heeled passengers from North America. Younger, active, environmentally conscious cruisers from a greater diversity of source markets are driving sustainable change and growth. New ships and new cruise lines sporting such features as active cruise experiences, hybrid energy plants and a more intimate cruising experience are making their debut. North American cruise passengers are being joined by the growing middle classes in Southeast Asia and India, with a growing number of ships being homeported (based) in these markets.

The growth in the number of ships is generating concomitant growth on land. Coastal hamlets, towns and cities are rapidly being added to the 1200 cruise ports currently operating. This growth is being driven both by the cruise lines and by potential destinations. The cruise lines continuously seek to refresh their itineraries by adding new destinations to attract repeat cruise passengers as well as new ones. Coastal communities of all sizes increasingly seek to partake in the revenues generated by cruise tourism. In fact, the interest in cruise tourism is so pervasive that there are many instances where the number of cruise passengers visiting a destination outnumbers the local population. Fundamental to this expansion is a destination's ability to offer interesting activities onshore, and at least a minimal level of amenities. However, passengers on small and exploration ships tend to want to pursue adventure and relatively unexplored destinations, thereby inspiring the addition of undeveloped and small ports.

There are no signs that this growth will abate. Even without the cruise lines' aggressive international expansion, only a quarter of the US population has ever taken an ocean cruise (Cruise Market Watch, 2018). Given that there are still a lot of people to attract to cruising, and as ships respond to consumer trends and implement the latest technologies to offer totally new cruising experiences, the future for cruising remains stellar.

## Notes

(1) The International Convention for the Safety of Life at Sea.
(2) The International Convention for the Prevention of Pollution from Ships.
(3) The International Convention on Standards of Training, Certification, and Watchkeeping for Seafarers.
(4) GRT refers to gross registered tonnage, the measurement by which a ship's internal volume capacity is measured (i.e. not weight).

## References

Artmonsky, R. (2012) *P&O: A History*. London: Bloomsbury.
Bosneagu, R., Coca, C.E. and Sorescu, F. (2015) Management and marketing elements in maritime cruises industry: European Cruise Market. *EIRP Proceedings* 10. Galati: Danubius University Press.

Branchik, B. (2014) Staying afloat: A history of maritime passenger industry marketing. *Journal of Historical Research in Marketing* 6 (2), 234–257.

Braudel, F. (1972) *The Mediterranean in the Age of Phillip II*, Vol. 1. Berkeley, CA: University of California Press.

Braun, B.M. and Tramell, F. (2006) The sources and magnitude of the economic impact on a local economy from cruise activities: Evidence from Port Canaveral, Florida. In R. Dowling (ed.) *Cruise Ship Tourism* (pp. 280–289). Wallingford: CABI.

BMJ (1899) The therapeutic value of ocean voyages. *British Medical Journal* 2 (2027), 1301–1302.

Chalkiti, K. and Sigala, M. (2006) Profiling Samuel Cunard: An assessment of his contributions to the contemporary cruise sector. *Journal of Hospitality & Tourism Education* 18 (3), 5–14.

Chapman, A. (1993) Watch the Port of Miami. *Tequesta* 53, 7–30.

Cruise Market Watch (2018) Capacity. See https://www.cruisemarketwatch.com/capacity/ (accessed 15 July 2018).

Fiji Hotel and Tourism Association (2017) *New Cruise Ship Reimagines What it Means to be Outdoors*. See https://fhta.com.fj/new-cruise-ship-reimagines-means-outdoors/ (accessed 4 August 2019).

F-CCA (2009) *Cruise Industry Overview – 2009*. Pembroke Pines, FL: Florida-Caribbean Cruise Association. See http://www.f-cca.com/downloads/2009-cruise-industry-overview-and-statistics.pdf (accessed 2 August 2018).

F-CCA (2018) *Cruise Industry Overview – 2018*. Pembroke Pines, FL: Florida-Caribbean Cruise Association. See http://www.f-cca.com/downloads/2018-Cruise-Industry-Overview-and-Statistics.pdf (accessed 2 August 2018).

Geels, F.W. (2002) Technological transitions as evolutionary reconfiguration processes: A multi-level perspective and a case-study. *Research Policy* 31 (8–9), 1257–1274.

Goldsbury, L. (2017) Cruise lines get adventurous with submersibles, go-karts, sky diving. *The Australian*, 25 February. See https://www.theaustralian.com.au/life/travel/cruise-lines-get-adventurous-with-submersibles-gokarts-sky-diving/news-story/99244807cce9c97bcdaddf2fcf8163f7 (accessed 2 August 2018).

Guttentag, D.A. (2010) Virtual reality: Applications and implications for tourism. *Tourism Management* 31 (5), 637–651.

Harpaz, B. (2018) Boom in small ship cruising: 'It's not about the slides'. *Financial Post*, 30 January. See http://business.financialpost.com/pmn/business-pmn/boom-in-small-ship-cruising-its-not-about-the-slides (accessed 2 August 2018).

Holland America Line (2018) *Cruise Atlas: April 2018–April 2019*. Seattle: Holland America Line. See https://digital.cenveomobile.com/publication/?i=419531&p=&pn=#{%22issue_id%22:419531,%22page%22:0&rcub; (accessed 3 August 2018).

IMO (1973) *International Convention for the Prevention of Pollution from Ships (MARPOL)*. London: International Maritime Organisation. See http://www.imo.org/en/about/conventions/listofconventions/pages/international-convention-for-the-prevention-of-pollution-from-ships-(marpol).aspx (accessed 3 August 2018).

IMO (1974) *International Convention for the Safety of Life at Sea (SOLAS)*. London: International Maritime Organization. See http://www.imo.org/en/About/Conventions/ListOfConventions/Pages/International-Convention-for-the-Safety-of-Life-at-Sea-(SOLAS),-1974.aspx (accessed 3 August 2018).

IMO (1978, 1995) *International Convention on Standards of Training, Certification, and Watchkeeping for Seafarers (STCW)*. London: International Maritime Organization. See http://www.imo.org/en/About/conventions/listofconventions/pages/international-convention-on-standards-of-training,-certification-and-watchkeeping-for-seafarers-(stcw).aspx (accessed 3 August 2018).

Karuk, M. and Chala, K. (2015) *History of Cruise Tourism*. Kiev: National University of Food Technologies.

Kester, J.G. (2003) Cruise tourism. *Tourism Economics* 9 (3), 337–350.

Kim, S. (2017) Cruise ship of the future: Facial recognition, fake views and virtual excursions. *Traveller*, 15 November. See http://www.traveller.com.au/cruise-ship-of-the-future-facial-recognition-fake-views-and-virtual-excursions-gzktxr (accessed 3 August 2018).

Koetse, M.J. and Rietveld, P. (2009) The impact of climate change and weather on transport: An overview of empirical findings. *Transportation Research Part D: Transport and Environment* 14 (3), 205–221.

Koteski, C., Dimitrov, N., Jakovlev, Z. and Petrov, N. (2016) Cruise as part of water transport-development and challenges. In *Proceeding Papers: Fifth International Scientific Conference 'Geographical Sciences and Education'*, Shumen.

Kwortnik, R.J. (2008) Shipscape influence on the leisure cruise experience. *International Journal of Culture, Tourism and Hospitality Research* 2 (4), 289–311.

Lane, F.C. (2001) *Ships for Victory: A History of Shipbuilding Under the US Maritime Commission in World War II*. Baltimore, MD: JHU Press.

London, W.R. (2012) Economic risk in the cruise sector. *Études Caribéennes* 18.

MacNeill, T. and Wozniak, D. (2018) The economic, social, and environmental impacts of cruise tourism. *Tourism Management* 66, 387–404.

Mason, D. (2017) Introduction to cruise ships. See http://www.san-shin.org/files/Introduction_to_Cruise_Ships.docx (accessed 10 August 2018).

Maxtone-Graham, J. (1972) *The Only Way to Cross*. New York: Barnes & Noble.

McCutcheon, C. (2009) *Port of Southampton*. Stroud: Amberley Publishing.

Miller, L. (1995) Five years in the rise of the modern cruise industry. *Hospitality Review* 13 (1), 33–40.

Miller, W.H. (2011) *The Last Atlantic Liners: Getting There is Half the Fun*. Stroud: Amberley Publishing.

Müller, D. K. and Wieckowski, M. (2017) *Tourism In Transitions: Recovering Decline, Managing Change*. Switzerland: Springer.

Napier, C.J. (1990) Fixed asset accounting in the shipping industry: P&O 1840–1914. *Accounting, Business & Financial History* 1 (1), 23–50.

Paris, C.M. and Teye, V. (2011) Segmenting the cruise market: An application of multiple correspondence analysis. *Tourism Analysis* 16 (5), 617–621.

Polat, N. (2015) Technical innovations in cruise tourism and results of sustainability. *Procedia – Social and Behavioral Sciences* 195, 438–445.

Princess Cruise Lines (2017) *Cruise Atlas: 2017–2018*. Santa Clarita: Princess Cruise Lines. See https://www.princess.com/downloads/pdf/hk_brochures/2017/2017-2018_EN_Cruise-Atlas.pdf (accessed 10 August 2018).

Sloan, E.W. (1992) Collins versus Cunard: The realities of a North Atlantic steamship rivalry, 1850–1858. *International Journal of Maritime History* 4 (1), 83–100.

Souza, B. (2016) Cruise ships of the future? *Cruise fever*, 27 December. See https://cruisefever.net/0120-cruise-ships-of-the-future/ (accessed 10 August 2018).

Stansfield Jr, C.A. (1977) Changes in the geography of passenger liner ports: The rise of the Southeastern Florida ports. *Southeastern Geographer* 17, 25–32.

Sulligoi, G., Bosich, D., Pelaschiar, R., Lipardi, G. and Tosato, F. (2015) Shore-to-ship power. *Proceedings of the IEEE* 103 (12), 2381–2400.

Twain, M. (1869) *The Innocents Abroad*. Hartford, CT: American Publishing Company.

Valenti, M. (1998) Luxury liners go green. *Mechanical Engineering* 120 (7), 72.

Wilkinson, P.F. (2006) The changing geography of cruise tourism in the Caribbean. In R. Dowling (ed.) *Cruise Ship Tourism* (pp. 170–183). Wallingford: CABI.

Witthaus, M. (2018) Introducing the first Cruise & Ferry green list. *Cruise&Ferry.net*, 2 April. See http://www.cruiseandferry.net/articles/introducing-the-first-cruise-ferry-green-list (accessed 11 August 2018).

# 10 From Muscles to Electrons: A Technological Look at the Futures of Energy, Transport and Tourism

Jonathan Hui

## Introduction

The history of tourism and human travel is often described in technological terms, in which developments in transportation gradually open up new forms of movement over larger and larger distances. In this chapter, I wish to build upon this history, yet with a focus on the development of energy systems and how energy is harnessed for moving goods, information and peoples around the world. Most importantly, I want to ask about what historical drivers there were for the technologies, the sociopolitical and economic contexts for the breakthrough, and how the advances changed society and paved the way for future innovations. The underlying interest is in transportation regimes and how they are reflected in landscapes, travel experiences and everyday technology.

Technology, to paraphrase historian Daniel Headrick (2009), consists of the many ways in which humans transform their environments to meet their needs. It opens doors to certain possibilities, while foreclosing others. These transformations have depended on different formations of the natural world, its resources and ecosystems, and of sociopolitical life as well. The range of possible starting points is numerous, from the evolution of bipedalism to the harnessing of fire for cooking and landscape control. This historical viewpoint will focus on developments since the emergence of *Homo sapiens* from Africa, with the bulk of the time spent on advances since the Industrial Revolution.

## Historical Turning Points

### Turning point 1: Sapiens, metabolism and movement

What are some of the historical drivers behind the evolution of human locomotion? This question is critical since tourism depends on human muscle power first and foremost, and represents the greatest technology we have yet evolved. The physiology and energy metabolism we have today is largely identical to that evolved by early humans two million years ago: an upright, bipedal, relatively hairless and slight-muscled hominid, in comparison to its great ape brethren. Like Neanderthals and *Homo erectus*, two of the closest relative species of humans, modern humans have a history of migration, tool-use and arguably language as well (Davis, 2018). What distinguishes *Homo sapiens sapiens* is the capacity for long-distance cooperative carnivorism, aided by a body that is incredibly efficient at sweating to disperse heat.

This remarkable adaptiveness is due, in part, to the development of muscle-enhancing technologies coupled with the ability to cooperate strategically in large groups. Hunting weapons such as spears, harpoons and bows greatly multiplied the amount of force applicable by muscles, all concentrated to a single point. Producing bows and arrows required sophisticated planning, specialization of skills and quality control, all competences advantageous to the later shift to settled agricultural life (Lombard & Haidle, 2012). Human metabolism, how much energy is expended through activity, is therefore an important facet and constraint of evolution and behavior. The ability to hunt and gather over diverse terrain and in groups meant that out human ancestors were born nomads, moving with the seasons and as food sources migrated.

### Turning point 2: Horse power

Domestication is the process of intentional and selective breeding of a species in order for them to fulfil a certain derived purpose. Wolves, more than 15,000 years ago, are considered as the first to have entered into this symbiosis with humans, transforming the horizons of possibility for animal–human partnerships (Fagan, 2015). Soon after came goats, pigs, sheep, cattle and horses, their grazing areas cordoned off and guarded by attentive dogs. The last two, cattle and horses, are of specific interest in changing the landscape of transportation and therefore tourism.

Oxen, castrated male cattle, were first domesticated around 4000 BC and were used primarily as draft animals on farms and for pulling heavy loads. Horses have been traced back to a similar period on the East Asian steppes, quickly spreading across the continent and into Europe for transportation and as working animals (Outram *et al.*, 2009). But what drove these people to consider domesticating horses? In terms of energy, a healthy horse consumes oats enough for six men, yet a day's work from a horse is equal in energy input to that of 10 men (Smil, 2017). The implications for the movement of people and goods from this change were wide ranging and gave groups able to tame them a dramatic advantage in war, trade and agricultural output. The Mongols under Genghis Khan were famed and feared for their prowess as horse archers, born and raised on the saddle and utilizing stirrups to fire arrows while retreating. Horses introduced into the Americas were similarly taken up by Native Americans, perfecting horse-riding and the use of guns (Headrick, 2010). Horses and oxen multiplied overland capacity for trade and movement, making land routes and long-distance journeys a new possibility.

### Pre-industrial travel

Travel in the Ancient World of Europe and Mesopotamia coincided with trade and pilgrimage. Perhaps best exemplified by the Roman Empire, vast network of roads were built across their territories for security and trade purposes, facilitating rapid movement of people, goods and ideas between Rome and its periphery provinces (Headrick, 2009). With the adoption of Christianity as the Roman state religion and the later emergence of Islam, pilgrimage became a dominant form of travel within the Judeo-Christian world. Spiritual and religious sites are major attractions and sites of prayer around the world, but it was the mass institutionalization of the practice by the Judeo-Christian religions that laid a foundation for modern tourism. The ancient Maya also developed networks of roads for trade and pilgrimage, making them a central part of their language and worldview. For them, walking along roads was akin to journeying along paths of history and spirit, representing 'not only a linear path from here to there, but also multiple, interwoven cycles of time that form the fabric of life' (Keller, 2009).

Pre-industrial tourism emerged from the Grand Tour of Europe among upper-class nobles and priests during the 16th century. It began with young, educated and landed men journeying to the great cities and sights of France, Italy and Germany, before declining as a cultural imperative with the rise of railways and steamboats. Two medieval inventions, the horseshoe and horse collar, greatly increased the amount of weight horses could pull, decreasing the costs of transport and allowing more widespread use of wagons and long-distance travel across Europe (Headrick, 2009). Key to this was the development of the post system, with horses kept at key intervals along a route for the highest efficiency. Extended through regular coach services in the 17th century and enhanced with new forms of capital contracts in the mid-18th century, vendors began to rent or sell carriages to tourists for the duration of their trips (Towner, 1985). Expansion systems of horse transportation meant that potential tourists could plan and create itineraries to a finer degree than ever before. City centers, like Rome and Naples, were transformed by seasonal influxes of grand tourists, with prices rising during high seasons and vice versa. By the 1830s, the Grand Tour was on the decline as young aristocrats turned to summer retreats while tourism was gradually opened up to middle-class people. However, in pioneering a mentality of self-development and solidifying travel routes of interest, it succeeded in setting the stage for the accelerated tourist industry to come.

### Turning point 3: Industrial innovations – steam power

First invented by Thomas Newcomen in the early 18th century, steam engines produce heat in order to boil water and use the resulting steam to power an assembly of pistons. They were dramatically improved and patented for commercial use by James Watt in 1769 before rising to wider prominence in the 19th century. Watt's engines were an order of magnitude jump in energy intensity, averaging five times the power output of watermills, three times that of windmills and 25 times that of a draft horse (Smil, 2004).

The advent of steam power ushered the Industrial Revolution, providing the necessary outlet for coal to be transformed into useful energy en masse. For the first time in human history, there was a challenge to the traditional primary energy sources of muscles, biomass (in the form of wood and charcoal), and wind and hydro power. However, the transition was a prolonged and uneven one, with draft animals, water wheels, windmills and steam engines all co-existing in Europe between the late 18th and mid-20th centuries (Smil, 2004). The distinctiveness of the steam engine lay in its sheer power output and modularity to different uses. Together with expansions in coal, iron and machine-making infrastructure, steam power found its most prominent uses in transportation such as trains and steamboats (Headrick, 2009). In 1807 US engineer Robert Fulton pioneered the steamboat on the Hudson River before starting a ferrying business, while two decades later British engineers George and Robert Stephenson would design a steam locomotive called the *Rocket*, capable of 30 mph and kicking off the growth of railways across the country.

### Turning point 4: Industrial innovations – combustion engines

Whereas coal became the key fuel for steam engines, oil and diesel became the choice for combustion engines. Whale oil was at first dominant for lighting in Europe yet, as industrialization accelerated, demand gradually outstripped the ocean's supply of sperm whales (Yergin, 1991). Two concurrent discoveries set the foundations for the internal combustion engine (ICE). In 1859 Edwin Drake became the first to successfully tap oil underground, combining a steam-powered wheel with an iron drill bit and setting off a worldwide hunt for the 'black gold' (Maugeri, 2006). Meanwhile, Karl Benz patented the four-stroke, one-cylinder gasoline engine and the first motorized carriage in 1886, developed in parallel with fellow German engine inventors, Gottlieb Daimler and Wilhelm Maybach (Smil, 2005). However, it would take the management prowess of Henry Ford, in pioneering the assembly line, 40-hour work week and high wages for workers, to make cars in enough volume to become affordable for the average consumer. First produced in 1908, the Model T would roll off the assembly line 15 million times before its discontinuation in 1927. The end of WWII would see the advent of mass travel and highway construction, as Europe was rebuilt and the US channeled its substantial wartime production back into its own growing cities, suburbs and economy.

### Turning point 5: Industrial innovations – gas turbines

The jet engine had its roots independently in England and Germany, seeing its first uses in military aircraft near the end of WWII. With the arrival of the Cold War came breakneck research in jet technology for both military and commercial purposes. The gas turbine found multiple uses such as for flexible industrial power generation, but its unique sequence of air compression followed by combustion allowed incredibly high temperatures to form within an engine. These temperatures could reach up to 1000°C, which then rotated the turbine and could be used for other forms of work or electricity (Smil, 2010). It is not overstating the case to say that the adoption of the jet engine was incredibly quick, with the first test flight in 1939 and first commercial flight by Pan Am occurring in 1958, a transatlantic flight from New York to Paris. The post-war arms race in terms of technological development, alongside exploding Baby Boomer populations, made for an easy sell regarding the use of these new engines.

### Industrial tourism

Modern tourism, as linked to liberalism, enlightened individualism and self-development, owes a debt of outlook to the Grand Tour. The Industrial Revolution, however, provided the material and technological foundations for a greatly expanded tourist experience in terms of speed, convenience and regularity. With advances in telegraph communications, this era of pre-WWI globalization rapidly shrunk the scope of the world for many. The application of steam power to the printing press helped ignite an explosion in publishing, with travel literature becoming a popular genre for those curious about distant cultures and lands.

Tourism and civilian transport were key drivers of this revolution. While military applications were quickly found, such as steamboats during the First Opium War, steam locomotives and railways dramatically linked urban centers together, cutting transit times by many orders of magnitude. Trains could reach speeds up to 10 times that of horse-drawn carriages, whereas transatlantic crossings by sailing ships in 1830 took over a week compared to under six days by steamship in 1890 (Smil, 2004). Massive infrastructure planning, capital investment and industrial production of steel laid the foundations for nation building, with the Trans-Siberian Railway, Canadian Pacific Railway and Transcontinental Railways all built by the early 20th century. Travelers came to depend on prompt departure and arrival times, driving the adoption of universal time zones linking what were previously disparate communities and clocks (GMT, n.d.).

The modern travel agency can also be traced back to the rise of the railways and coal. British cabinet maker Thomas Cook began his enterprise by organizing travel packages and arrangements in 1841, taking 500 people on a round trip of 12 miles for the price of a shilling (Cook, n.d.). His company continues to provide leisure travel services to this day, having also pioneered destination guides and hotel coupons along the way. The confluence of industrialization, urbanization and new classes of wealth concentrated labor and populations geographically, providing a convenient audience for Cook's project. By applying these new forms of transport and energy systems not simply to the movement of goods and commodities but to people, mass tourism and the development of destination-specific tourist economies reached new heights.

While urban road-based transportation would remain tied to horse-drawn carriages into the beginning of the 20th century, automobiles and oil together quickly combined to replace them, transforming travel and cities in the process. Since its earliest days, the motor car has been associated in travel literature with adventure and freedom, freeing its riders from the well-defined tracks of railways. The early 1900s saw the rise of the motor tour and hinted at the sociocultural implications of this new form of transport, with many pioneering women traveling independently and documenting their experiences (Speake, 2003). Similar to horse transport, an ecosystem of industries and jobs grew to accommodate this new form of travel, from detailed maps and roadside motels to urban planning and car rental hubs. The internal combustion engine has transformed landscapes as well, with roads transecting ecosystems and bringing financial support to national parks by creating 'windshield wilderness' (Louter, 2006). In 2015, 36% of Europeans used rented or private cars for travel or vacation within the continent, a close second behind air travel (Eurostat, n.d.). Whereas steam power and internal combustion engines set the stage for a fossil-fueled dominance of land and sea transport, it would take a different trajectory of technological research to bring travel by air to the masses.

Modern tourism owes much to the concentrated generative capacity of gas turbines. In terms of global annual passenger kilometers, people traveled 40 billion km in 1950 versus 3 trillion in 2000, an increase of nearly two orders of magnitude (Smil, 2017). Infrastructure to accommodate its growth has exploded as well, including airports, jet fuel production, climate and weather modeling, air traffic control unions, and international aviation bodies such as the ICAO. The experience of time and space has been reshaped by the advent of mass jet travel, where nearly any point on the globe is now accessible within 24 hours. Pre-jet era pioneers of air travel like Charles Lindbergh and Amelia Earhart painted their experiences through personal musings and autobiography, emphasizing a sense of nostalgia in anticipation of mass commercialization. Lindbergh's observation that 'planes of the future will cross the ocean without any sense of the water below' eerily foretold of the hyper-connected and invisible nature of modern mass air travel (Speake, 2003). As with other forms of travel, the cultural novelty of planes eventually wore off with the proliferation of air routes and airports, becoming simply a vector to reach the destination. If industry projections that passenger demand will double over the next 20 years are any indication, the era of the gas turbine looks set to expand.

## Futures Direction

I have attempted to describe several historical junctures in the adoption of different prime movers. These technologies have fundamentally changed the form and experiences of travel, beginning with their roots in wind and muscle power, and more recently with the unearthing of fossil fuels. Turning to the future, what kinds of changes in transportation can be forecast? One primary trend is the move towards the electrification of transport, in line with greening economies. It is important to note that transitions in energy or infrastructure can take several decades, the effects of which are unequally distributed in time and geography. For example, horse-drawn carriages, steam-powered trains and muscle-powered humans all coexisted at the turn of the 20th century and the same will be true of the rising renewable-powered and existing fossil-fueled infrastructure. Having focused on the changes brought by transitioning to fossil fuels, I will now center on forecasting the trend towards electrification and different zero- and low-carbon forms of transport.

### Future turning point 1: Electric vehicles and energy storage

The electric car has a parallel if underappreciated history with that of the ICE automobile. Its competition with ICE automobiles dates back to its 'golden age' in the 20th century, where one-third of cars in US cities were electric (Hoyer, 2007). Breakthroughs in electricity transmission, motors and generation were taking place and to pioneers like Thomas Edison it was only a matter of time before electric cars would win the battle of the engines. Advances on the side of the combustion engine would eventually push the debate in its favor, namely in terms of its affordability, flexibility and ease of use (Smil, 2005). Cost-effective, high-density batteries have been slow to develop, characterized in the media and by consumers with the term 'range anxiety'. Whether they use early 20th century lead-acid batteries or the lithium-ion battery packs on the Tesla Model S, electric vehicles (EVs) are dependent on the resilience and responsiveness of the electric grid, just as gasoline cars depend on networks of pipelines, refineries and extraction infrastructure. Therefore, a major challenge for utilities will be not just the transition away from fossil fuel generation sources, but also handling higher peak power demands as more electric cars hit the roads. Primary research, subsidies and industry standards will remain indispensable into the foreseeable future, especially for a market that still only makes up 0.2% of all light-passenger vehicles (IEA, 2017). Despite this slow progress, political pressure and activism around issues like air pollution, carbon emissions and gasoline prices support the trend towards electrification.

In terms of traveler experience, electric cars are quieter, have smoother acceleration and generally allow for more precise digital sensors and integration than gasoline-powered cars (Morris, 2015). EVs therefore become natural extensions of other electric-powered systems such as the internet and AI for autonomous driving. In other words, car culture could shift away from its twin 20th century pillars, both car making (or maintaining) and car owning, and towards reframing the car as one of a panoply of digital devices and appliances for use on demand (Smil, 2005). Trends in this direction include car-sharing, the push towards self-driving fleets and the design of smart cities integrated with GPS, cameras, algorithms and social media. Charging times are also a factor, with household plugs charging an EV from 0 to 100% in 8–20 hours while DC Fast Chargers can charge from 0 to 80% in 30 minutes. This need to plan trips accordingly around charging station types, battery levels and wait times adds a level of complexity to EV travel that is less prevalent when driving fill-and-forget gasoline cars. It also means that EV uptake will be greatest in urban areas, where charging station densities will be highest and transmission infrastructure the most developed.

For tourists, EV adoption can seem an invisible trend, as taxis and ride-sharing and car-sharing services increasingly swap combustion engines for electric motors. According to the IEA, the stock of EVs is slated to grow from 2 million cars globally in 2016 to 40–70 million by 2025 (IEA, 2017). Within the United States, states and cities are moving forward with EV tourism, with plans in New York State to support 'EV tours' – inclusive train ticket and EV rental packages – and promoting charging infrastructure around high-traffic destination areas (Harding, 2015). The greatest changes thus will be in how cities and their digital-electrical infrastructure are transformed by this growing demand for electricity, efficiency, automation and connectivity, and how this will shape the spaces the tourists of the future will come to inhabit and explore.

### Future turning point 2: Cycling and scooters

Bicycles seem ubiquitous now, yet it was not till 1886 that designs with 'equal sized wheels, direct steering, and diamond frame of tubular steel' were released and adopted (Smil, 2005). Relying on paved roads, regular maintenance and traffic rules, bikes have had a parallel and intertwined history with cars since their invention. Motorcycles and scooters soon followed in bringing together combustion engines and bicycles. Scooters, due to their lower speeds and affordability, have become popular in south and southeast Asia, while bicycles are more common in advanced economies and in total outnumber cars around the world (Poushter, 2015). Scooters face headwinds in many developing countries due to congestion, inadequate driving etiquette and contribution to air pollution, having been banned from Hanoi, Beijing and Shanghai (BBC, 2016). Electrification is underway for two-wheeled vehicles, although many of the same challenges as discussed for electric cars apply, such as charging infrastructure and battery capacities. However, a push towards bicycles is gaining traction in cities for emissions, air pollution, density and health reasons, with ideas like dedicated bike lanes, intersections and even highways (Bliss, 2017).

What impacts might two-wheeled vehicles have on tourism in the future? While cars have the added design pressure of requiring flexibility in trip distances, scooters and bikes are generally geared towards urban travel and relatively short trips. Recent research from the EU has found scooters to be a major contributor to air pollution and emissions per fuel unit, due in part to a lack of adequate emissions standards for the vehicles (Platt *et al.*, 2014). Both bikes and scooters enable a flexibility and freedom not open to cars, as cities become denser and more heavily trafficked, and as air pollution rises in the public consciousness. For example, China's electric bike production grew from 40,000 in 1998 to 10 million in 2005, a trend that has continued until today (Xinhua, 2017). Issues around regulation and insuring these riders has caused the government to crack down on the bike- and scooter-sharing industries in Shanghai and Beijing, with temporary bans on these services being implemented (Xinhua, 2017). In the long term, electric bikes and scooters seem well positioned to benefit from many of the same social, political and economic tailwinds as electric cars. However, they also reshape city traffic and the tempo of city life, as greater access to electricity and affordability allows more and more people to 'upgrade' from bikes to electric or motorized scooters. As cases in the Netherlands and China show, smaller vehicles can often defy regulation and open new horizons to people of many levels of training and driving experience, posing challenges to policy makers and potential disruption to both local people and tourists (Bicycle Dutch, 2013).

### Future turning point 3: Buses and high-speed trains

Trains and buses have an interrelated history despite relying on different sets of infrastructure. In some ways they compete, yet they are bound by similar functional areas when it comes to passenger use. The origin of the word bus derives from *omnibus*, meaning 'transport for all', and was applied to transporting large numbers of passengers in order to benefit from economies of scale (Speake, 2003). In that sense, trains similarly attempt to capitalize on transporting high volumes of goods and passengers, although on fixed tracks as opposed to paved road infrastructure. Both have their roots in horse power – trains and carriages having originally been pulled by teams of horses before the advent of steam and fossil fuels in the 19th century. From these origins in communal, bulk capacity transport, the futures of buses and trains will be highly dependent on the economic and environmental characteristics of high-density centers such as cities, in which such means of transport best fit.

The futures of rail and bus transport are happening within a wider re-imagining of the role of cities and the changing nature of the energy infrastructure. Amid greater emphasis on renewable energy, smart cities and the prospect of autonomous EVs, some voices in the tech world view car-sharing as a future replacement for public transit, with door-to-door service instead of bus stops. However, the fact remains that some routes will tend to be more heavily trafficked than others, thus making buses or trains viable in an interconnected city of autonomous vehicles (Marshall, 2017). Further, if people continue to own autonomous cars as private property and eschew car-sharing, then current problems around congestion and traffic could continue to worsen. Outside and between cities is perhaps where the line between resident and tourist has blurred the most. Long-distance high-speed rail in Japan and China has promoted greater regional interconnection of people and economies along such corridors as Tokyo–Osaka and Beijing–Shanghai. Greater regional integration has already been demonstrated through the Trans-European rail network and is forecast to continue under the Chinese One-Belt-One-Road initiative connecting China, the Middle East and Europe with railways, ports and other infrastructure (Cai, n.d.). With greater regional interconnection of transit infrastructure, the tourism sector stands to benefit as cities become more closely linked and networked electrically.

## Post-Industrial Tourism and Futures

The foundations of transportation lie in how work is done, in expending energy from one form to another. At first the human body presented the sole vehicle for movement, especially efficient over long distances for hunting prey. One of these species of prey was the horse, admired for its strength and speed in comparison to our own. The domestication of horses opened up vast new possibilities for trade and movement of surplus through their greater energy returns and power. The Industrial Revolution triggered investment and innovation around new engines of steam, internal combustion and diesel, alongside the worldwide hunt for coal and oil reserves. These engines spurred the acceleration of transport routes over land and water, but it would take until the 20th century for the gas turbine to develop into global jet travel. The result is what Jan Aart Scholte has called the 'transplanetary' character and 'supraterritoriality' of social space, in which previous notions of territory, place and identity are being reshaped by the instantaneity of connection (and disconnection) and dizzying complexity of the system that is being formed (Scholte, 2005). Alongside the electronic travel of ideas and images, tourism now encompasses not just the energy required for transportation infrastructure, but also the massive computational resources necessary to maintain networks of Instagram accounts, blogs, review sites, payment providers and hospitality platforms.

In forecasting futures, I have attempted to look at technological shifts in how energy is harnessed, primarily regarding renewable energies, electrification and autonomous sensory regimes. Geologists and social scientists have recently proposed the 'Anthropocene' as the name for the new geological epoch in which humans present a new large-scale driver of global environmental change. Amid challenges like climate change, biodiversity loss, resource overconsumption and demographic shifts, the futures of tourism will need to navigate the narrow strait between growing tourist bases and the subsequent contributions to these problems. Technologically, these are in part being addressed through EVs and expanded grids, batteries and renewables across road, rail and sea. These interventions are also being mirrored by calls to expand the efficiency and connectedness of public transit, in order to minimize the disruptions of ongoing urbanization. For the experience of tourism, my argument has been that it will increasingly mirror the speed, consistency and fickleness of electricity, just as the industrial era relied on the explosiveness, noise and fumes of fossil fuels. As the long transition away from fossil fuels advances, the futures of transportation and tourism will continue to evolve into an era of plural and competing energy sources, systems and vehicles.

## References

BBC (2016) Beijing bans electric scooters and segways on roads. *BBC website*, 31 August. See https://www.bbc.co.uk/news/business-37227562 (accessed 15 January 2019).

Bicycle Dutch (2013) The moped menace in the Netherlands. *Bicycle Dutch*, 23 February. See https://bicycledutch.wordpress.com/2013/02/23/the-moped-menace-in-the-netherlands/ (accessed 15 January 2019).

Bliss, L. (2017) This Dutch superhighway was built strictly for bikes. *CityLab*, 22 June. See https://www.citylab.com/transportation/2017/06/cruising-a-superhighway-built-for-bikes/531246/ (accessed 9 October 2018).

Cai, P. (n.d.) Understanding China's Belt and Road Initiative. Sydney: Lowy Institute. See https://www.lowyinstitute.org/publications/understanding-belt-and-road-initiative (accessed 9 October 2018).

Cook, T. (n.d.) Thomas Cook History. See https://www.thomascook.com/thomas-cook-history/ (accessed 9 October 2018).

Darling, J.A., Erickson, C.L. and Snead, J.E. (eds) (2011) *Landscapes of Movement: Trails, Paths, and Roads in Anthropological Perspective*. Philadelphia, PA: Penn Press, University of Pennsylvania.

Davis, N. (2018) Homo erectus may have been a sailor – and able to speak. *Guardian Online*, 20 February. See https://www.theguardian.com/science/2018/feb/20/homo-erectus-may-have-been-a-sailor-and-able-to-speak (accessed 16 January 2019).

Eurostat (n.d.) Tourism statistics – intra-EU tourism flows. *Eurostat Statistics Explained*. See https://ec.europa.eu/eurostat/statistics-explained/index.php/Tourism_statistics_-_intra-EU_tourism_flows#Aeroplane_most_common_for_travelling_to_another_EU_country (accessed 9 October 2018).

Fagan, B.M. (2015) *The Intimate Bond: How Animals Shaped Human History*. New York: Bloomsbury Press.

GMT (n.d.) Railway time. See https://greenwichmeantime.com/info/railway/ (accessed 9 October 2018).

Harding, E. (2015) How tourism and travel will increase the use of electric vehicles. See https://www.energydigital.com/sustainability/how-tourism-and-travel-will-increase-use-electric-vehicles (accessed 30 September 2018).

Headrick, D.R. (2009) *Technology: A World History*. The New Oxford World History. Oxford and New York: Oxford University Press.

Headrick, D.R. (2010) *Power over Peoples: Echnology, Environments, and Western Imperialism, 1400 to the Present*. The Princeton Economic History of the Western World. Princeton, NJ: Princeton University Press.

Hoyer, K.G. (2007) The battle of batteries: A history of innovation in alternative energy cars. *International Journal of Alternative Propulsion* 1 (4), 369. doi.org/10.1504/IJAP.2007.013330

IEA (2017) *Global EV Outlook 2017: Two Million and Counting*. Paris: International Energy Agency. See https://www.iea.org/publications/freepublications/publication/GlobalEVOutlook2017.pdf (accessed 19 January 2019).

Keller, A.H. (2009) A road by any other name: Trails, paths, and roads in Maya language and thought. In J.A. Darling, C.L. Erickson and J.E. Snead (eds) (2011) *Landscapes of Movement: Trails, Paths, and Roads in Anthropological Perspective* (pp. 133–157). Philadelphia, PA: Penn Press, University of Pennsylvania.

Lombard, M. and Haidle, M.N. (2012) Thinking a bow-and-arrow set: Cognitive implications of Middle Stone Age bow and stone-tipped arrow technology. *Cambridge Archaeological Journal* 22, 237–264.

Louter, D. (2006) *Windshield Wilderness: Cars, Roads, and Nature in Washington's National Parks*. Seattle, WA: Weyerhaeuser Environmental Books, University of Washington Press.

Marshall, A. (2017) Don't look now, but even buses are going autonomous. *Wired*, 2 May. See https://www.wired.com/2017/05/reno-nevada-autonomous-bus/ (accessed 20 January 2019).

Maugeri, L. (2006) *The Age of Oil: The Mythology, History and Future of the World's Most Controversial Resource*. Westport, CT: Praeger.

Morris, D. (2015) Tesla veteran explains how electric motors crush engines. *Fortune*, 17 November. See http://fortune.com/2015/11/17/electric-motors-crush-gas-engines/ (accessed 9 October 2018).

Outram, A.K., Stear, N.A., Bendrey, R., Olsen, S., Kasparov, A., Zaibert, V., Thorpe, N. and Evershed, R.P. (2009) The earliest horse harnessing and milking. *Science* 323, 1332–1335.

Platt, S.M., Haddad, I.E., Pieber, S.M., *et al.* (2014) Two-stroke scooters are a dominant source of air pollution in many cities. *Nature Communications* 5. See https://doi.org/10.1038/ncomms4749 (accessed 20 January 2019).

Poushter, J. (2015) Car, bike or motorcycle? Depends on where you live. *Pew Research Center FactTank*, 16 April. See https://www.pewresearch.org/fact-tank/2015/04/16/car-bike-or-motorcycle-depends-on-where-you-live/ (accessed 20 January 2019).

Scholte, J.A. (2005) *Globalization: A Critical Introduction* (2nd edn). Basingstoke and New York: Palgrave Macmillan.

Smil, V. (2004) World history and energy. *Encyclopedia of Energy* (pp. 549–561). See https://doi.org/10.1016/B0-12-176480-X/00025-5 (accessed 20 January 2019).

Smil, V. (2005) *Creating the Twentieth Century: Technical Innovations of 1867–1914 and Their Lasting Impact*. Oxford and New York: Oxford University Press.

Smil, V. (2010) *Two Prime Movers of Globalization: The History and Impact of Diesel Engines and Gas Turbines*. Cambridge, MA: MIT Press.

Smil, V. (2017) *Energy: A Beginner's Guide* (2nd edn). Oxford: Oneworld.

Speake, J. (ed.) (2003) *Literature of Travel and Exploration: An Encyclopedia*. New York: Fitzroy Dearborn.

Towner, J. (1985) The grand tour. *Annals of Tourism Research* 12 (3), 297–333.

Xinhua (2017) Shanghai bans electric bicycle sharing. *XinhuaNet*, 10 November. See http://www.xinhuanet.com/english/2017-11/10/c_136740746.htm (accessed 9 October 2018).

Yergin, D. (1991) *The Prize: The Epic Quest for Oil, Money, and Power*. New York: Simon & Schuster.

# Part 4

# The Hotel

# 11 Hotel History

Kevin James

## Introduction: Turning Points in Material and Cultural Configurations of the Modern Hotel

The hotel is indisputably integral to modern travel. Yet identifying specific attributes to define a continuous material form is a challenging task, if not a fool's errand, given that one of the hotel's historical characteristics is mutability. To add to the confusion, the contemporary nomenclature of commercial lodgings – take 'Holiday Inn' or 'Travelodge' as examples – is a lexical hotchpotch, reflecting the dissolution of a vocabulary that historically denoted distinctive institutions (Walton, 2011). Tracing the hotel's institutional origins requires an imaginative intellectual apparatus. It demands an approach that accounts for the exceptional diversity of hotels' manifestations – places of various scales and built forms, offering different services to diverse clienteles – while acknowledging the dominance of historically specific discursive constructions of the hotel. These conceptual anchors were elaborated in discussions of travel and lodging, as well as in wider appraisals of technology, urban life and modernity. By paying close attention to formulations and reformulations of the hotel as an idea as it charts the profusion of hotel forms, this analysis does not present a determinative account of hotel development as much as it underlines how 'hotel history' has centred on historically contingent imaginative configurations.

Writers have identified a number of functional and physical properties intrinsic to the archetypical hotel as it evolved from the 18th century. It was seen as characteristic of a new, modern institution whose 'rise' variously elicited wonder, fear and scorn. Broadly speaking, it was a purpose-built building type which, while evincing regional and national inflections, was regarded as an essentially transnational institution – born in the West and transplanted to serve the needs of Westerners in colonies and other places of travel and settlement. It was erected on a notably larger scale than had been hitherto associated with commercial lodgings. It tended, over time, to adopt high levels of technology (especially in the United States). In addition to these physical, functional and locational features, the archetypal hotel fostered a particular set of sociabilities, and filtered 'hotel life' downwards to aspirational sections of the middle classes. These criteria were elaborated in genealogies that identified the modern hotel's ancestor as the aristocratic home. Its culture was reproduced as this new category of commercial accommodation developed in the 18th and early 19th centuries. Yet that evaluation obscured many nuanced developments on the ground.

## The 'Birth' of the Hotel: Precedents and Contexts

There was no dearth of lodgings for travellers in the centuries before the idea of the hotel as a modern invention gained cultural currency. Indeed, other forms of putatively more 'primitive' accommodation became foundational to narratives of the hotel's roots, and also to expositions on its originality. Commercial accommodations servicing traders, pilgrims and other travellers existed for centuries – millennia, even. Outside Europe, the caravanserais of the Islamic world developed along key trading roots: they were often invoked when writers searched for an ancient institution comparable in scale and function to the 19th century hotel (Griffiths, 1881: 36).

In Europe, too, expansive networks of commercial lodgings were well established along commercial and pilgrimage routes – indeed they were indispensable to systems of trade (Hunt & Murray, 1999: 62–64). The kinds of establishments to which travellers turned depended very much on their socio-economic status. It was common practice, for instance, for travelling elites to be accommodated in the private houses of their peers, upon presentation of letters of introduction to those hosts with whom they were not personally acquainted (Stobart, 2017). At the opposite end of the socio-economic spectrum, many travellers who were unable to pay the costs of inns turned to extended kin and others who hosted them *gratis* (Heal, 1990). Travellers of various stations also sought access to sites such as monasteries for short-term hospitality (defined here as the core functions of shelter, board and protection), especially while participating in acts of devotional travel. That hospitality depended upon mutual, often tacit, knowledge of the codes that governed interactions between hosts and guests, from the conventional period of a stay to the extent of physical separation between them. These practices took travellers outside the realms of commercial accommodation (the central role that unpaid lodging plays in travel persists today in the identification of a whole category of travellers – 'VFR' or 'Visiting Friends and Relatives' – who often make limited, if any, use of such accommodation). Yet there was also extensive resort, over many centuries, to various forms of commercial accommodation, from paid use of private dwellings to purpose-built lodgings (supplying room and board for both people and beasts), along the main lines of travel. Critically, in the early modern period, one particular lodging form came to be associated with the nascent 'hotel'. It had its coordinates in the Grand Tour.

## Historical Turning Points

### Turning point 1: The country house's commercialization

The first historical turning point which was crucial to the elaboration of the hotel form was nourished by socio-economic and cultural developments that included the widening compass and constituency of travellers in 18th century Europe, the growing accessibility and popularity of the Grand Tour as a template for middle-class travel, and an imbrication of the private home and the commercial lodging.

The Grand Tour was conceived of as a masculine exercise in early-modern educational travel associated with the British and northern European elite (Black, 2003; Chard, 1999; Sweet, 2012). It involved extensive travel on the continent and networks of commercial lodgings frequented by cosmopolitan society. As commercial institutions grew to meet the expanding ranks of Grand Tourists, including women and families, self-styled 'hotels' often took the form of such high-status lodgings. Many tended to replicate the amenities and atmosphere of the aristocratic abodes with which these elite travellers were familiar. Grand Tourists were often compelled to settle for more rustic accommodation in their travels between major cities. But Rome, Venice and other urban centres furnished lodgings that were celebrated for their comforts and for the personal attendance of their landlords – perhaps most famously the Englishman Charles Hadfield's hostelry in Florence (Black, 2003: 69–70). Indeed, the 'character' of such lodgings was a preoccupation in diaries of the continental tour, in which amenities and comforts were often extensively appraised and compared. The extension of this accommodation network was accompanied by the expansion of many prestigious, historical European resort accommodations. They catered to a growing market of patrons of fashionable national and regional spa centres: previously, their elite patronage was matched by a correspondingly limited provision of commercial accommodation.

The presence of establishments servicing an international coterie of travellers influenced the development of a lodging form that incorporated the ethos of a non-commercial site of accommodation – the country house to which many elite travellers had historically turned (Riley, 1984). In French, the word 'hôtel' historically denoted a style of private accommodation that was also used for large-scale public edifices, such as the 'hôtel de ville' ('city hall'). When it was adopted in the late 18th century Anglophone world to christen a particular form of public, commercial accommodation, it denoted a place in which styles and features associated with the aristocratic home were replicated: opulent décor; personalized, genteel and attentive service; abundant, fashionable and often foreign fare. Although many establishments rushed to re-christen themselves, the elite ethos of the archetypal hotel was difficult to replicate. Its guests possessed material resources to access it, and the cultural capital to navigate its codes.

Other institutions served as a foil in elaborations of this hotel ideal. In terms of its amenities and its ineffable 'culture', the hotel was seen as distinct from the more rustic, familiar, indigenous and historical forms of commercial accommodation. While these other institutions bore various names in different regions – the *pensione*, *Gasthof* and *auberge*, for example – English-speaking travellers often aggregated them under the rubric of the 'inn'. Their features, from fare to styles of attendance, were regarded as comparatively intimate and simple. They were often identified with rural spaces, and were superintended by 'locals' who were as much of the soil as the food they proffered guests (Riley, 1984). In reality, in architectural form and in amenity, there was often comparatively little difference between the hotel and these other lodgings – after all, there were large-scale urban inns and many small and comparatively rudimentary establishments that advertised themselves as hotels (Maudlin, in James *et al.*, 2017). However, the distinction in nomenclature was salient to elaborations of an *idea* of the hotel as a comparatively new, cosmopolitan and high-status institution (Pevsner, 1976: 169–192). That idea was central to validating its 19th century incarnations, as the institution retained associations with specific physical forms and refined hospitality.

## Turning point 2: Transport innovation and industrialization

A second critical turning point which influenced the diffusion of hotels and consolidated the cultural influence of one specific hotel form is associated with socio-technological changes driven by steam technology in the transport sector. These were especially salient with the rise of the steamboat and the steamship in the first decades of the 19th century, and the railway after the 1820s. They also were enfolded within wider processes of industrialization, class stratification and the emergence of a commercial tourism infrastructure in the first half of the 19th century. New forms and scales of paid, public accommodation evolved to meet growing demand. The increasing scope, ease and affordability of travel and the development of new means of locomotion marked an era of rapid industrialization. This was led by the UK, and followed by America and by industrializing states in Western Europe. A proliferation of commercial lodging types accommodated these mobile populations, even as the 'grand hotel' was elevated as the institution's dominant, valorised, imaginative form.

Framing this turning point as a decisive break from the past risks understating the role of earlier travel networks, including roads, canals and lodgings, which enabled mobility before the steamship and railway. Moreover, transport networks in the Age of Steam knitted together railways, horse-drawn carriages, practices of pedestrianism and, from the later 19th century, the bicycle. Many lodgings were accessible through a combination of these modes of transport. The railway may have increased the speed of travel and democratized travel by providing fare structures that serviced greater numbers of the population, but this did not alone supply an impetus to commercial lodging. A critical factor was rising real incomes, which made hotels accessible to a wider range of the travelling public and promoted the expansion of the range of accommodation products.

Market expansion and segmentation produced an array of paid lodgings to meet different levels of demand for different forms of accommodation. Social stratification associated with industrial society was expressed in differentiation in lodging types. The rhythms of industrial society also produced new travel practices, including Cook's famous tours. For many workers in sectors of the economy that were based on mechanized, centralized production, the logic of industrial capitalism introduced sharper distinctions between 'work' and 'leisure', and between 'factory time' and 'family time'. 'Family hotels' catered to units of kin, many of whom were engaged in leisure travel over longer distances than in previous decades (Sandoval-Strausz, 2007). Other travellers were content with renting rooms in seaside homes. The growth of resort hotel facilities with comparatively modest amenities responded to the desire of a new market to reproduce elite leisure patterns, at a lower cost. 'Commercial hotels' became deeply embedded within work regimes for travellers, too (French, 2005; Hosgood, 1994). One commercial traveller commented on the remarkable changes that had occurred within a short span of time, with a generation still alive in 1885 that recalled the age of 'horse and saddlebags' (Anon, 1885: 243). Some hotels furnished the commercial traveller with sample rooms; others operated on a much more modest scale. In smaller communities, inn signs were buffed and hung on establishments that now proclaimed themselves to be hotels. In Britain and other jurisdictions where British legal traditions prevailed, the Law of Innkeepers, which had initially grappled with the putative differences between the hotel and inn, largely abandoned such efforts by the mid-19th century.

Yet the expansion of accommodation types did not tend to dilute the idea of the hotel as a fundamentally different creature, or diminish the influence of the grand hotel idea: it supplied a yardstick against which commercial lodgings were measured. In colonial environs, such hotels were appraised as markers of European civilization, and they often telegraphed an aura of exclusivity and pampered luxury even as they served as sites of intercultural encounter and assumed hybridized forms (Peleggi, 2012). In America, hotels began to be discursively positioned as distinctive variants on the original, European form – institutions that were especially well suited to serve the needs of a highly mobile population, and whose demotic ethos aligned with the egalitarian spirit of the New Republic. American hotels supplied public squares and contrasted with the enduring snobbishness and exclusivity of their European counterparts. Practices of meal service, the uses of public spaces and the greater preference among Americans to adopt the hotel as a permanent residence were signals of institutional divergence across the Atlantic. The idea of the hotel bifurcated along national lines was something that many observers, from Anthony Trollope (1862) to Henry James (1907), affirmed in the 19th and early 20th centuries.

### Turning point 3: Innovations in business structure

A third turning point in the development of commercial accommodation was grounded in mid-19th century legal developments, involving new (and the revival of old) means for aggregating capital and mitigating risk as hotels attracted greater investment, often as part of wider business ventures. Hotels, for example, were often incorporated within railway enterprises, through vertical integration. The demands for new legal mechanisms to pool capital at new scales while mitigating risk in railways and hotels mirrored the requirements of industry, and governments responded by developing legal frameworks for capital accumulation and investment in the sector.

Limited liability was an important principle of business organization that came to be enshrined within the law, especially in the second half of the 19th century (Kindleberger, 1984: 202–209). This principle limited the liability of an investor in a business to the sum actually invested, and protected the investor from the seizure of personal assets to satisfy a debt incurred by a company. It had a critical relationship to incorporation, which in Britain was greatly facilitated by the *Joint Stock Companies Act* in 1844. It created a framework for investors to engage in business through an institution with a separate legal personality. The doctrine of limited liability was adopted in a restricted form in New York State in 1811, and under Acts of Parliament at Westminster in 1855–1856. This UK legislation enshrined the principle and enabled businesses to incorporate without prior parliamentary consent; its provisions were expanded by further legislation in 1862. Thereafter the principle was adopted in other Western European jurisdictions.

The implications of the adoption of limited liability and incorporation were profound for the hotel sector. Incorporation under the limited liability principle was not the exclusive province of the grand hotel – it was also adopted by smaller establishments (Ball & Sutherland, 2001: 154–156). But the heavy investment by railway companies and other actors in the hotel sector reflected the influence of the new legal framework on the sector's landscape, as well as a pattern of vertical integration that was to be realized in the 20th century with airlines, too. Innovations in construction and design, especially the use of steel frames in American hotels in the 1880s (and later elsewhere too), along with the invention of the lift, enabled the largest hotels to rise even higher. Greater capital and technical and technological improvement contributed to aggrandizing even the grandest of hotels. The precocious adoption of these technologies in America nourished the idea that the British and European hotel was a comparative laggard. And as the hotel's scale grew, so too did the chorus of anxiety centred on its essential impersonality.

Both incorporation and the public limited liability model were invoked in a developing late 19th century discourse that castigated the modern 'Hotel Company, Limited' – a steel-framed, city block sized, urban establishment. Its foil was the time-honoured inn, whose scale fostered intimate connections. This binary, which inverted some of the previously dominant valorizations of the grand hotel, also promoted the idea of the hotel as a peculiarly modern place. Here, the social configurations characteristic of modernity were in stark evidence, especially in its lobby, where bustling, anonymous crowds passed each other, where guests checking in were assigned a room number in place of a name, and where in a 'city within a city' people dined, shopped, slept, met and inhabited spaces that were highly specialized and technologized. Although the majority of American and European hotels operated at a much more diminutive scale, this dominant idea of the sprawling urban hotel found its epitome in American cities such as New York (Berger, 2011; Fick, 2017). There, observers marvelled at the scale, labour-substituting technologies and openness of the hotel, and continued to contrast it with the purported stuffiness of its Old World counterpart (Paul, 1896).

### Turning point 4: Spatial efficiencies and global expansion

The fourth turning point in this analysis relates to the application of models of efficiency associated with systems of industrial production to the spatial organization and operations of the hotel, and also to the expansion of hotel 'chains' and brands on national and global scales. Both processes dealt with hotel space – its internal organization and its external reach.

Large-scale hotels tended, over time, to develop an elaborate spatial matrix in which lodging and the provision of food and drink were elements in an institution with functional specialization that included the development of customized retail and service outlets, commercial rooms, and a spatial logic that reflected gender ideologies. A crucial turning point in refining and extending the rational spatial logic in the hotel centred on the adoption of new technologies in large and high-status hotels, such as sophisticated plumbing. Innovations in the application of technology and the organization of space were accompanied by innovations in business forms and in marketing that tapped the reputation of particular figures and created 'brands' from them, often underscoring the quality and consistency of their amenities through the christening of multiple establishments with the same name.

Despite the celebrity that later accrued to Ellsworth Milton Statler and other Americans, the most famous early European figure in the sector was César Ritz, a Swiss hotelier who boasted a record of entrepreneurialism and hotel management of such places as London's Savoy, and co-established the Ritz hotel syndicate in 1896, lending his name to hotels in Paris, London and

Madrid (Denby, 1998). Britain's Gordon Hotels Company Limited, incorporated in 1890 to acquire hotels that had each been separately held under limited companies, was a prominent chain in the late 19th and early 20th centuries. On the death of founder Frederick Gordon in 1904, it was declared to hold 'the most valuable hotel property belonging to any single concern in the world' (Anon, 1904: 3). But the development of a chain based on a standardized 'product' and a consistent branding strategy was strongly associated with Americans such as Statler and his eponymous hotel chains. His company was also critical in developing processes through which hotel space was rationalized and subjected to new efficiencies – standardized room sizes, for instance – and, through such processes, by democratizing the grand hotel form and extending it across the nation (Pfueller Davidson, 2005a, 2005b). These developments in business organization strengthened the idea of the 'American hotel' as a distinctive, and advanced, form of an institution whose Old World origins were increasingly removed from the ingenuous New World Fordist efficiencies.

In the interwar period, Conrad Hilton developed a business model which, after WWII, extended the American hotel into the world in ways that reinforced America's status as the sector's dominant global actor (Wharton, 2001). Creative forms of business organization, such as franchising and the separation of property ownership and hotel management functions, made American companies especially nimble actors in the internationalized hospitality sector. Long associated with innovative technological applications and inventive business organizations, the American hotel was now the undisputed epitome of modern hospitality.

### Turning point 5: Automobility

A fifth critical turning point in the hotel's historical development, like the second, is associated with new transport technologies – namely the application of internal combustion to locomotion in the form of the automobile, and the emergence of types of commercial accommodation that were designed around automobile travel. In particular, the development of the auto-camp in America and, from it, the motorcourt and motel as distinctive architectural forms with distinctive locational features, responded to new transport modes and travel preferences that were apparent in the interwar period and accelerated rapidly in the 1950s and 1960s (Jakle *et al.*, 1996). The distribution and forms of commercial lodging often favoured sites outside major city centres, to accommodate automobile travellers. At the same time, many new hotels incorporated design features for the automobile traveller, such as the car entrance, while other hotels erected before the automobile were materially adapted. Often overlooked in studies of the age of automobility, the coach also became an important mode of collective travel – undermining the strong association between the internal combustion engine and individualized travel practices, and affirming continuities between mass transport and road travel that stretched back centuries (Walton, 2011).

The consequences of the evolution of car-based travel on hotel choice were complex: in some cases the automobile generated new demand for lodgings that were believed to have been superseded by the urban hotel in the 19th century. Rustic accommodation emerged in a materially and culturally reconfigured form (Bates, 2003) – ironically responding to the enhanced mobility associated with the automobile, which allowed travellers to access more remote landscapes and 'discover' the country inn. There was a concomitant recasting of the historical urban grand hotel with respect to the changing uses of the downtown business districts with which it had been associated. In some cases, 'grand hotels' saw their cachet enhanced as heritage institutions as they adapted to the fragmenting tourism market by developing narratives that capitalized on a storied past, but others were seen as poorly adapted to the exigencies of suburban, automobile-oriented culture. These developments marked the diminishing influence of the grand hotel ideal type, although it retained a strong position in popular consciousness, through novels and film, the various eulogistic coffee-table enterprises that fetishized it, and efforts to preserve or restore individual establishments. The cultural representation of the hotel, however, shifted in notable ways away from a site that exemplified modern life to one whose historically esteemed form was now cast as obsolete. Nostalgic representations of the 'grand hotel' as a hallmark of a world whose apogee was in the interwar period proliferated in the three decades after the end of WWII, and repositioned it within a new discourse in which it was no longer hailed as the epitome of refined travel culture, but rather a holdover from a romantic age now in an eclipse as definitive as the far-flung European empires where it boasted long-established outposts.

**Future Turning Points**

**Future turning point 1: Redefining the hotel form**

The hotel's cultural moorings in the ethos of the aristocratic abode, the dominance of the grand hotel in popular understandings of the institutional form and the representation of distinctive American and European forms waned in the last decades of the 20th century, leaving the very label 'hotel' as a generic descriptor for a variety of lodging types, and opening the door for continuing and future transformations in the sector. These developments have provided the context for the first future turning point in the hotel's evolution, which extends and amplifies historical patterns, and also adopts new directions: innovative uses of 'history' in branding and re-branding imaginative repositionings of the hotel through historical narratives, and the creative adaption and repurposing of historical built spaces.

As we have seen, the elaboration of new scales and types of commercial lodging is part of a historical pattern. Important aspects of the hotel's future relate to how the sector will respond to the decline of some historical (and historic) hotels, to the increasing fragmentation of tourist markets and to the intrinsic diversity of the hotel's built forms. One strategy, discussed above, will be to continue to organize grand hotel preservation and 'restoration' around nostalgic dispositions, through hotel heritage projects. The grand hotel form may take new forms in places such as themed vacation resorts and fast-growing cities where there is no preceding 'heritage' building, but demand for a 'heritage' product. Where the prospects for such projects are dim, radical reuses of former hotel buildings will reflect the adaptability of the built form for a variety of institutional purposes. Already there has been creative repurposing of former hotels – and many will continue to assume an afterlife in the form of hospices, apartments and other institutions, as the idea of the hotel as a purpose-built edifice becomes less and less strongly associated with its identity as an institution, as the nature and geography of demand for accommodation evolves, and as the suitability for hotel decommissioning, sometimes for urgent uses such as temporary refugee shelters, becomes more apparent.

A related development will be the accelerated adaptation of other buildings as hotels, as a response to the increasing interest in lodgings as part of a more immersive experience – again reflecting the waning idea of the purpose-built, insular grand hotel as a paragon (Lee & Chhabra, 2015). Castles, plantations and stately homes, as well as asylums, prisons and other institutions, will continue to be adapted within the tourism sector, with commercial accommodation in the leisure sector incorporated within customized 'tourist packages'. The development of prestige establishments and lodging types whose built structure and ethos are in some respects erected in opposition to historical forms strongly associated with the institution – large-scale, purpose-built edifices, for example – is already in strong evidence in the form of boutique hotels, which offer bespoke services in more intimate environments, often occupying repurposed buildings. Treetop hotels, huts, yurts and other novel 'experiential' accommodation types will continue to attract diverse clienteles and demonstrate the adaptiveness of the sector. This will be paralleled by a final, critical process that is already very much in evidence – the intensified imbrication of home and hostelry (a historical feature of the sector) through the accessibility of home stays, farm stays and other paid lodging (Lynch *et al.*, 2009), much of which is made available through online brokers.

### Future turning point 2: 'Escaping' the hotel

Another, second critical future turning point marks not so much a rupture with past practices as the adaptation of technology to expand the availability of, and access to, a range of accommodation products, drawing in a pool of new users, both consumers and suppliers, and an entrenchment of the imaginative association of 'home-based' accommodation as more integrated within localities and therefore better able to deliver an 'authentic' experience to users. Increasing small-scale entrepreneurialism in the sector, enabled by technology, the division between property ownership and virtual brokerage platforms and an enduring, historical quest for alternative accommodation that has long been manifested in the popularity of bed & breakfasts, farm stays, self-catering facilities and non-commercial options such as staying with friends and family, is boldly expressed by the expansion of putatively collaborative 'peer-to-peer', web-based companies such as Airbnb. The idea of escaping the tyranny of the hotel is by no means new – indeed the development of historically specialized 'non-hotel' lodgings, such as self-catering accommodation, was a response to guests' encounter with durable barriers to full participation in the culture of the hotel (Wood, 1994). Yet whereas there has often been a marked demographic market segmentation among forms of accommodation – witness the comparatively durable association of hostels with youth markets – in the case of Airbnb and similar companies, youth will play the role of early adopters. The association of the large hotel with both danger and surveillance will motivate new behaviours around product marketing and choices. The idea of the hotel as a site of anonymity has always been more imagined than real, but the prospect of the hotel-as-panopticon will be reinforced by its increasing securitization and technologization – especially through biometrics. The flight by certain travellers seeking greater 'privacy' will be matched by prospects for 'authentic' local integration. It will be enabled by pairing clients with proprietors of accommodation through brokers such as Airbnb. Hotel companies will respond by seeking to adapt their own brands to these markets, even as jurisdictions promote more formal surveillance over the informal paid accommodation sector.

The sectors' own systems of classification – most famously reflected in 'stars' – will continue to adapt to demands for a novel 'local' experience, but also to the challenges associated with disruptors such as Airbnb where online reviews are the exclusive form of evaluation, which claims to be built on a 'new' relationship of trust but which, for a variety of reasons which include institutional policies and less clearly defined criteria, also generate few meaningful distinctions among property ratings. Hotels will therefore strike a delicate balancing act, through the proliferation of their brands, between those that have strong local and regional identities and those that promise a well-recognized national and international standard of amenity and service. It may deliberately contrast with the business model behind many vacation brokerages, which often have more restrictive payment and cancellation policies as well as fewer amenities, and a comparatively loose internal rating system. The environment that hotels face is not new – the history of the commercial lodging sector has been characterized by ever-greater provision of distinctive accommodation products. But the increasing capacity for travellers to satisfy specific, personal preferences will spur adaptations in hotels' business strategies in years to come.

## Conclusion

The history of commercial accommodations presented here stresses technologies of transport and construction, innovations in legal frameworks and the proliferation of lodging types since the late 18th century as factors which, far from forging the hotel as an institutional form, have tended to highlight its intrinsic diversity. This chapter has underscored the historical dissonance between dominant discursive configurations of the hotel in particular eras and the concomitant expansion of accommodation products. During more than 200 years, the commercial accommodation sector has developed a remarkably broad and heterogeneous range of institutions and lodging types in the context of changing travel cultures and practices, innovations in transport, new frameworks for business organization and the expansion of the ease, efficiency and scale of travel generally. While boarding houses, guest houses, lodges, inns and other types increased in number, a 'hotel' until recently referenced a durable popular construct distinguished by social tone, building style, scale, regimes of technology and, in the 20th century, efficient spatial organization, managerial innovation and American entrepreneurialism. To understand the future of the hotel and its historical relationship to travel, we must understand how processes of product differentiation have continued, and how the dominant 'hotel idea' was erected and then progressively dislodged in popular culture, resulting in only the faint influence of the grand hotel as the hospitality sector looks to a future in which disruptions to hospitality practices, brands and markets will accelerate.

## References

Anon. (1885) Commercial travellers. By one of them. *The Leisure Hour* 35 (April), 242–247.
Anon. (1904) Death of Mr Fredk. Gordon, formerly of Ross. *The Ross Gazette* 31 March, 3.
Ball, M. and Sutherland, D. (2001) *An Economic History of London, 1800–1914*. London and New York: Routledge.
Bates, C. (2003) Hotel histories: Modern tourists, modern nomads and the culture of hotel-consciousness. *Literature & History* 12 (2), 62–75.
Berger, M.W. (2011) *Hotel Dreams: Luxury, Technology, and Urban Ambition in America, 1829–1929*. Baltimore, MD: Johns Hopkins University Press.
Black, J. (2003) *Italy and the Grand Tour*. New Haven, CT and London: Yale University Press.
Chard, C. (1999) *Pleasure and Guilt on the Grand Tour: Travel Writing and Imaginative Geography, 1600–1830*. Manchester: Manchester University Press.
Denby, E. (1998) *Grand Hotels – Reality and Illusion: An Architectural and Social History*. London: Reaktion Books.

Fick, A. (2017) *New York Hotel Experience: Cultural and Societal Impacts of an American Invention*. Bielfeld: Transcript Verlag.
French, M. (2005) Commercials, careers, and culture: Travelling salesmen in Britain, 1890s–1930s. *Economic History Review* 58 (2), 352–377.
Griffiths, A. (1881) Mine ease at mine inn. *Time: A Monthly Miscellany of Interesting & Amusing Literature* 5, 36–41.
Heal, F. (1990) *Hospitality in Early Modern England*. Oxford: Oxford University Press.
Hosgood, C.P. (1994) The 'Knights of the Road': Commercial travellers and the culture of the commercial room in late-Victorian and Edwardian England. *Victorian Studies* 37 (4) 519–547.
Hunt, E.S. and Murray, J.M. (1999) *A History of Business in Medieval Europe, 1200–1550*. Cambridge: Cambridge University Press.
Jakle, J.A. Sculle, K.A. and Rogers, J.S. (1996) *The Motel in America*. Baltimore, MD: Johns Hopkins University Press.
James, H. (1907) *The American Scene*. London: Chapman & Hall.
James, K.J., Sandoval-Strausz, A.K., Maudlin, D., Peleggi, M., Humair, C. and Berger, M.W. (2017) The hotel in history: Evolving perspectives. *Journal of Tourism History* 9 (1), 92–111.
Kindleberger, C.P. (1984) *A Financial History of Western Europe*. London: George Allen & Unwin.
Lee, W. and Chhabra, D. (2015) Heritage hotels and historic lodging: Perspectives on experiential marketing and sustainable culture. *Journal of Heritage Tourism* 10 (2), 103–110.
Lynch, P.A., McIntosh, A.J. and Tucker, H. (eds) (2009) *Commercial Homes in Tourism: An International Perspective*. London and New York: Routledge.
Paul, H. (1896) The new hotels of New York. *The Caterer and Hotel-Keepers' Gazette* 19 (215), 16 March, 112–114.
Peleggi, M. (2012) The social and material life of colonial hotels: Comfort zones as contact zones in British Colombo and Singapore, ca. 1870–1930. *Journal of Social History* 46 (1), 124–153.
Pevsner, N. (1976) *A History of Building Types*. London: Thames & Hudson.
Pfueller Davidson, L. (2005a) 'A service machine': Hotel guests and the development of an early-twentieth-century building type. In K.A. Breisch and A.K. Hoagland (eds) *Building Environments: Perspectives in Vernacular Architecture*, Vol. 10 (pp. 113–132). Knoxville, TN: University of Tennessee Press.
Pfueller Davidson, L. (2005b) Twentieth-century hotel architects and the origins of standardization. *Journal of Decorative and Propaganda Arts* 25, 72–103.
Riley, M. (1984) Hotels and group identity. *Tourism Management* 5 (2), 102–109.
Sandoval-Strausz, A.K. (2007) *Hotel: An American History*. New Haven, CT: Yale University Press.
Stobart, J. (2017) Introduction. In J. Stobart (ed.) *Travel and the British Country House: Cultures, Critiques and Consumption in the Long Eighteenth Century*. Manchester: Manchester University Press.
Sweet, R. (2012) *Cities and the Grand Tour: The British in Italy, c. 1690–1820*. Cambridge: Cambridge University Press.
Trollope, A. (1862) *North America*. New York: Harper & Brothers.
Walton, J.K. (2011) The origins of the modern package tour? British motor-coach tours in Europe, 1930–70. *Journal of Transport History* 32 (2), 145–163.
Walton, J.K. (2014) Family firm, health resort and industrial colony: The grand hotel and mineral springs at Mondariz Balneario, Spain, 1873–1932. *Business History* 56 (7), 1037–1056.
Wharton, A.J. (2001) *Building the Cold War: Hilton International Hotels and Modern Architecture*. Chicago, IL: University of Chicago Press.
Wood, R.C. (1994) Hotel culture and social control. *Annals of Tourism Research* 21 (1), 65–80.

# 12 Historical Employment Relations in the New Zealand Tourism Hotel Sector: From a Collective Past to an Individual Future

David Williamson

## Introduction

As in many countries, tourism has become a major economic driver in New Zealand, overtaking the dairy sector to become the nation's largest earner of export dollars. It is well on track to achieve the Tourism Industry Association's 'aspirational' goal of NZ$41bn in revenue and 4.5 million visitors by 2025 (Stuff, 2016; Tourism Industry Association New Zealand, 2014, 2015). Yet despite its economic importance, labour in this sector demonstrates all of the challenges commonly associated with the international tourism workforce: low wages, high turnover, high levels of casualization, skills shortages and a dependence on migrant workers (Lincoln University, 2007; Ministry of Business, Innovation & Employment, 2013; Tourism Industry Association New Zealand, 2015). There is a growing concern that labour issues may be a major limiter on future tourism growth and development. This chapter will address the question: How did we get here and what does this mean for the future?

This chapter takes a critical, historical employment relations approach, drawing on interviews and archival research to describe a cascading series of impacts on employment in the hotel sector. The chapter argues that the key employment relations turning points discussed below were greatly influenced by, or were the direct result of, dramatic post-war changes in both international and New Zealand political, economic and social policies. It will highlight New Zealand's transformation from a post-war corporatist nation to a post-1980s example of neoliberal orthodoxy. Corporatism refers to the sociopolitical organization of society into major interest or 'corporate' groups (business, labour, military or ethnic groups) based on common interests (Wiarda, 1997). Corporatism has been widely used to interpret the results of international government responses to the oil crisis driven economic upheavals of the late 1970s and 1980s and provides rich analysis of the New Zealand post-war experience (Molina & Rhodes, 2002; Schmitter, 1989; Schmitter & Grote, 1997; Wiarda, 1997).

The chapter will initially present four turning points, in matching pairs. The first two turning points highlight conditions under post-war corporatism and present the rise of the Hotel Workers Union and the birth of the Tourism Hotel Corporation (THC). The last two turning points trace the changes wrought under the post-1984 neoliberal revolution, highlighting the fall of the Hotel Workers Union and the demise of the THC. The chapter will conclude by suggesting two dramatically different scenarios for the future of the tourist hotel workforce.

## Who are the voices in this narrative?

Table 12.1 Key voices in the narrative

| | |
|---|---|
| Julian Bugledich | Julian was Regional Manager, New Zealand, of Southern Pacific Hotels (SPHC) from 1990 to 2000. He oversaw the purchase of the THC assets by SPHC in 1990 and was head of the first major multinational corporate hotel group in New Zealand. |
| Denis Callesen | Denis joined the THC in 1970 as a trainee manager and worked his way through many properties before achieving Deputy CEO of the THC by the age of 28. Denis continued to work for the SPHC after 1990. He managed the Hermitage Hotel for many years and is closely associated with this iconic property. |
| Tim DiMattina | Tim joined the THC as an accountant in 1973; he was Financial Controller of the THC by the age of 25. Tim oversaw the financial management of the THC until 1982 and then later worked for the SPHC on the purchase of the THC. |
| Nigel Harper | Nigel arrived in New Zealand in 1968 and worked for the Intercontinental Hotel and THC Wairakei Hotel before becoming General Manager at the Hermitage Hotel. Nigel has been area manager for hotels in Asia and Australia and has worked for Accor. |
| Rick Barker | Rick joined the Hotel Workers Union in 1976. He is a past Secretary of the Northern Union, formed the SFWU, and is a former Labour Party MP. |

## Historical Turning Points

### Turning point 1: The Hotel Workers Union and rising wages

New Zealand had developed a particularly rigorous form of corporatism over the period of WWI, the Great Depression and WWII (Belich,

2001; Williamson, 2017a). These three disasters had forged a widely held consensus that collective, corporatist cooperation was the best way for the country to 'survive and thrive' (King, 2003). The attributes of this 'kiwi' corporatist system included an expansive, interventionist state that granted monopoly rights to industries and unions, while ensuring full employment, high wages and a 'cradle to the grave' welfare system (Belich, 2001). From education to wage and price levels, the state involved itself everywhere, by using public ownership of key industries and services and highly regulative legislation. As part of this consensus, the relationships between heads of leading industries, the union leadership and the government were very close. Employment relations in New Zealand were no exception to this corporatist approach and resulting employment legislation embodied the principles of collectivism.

The Industrial Conciliation and Arbitration Act (1894) can be seen as a classic corporatist approach, which determined New Zealand employment relations for almost a century. By 1955 this Act included compulsory unionism and a fully restored arbitration system, in addition to longstanding support for industry-wide collective agreements. For hotel workers, the main impact of this legislation was its powerful support for unions; the advantages of the IC&A Act to the Hotel Workers' Union, specifically, were profound. The industry that this union represented was typically made up of many small businesses – hotels and restaurants, cafes and tearooms – most having fewer than 20 employees. The ability to exclusively enforce an award across all workplaces in an entire industry, along with compulsory union membership, gave the union the ability to organize an otherwise very unstable and atomized workforce. This collectivist approach to employment relations reflected the wider social consensus of the time that the prosperity and social cohesion of New Zealand was well served by corporatist collective action (Kelsey, 1995; King, 2003; Roper, 2005; Trotter, 2007).

Thus, post-war employees in New Zealand tourist hotels found themselves protected by an industry-wide collective agreement which was negotiated by a very powerful and secure union. This was backed by a government that was committed to full employment and high wages. In this environment, hotel union membership grew from just over 20,000 in 1955 to over 60,000 members by 1990 (Williamson, 2017a). By enforcing a well-established collective agreement with multiple penalty rates and allowances, the Hotel Workers Union worked hard to ensure steadily rising wages. Data from Statistics New Zealand yearbooks show both the national wage and the hotel wage steadily growing in real value from 1957 to 1974 (+57% for the national average rate and +60% for the hotel rate; Williamson, 2017a). Both rates gain considerable value compared to the minimum wage during these years, with the hotel wage moving from being a modest 20% higher than the minimum wage in 1957 to double the minimum wage by 1979 (Williamson, 2017a). This 'state-led' form of corporatist collectivism not only resulted in the rise of the Hotel Workers Union (with consistently rising wages for hotel workers), but it also resulted in a new form of employer organization in the accommodation sector. The government sought to further develop tourism by creating a strong, state-owned hotel corporation. This new 'corporatist corporation' would have lasting impacts on training, skill development and career paths in the hotel industry.

### Turning point 2: The birth of the Tourist Hotel Corporation – training, skills and careers

In 1950 the magnificently titled 'Conference of New Zealand Tourist and Travel Interests and Expand the Dollar-spending Trade' reviewed the state of tourism and hotels in New Zealand. This conference can be seen as the birthplace of the modern tourism and hotel industry (McClure, 2004). Speakers described an industry that had struggled to make private capital work, particularly in the development of tourist hotels. The previous 60 years had seen entrepreneurs like Rodolph Wigley (founder of the Mount Cook Motor Car Service, eventual leasee of the Hermitage Hotel and builder of the Chateau Hotel, Tongariro) struggle to survive. By 1955, the Department of Tourist and Health Resorts (DTHR) had gathered 10 of these enterprises under its wing, mostly elderly, costly and remote tourist hotels. The DTHR was faced with rapidly growing numbers of inbound tourists (as wartime travel restrictions ended), but in addition to aging hotel stock, they faced disrupted capacity due to hotel fires, volcanic eruptions and tight government restrictions on building supplies, imports, labour, liquor and pricing (McClure, 2004).

The conference resulted in the formation of the Travel and Holiday Association, which began promoting the idea of a specialized hotel management corporation to advance New Zealand tourism. Intensive lobbing of National MP Eric Halstead led to the government presenting the Tourist Hotel Corporation Act for its first reading on 24 August 1955. The Act established a 'state owned enterprise', a government-owned corporation, to be managed by experienced hoteliers and businessmen. The THC would be tasked with achieving two things: stem the growing losses associated with running the hotels and crucially, *encourage the development of the tourist hotel industry in New Zealand*. The THC was owned by the state, tasked with developing the tourism industry, and went about its business in tight cooperation with the Hotel Workers Union (Williamson, 2017a). The impact of a 'development focused' THC on front-line workers was immediate.

Front-line hotel workers enjoyed significant training and development through the THC. In addition to union-based chef training, the THC quickly gained a reputation as the pre-eminent hotel group in terms of service provision and development of their staff: 'You've got to remember that the THC was unique in that the trainee programme was world class at that time' (T. DiMattina, personal communication, 27 June 2014). The THC quickly became the 'gold standard' for hotel service delivery in New Zealand: 'The THC was the leading trainer of staff in hotels. They saw themselves as the premier hotel [group] and if you got an apprenticeship at the THC as a chef, well then you would be considered well trained. If you worked in the waiting or bar staff ... you were like an A grade if you had worked your time at the THC' (R. Barker, personal communication, 30 July 2014). From its foundation in 1955 through to its sale in 1990, the THC channelled government investment into raising service standards nationwide. This provided much of the foundation for the subsequent growth of the New Zealand tourism sector (Williamson, 2017a).

Managers also greatly benefited from this THC commitment to career progression and training. Many senior general managers attribute much of the success in their careers to the start the THC gave them. 'I was general manager when I was 26 ... so the opportunity was there ... [the THC] launched my career' (N. Harper, personal communication, 23 June 2014); 'I was sent to the Rotorua International Hotel as second in command, I was 21 years old! That set the scene of my career for the next 42 odd years' (D. Callesen, personal communication, 14 May 2014). The rapid promotion of talented young managers established an entire generation of future leaders in the hotel sector. Tim DiMattina describes a similar experience of easy entry, high pay and a fast-track career: 'I went in and the guy who interviewed me ... was clearly in a total shambles [but] it was well paid and interesting ... within three or four years I was finance controller [of the entire THC] ... I was 25 years old' (T. DiMattina, personal communication, 27 June 2014). The point must be stressed that all of this talent development and capacity building was built on government commitment and taxpayer investment as part of the corporatist consensus.

### Summary of the post-war corporatist period

The post-war corporatist period drove real wages up and developed infrastructure, front-line employee skills and managerial careers which underpinned the future growth of New Zealand's tourism industry. To a large extent, the debt the New Zealand tourism industry owes this collectivist, corporatist period has been forgotten. However, by 1975, the corporatist consensus was failing and a new neoliberal philosophy sought to create an entirely new approach to work in the hotel sector which would have a destabilizing impact on hotel wages, training and careers.

### Turning point 3: The neoliberal revolution – the failing union and falling wages

Politically and economically, the 1970s saw the 'beginning of the end' of highly interventionist, corporatist government policies and the arrival of a series of nasty economic shocks. Pressure to change the fundamental approach of political and economic management in New Zealand was growing in a number of quarters. UK 'Thatcherite' and US 'Reaganite' policies of neoliberal, deregulated, free-market economics were promoted by powerful groups within New Zealand, including the Treasury, the Reserve Bank, the Business Roundtable (a prominent pro-business lobby group), the Employers' Federation and an influential faction within the Labour Party (James, 1986, 1992; Kelsey, 1993, 1995; Rasmussen, 2009; Roper, 2005; Walsh, 1991). Upon winning the 1984 Election, the Fourth Labour Government undertook a radical series of deregulatory reforms based on neoliberal, free-market philosophies. These reforms included a dramatic restructuring of the public sector, the selling of state assets and the comprehensive deregulation of the economy (Kelsey, 1993; King, 2003). The pace and scale of these reforms led to this period being commonly described as a new right or neoliberal revolution (Belich, 2001; Kelsey, 1993, 1995), and arguably, 'despite the controversies generated by Labour's 1984–90 reforms, few of them were reformed by subsequent administrations' (King, 2003: 495).

This movement from collectivist and corporatist approaches to individualist and neoliberal policies expressed itself in a gradual weakening of employment legislation protection for unions and workers. By 1991, the Employment Contracts Act fully expressed the neoliberal philosophy, seriously undermining remaining collectives by strong encouragement for individual contracting and removing compulsory unionism. The outcomes for hotel workers were dramatic. Across all sectors in New Zealand, multi-employer collective agreements dropped from 59% in 1991 to 8% by 1992 (Rasmussen, 2009: 87). Total union membership fell by half over the same period. However, the hospitality sector was disproportionately affected by these changes, with hospitality union membership falling by 81%, collective agreements being rapidly replaced by individual contracts and the resulting wholesale removal of penalty rates (Rasmussen, 2009; Williamson, 2017a, 2017b). This disproportionate effect was partly caused by the effect on the union of high labour turnover in an industry made up of multiple small to medium-sized employers; without the protections of compulsory unionism and collective agreements, the Hotel Workers Union virtually collapsed.

The significant weakening of the union, the reduction of collective agreements, the removal of penalty rates and rising unemployment all drove real hourly wages (adjusted for inflation) down by 24.5% from 1975 to 2000 (Williamson, 2017a). Thus, 30 years of steady rises in real hotel wages under the corporatist environment gave way to 25 years of significant reductions in real wages under the neoliberal consensus. The Employment Relations Act of 2000, while signalling the end of further weakening of employment legislation, failed to make a significant impact on union membership, wage rates or conditions in the hotel sector. The argument can be made that strong growth in the hospitality and tourism sector, combined with the collapse of wages and conditions discussed above, has resulted in a steady rise in the use of temporary migrant labour since the mid-1970s. The industry has increasingly struggled to fill vacancies since de-corporatization (Cropp, 2016; New Zealand Tourism Industry Association, 2015; Williamson, 2017b).

### Turning point 4: The end of the Tourism Hotel Corporation – the stagnation of training and development

It seems remarkable that in 1991 both of the monolithic organizations that had dominated the hotel sector for the last 35 years ceased to exist. In a bid to survive, the Hotel Workers Union had amalgamated with other unions and formed the Service Workers Union, and the THC was sold to the Southern Pacific Hotel Corporation (SPHC). The sale of the THC marked a dramatic change in the strategies, aims and employment relations approaches of this hotel management group. The hotels were now owned by a Chicago-based multinational hotel corporation, rather than the THC, a New Zealand based state-owned enterprise. The focus was no longer on the development of the New Zealand tourism sector and its human capital, but rather profit and shareholder value. This immediately altered management aims and the tone in which they were expressed:

> We basically ripped the crap out of the company and just kept the jewels. We wanted to sell the hotels. Our philosophy was, we were not hotel owners, we were hotel managers. If we could dispose of the properties and maintain management agreements, that was the objective. We sold about $90 million worth of property and still had ... the management rights. So we recouped ... the [initial] funds basically and were still earning good money. (J. Bugledich, personal communication, 21 June 2014)

SPHC were the first New Zealand hotel group to establish the now common practice of separating the physical ownership of hotel property (land and buildings) from ownership of ongoing management rights. The shedding of physical property ownership freed up capital for the new hotel group to develop their marketing and brand strengths, to concentrate on distribution channels and to focus on cost-reducing management of the hotel labour force. Realizing the freedoms that the new Employment Contracts Act (1991) granted them, senior SPHC managers immediately set about removing the penalty rates from new agreements. However, senior management realized that under the ECA, not only the penalty rates but perhaps even the union itself could be disposed of. The SPHC head Julian Bugledich and Regional HR Manager Terry Hiltz engaged in a year-long campaign to replace the old award with the new agreement:

> With Terry we ... embarked on a road trip ... convincing employees to come into this agreement. We [moved] into individual employment

agreements, but we had a lot of convincing [to do], we had about 1,800 union members. The union didn't want to take part ... so we didn't increase their [the employees in the union agreement] salaries. The union didn't want a bar of it, as a result, they were excluded from the bargaining process, which was quite a big shift industrially ... it took away a lot of the negotiating ability of the union. We went from 1,800 [union members] down to about 120. (J. Bugledich, personal communication, 21 June 2014)

While some senior managers saw the 1990s as a triumphant period of union defeat, business success and employee empowerment, others are more circumspect about the impacts on employees: 'It was better for business, it was a huge leap backwards for the workforce themselves' (D. Callesen, personal communication, 14 May 2014). The arrival of the SPHC approach raised questions with some hotel leaders. According to Tim DiMattina, the modern hotel industry, with its high staff turnover, low wages and a dependence on short-term visa based migrant labour, was the result of a 'double whammy' in the 1990s: 'I thought at the time the legislative changes had the greatest impact ... but then **that** was swamped by the internationalisation of the industry' (T. DiMattina, personal communication, 27 June 2014). DiMattina goes on to argue that the combination of 1990s legislative change and the arrival of an international, almost entirely profit-focused management approach has resulted in the current poor conditions of employment in New Zealand hotels: 'The industry has done it to itself ... you know, with cost savings. You've got to understand, the profit motive drives everything' (T. DiMattina, personal communication, 27 June 2014).

The legacy of the THC transfer of assets to the Southern Pacific Hotel Corporation proved to be mixed. Expectations were that the SPHC, freed from the capital restraints of property management and freed from the restraints of 'rigid' union-based collective negotiation, would prosper. This proved to be the case, but it did not end up being of advantage to the New Zealand hotels, managers or employees: 'SPHC ... New Zealand were doing well, [but] SPHC Australia was a disaster' (D. Callesen, personal communication, 14 May 2014); 'We were sending back [to Australia] about five or six million dollars a year at that stage' (J. Bugledich, personal communication, 21 June 2014). Denis Callesen argues that in the 1990s, the SPHC actually oversaw a large step backwards in service and hotel quality:

> All the cash profits were going to Australia, so [there was] no capital expenditure, a sinking lid type [of management] ... I think it took quite a leap backwards. THC's level of service, training, management, quality was far better than what SPHC was [offering]. (D. Callesen, personal communication, 14 May 2014)

Thus, replacing the long-term, developmental focus of the THC with the short-term, profit-focused approach of the SPHC resulted in the disruption of training and development, falling skills and service levels and the ultimate failure and sale of the new hotel chain.

## Future Scenario One: Dystopia

### Future turning point 1: Ongoing neoliberal consensus and the 'individual' future

In this scenario, the progress of the New Zealand tourism and accommodation industry remains uninterrupted by major international disasters and the current social, economic and political neoliberal consensus remains intact. The impacts from the milestone identified above continue on their current trajectory. In this setting, the political and economic goals remain growth, free-market openness and the favouring of business success/profit/shareholder return over the protection and prosperity of the average local citizen. The tourism sector continues to grow predominantly through increasing volume rather than value, creating a bifurcated industry with a few high-end opportunities but a majority of undifferentiated, 'mass-service factory' hotels. Current disruptions by online sales platforms continue to expand, driving increasing price pressure on hotels, which in turn increases the pressure for cost savings in labour. In this scenario the following turning points could be expected.

*Collectivism, unions and pay*

The continuation of the neoliberal consensus results in little appetite for a return to strong unions, collective negotiation or industry-wide collective agreements. As a result, unions remain weak in the sector and there is little major change in employment legislation to substantively alter the conditions of hotel workers. The average hotel wages continue to hover just above the minimum wage rate, while in other industries average pay rates continue to rise, slowly pulling further and further away. This pay differential continues to be a major driver of labour turnover as other industries offer more attractive wages and conditions than hotels. The already acute lack of local employees in the hotel sector becomes accentuated as relative pay and conditions in other sectors become increasingly attractive to prospective New Zealand workers. This, combined with ongoing growth in the sector, feeds an insatiable need for migrant labourers to fill the gaps. Political tensions that arise from the required high levels of temporary migration would have to be managed by emphasizing the neoliberal foundations of international free movement of people and goods.

*The industry, managers and development*

The continuation of current settings results in little change for future industry operations. Concentrated international ownership remains the standard for the hotel industry with very little participation of local capital in the sector. Management continues to pursue a mostly unitarist, short-term profit focus, prioritizing the return of superior shareholder value above considerations of local development or reward. Downward price pressures from online aggregators accentuate long-term trends of focusing on efficiency, productivity and replacing humans with technology wherever possible. This in turn continues to result in low pay, intensive work conditions and high labour turnover in the majority of hotels. The hotel sector could be expected to invest heavily in supporting political parties that continue the neoliberal consensus and would be very active in lobbying for ongoing access to temporary migrant labour. Growth in 'super-high end' properties could lead to opportunities for highly skilled employees to earn above average wages and enjoy superior training and development, but the majority of the industry would be geared around the current mass-service provision and high churn labour approaches. Local workers cease to seek training or education for hospitality as the sector continues to become less and less attractive compared to other industries.

*What work in the future industry looks like*

Individual workers struggle to build careers in this future hotel industry, fighting to gain roles in the minority of upmarket operations that invest in training and above-average wages. Unions are too weak to drive change. In the majority of operations, a mostly international workforce of temporary migrant labourers exist on minimum wages, and just-in-time training ensures the basic delivery of services for bargain price accommodation. Customers will continue to remark on the lack of locals working in the sector and the poor service standards; only those who can afford the luxury class hotels will find New Zealand workers and decent service. An increasing poorly paid and pressurized managerial class continue to use technology to replace people wherever they can and otherwise drive efficiency and productivity to maximize profit.

**Future Scenario Two: Utopia**

**Future turning point 2: Disruption drives change – a collective future**

In this scenario, the New Zealand tourism and accommodation industry is disrupted by major economic challenges and the current social, economic and political neoliberal consensus is significantly revised. Imagine major international economic disruptions hitting New Zealand in the near future. International trade, share markets and national economies tremble. International tourism is severely disrupted. As a result, serious questions are asked about the political and economic direction the country has taken in the past 30 years. In this setting, a national debate rekindles the lessons of previous disasters, the great depression and the world wars. A strong political movement calls for a return to more collectivist, nationalist protection of New Zealand citizens over international shareholder returns. A new government bases recovery for the country and the tourism sector on increasing volume and value equally. As part of a generally more protectionist and interventionist approach, this government enacts a series of legislations that limit completion in the hotel market, fix minimum room prices and strongly encourage collective bargaining in the industry. A major drive begins to build a sustainable, high-value accommodation sector which can be mostly staffed by well-paid and well-trained local employees. In this scenario the following turning points could be expected.

*Collectivism, unions and pay*

As part of a general move back to a more collectivist, interventionist approach, the government enacts employment legislation with strong support for unions and compulsory, industry-wide collective agreements that stipulate a comprehensive set of minimum conditions. Using the collectives as a vehicle for lifting wages and other conditions in the hotel sector, the government convinces employers to participate by limiting completion and driving up room rates through minimum price controls. The legislation requires hotel employees to belong to a union in order to enjoy the protections of the collective agreement. This increases union membership dramatically, as the conditions in hotels under the collective agreement quickly become more attractive than those under individual agreements. Reverting to a neocorporatist approach, the state requires unions and employers to work together on a series of strategic projects to develop the tourism and hotel sector. Backed by considerable investment by the government, long-term educational, training and professional development systems are coordinated and integrated into the unions and hotel groups, in conjunction with other stakeholders. The goal is to create clear, visible and rewarding career paths in tourism and hospitality, from school level to senior management. The collective agreements strongly support the hiring and development of local, full-time employees, rather than temporary and casual staff. There is a structural move away from relying on migrant labour towards attracting local workers.

*The industry, managers and development*

Faced with considerable economic challenges, hotel organizations reluctantly accept the neocorporatist shift in politics. Accepting a trade-off between government-protected room rate increases and rising wages and conditions, employers conclude that adapting to the new, long-term and cooperative approach 'inside the government tent' will be better than trying to survive 'outside' and alone. Managers are now faced with having to build a local, stable, long-term, well-trained and well-rewarded workforce, rather than being reliant on high-turnover temporary migrant labour. As a result, these managers accept the fact that, in order to build this workforce, real engagement and genuine care will be required. With the buffer of government support and minimum prices, the hotels are partially relieved from a relentless drive for increased efficiency, instead focusing on building value through developing highly skilled and highly productive work teams. As the government invests more heavily in local skill development and building tourism and hotel infrastructure, both managers and front-line staff can see long-term futures in the New Zealand hotel industry.

### What work in the future industry looks like

Local workers are attracted to a vibrant hotel industry by a well-planned network of career paths which start in school courses and continue all the way to senior career development learning. Jobs in hotels are well paid and secure and guarantee significant training and development to ensure exciting career opportunities. Workers know that a strong union has their back, but also that the union, employers and government are working together to develop training, development and rewards initiatives to ensure the sustainability of the labour market. By focusing on value as well as volume, the New Zealand tourism and hotel sector has the fastest rising levels of service excellence in the world; the training systems and management approaches are considered world leading. Tourists engage with highly trained, motivated and happy locals working alongside talented international staff. While visitors acknowledge the higher prices of New Zealand hotels, they also value the higher service standards and general excellence of the product. New Zealand hotel staff are considered 'A' grade by international recruiters. Slowly, hotel work becomes viewed as an attractive and rewarding career choice, no longer consider menial, part-time or second rate.

## Conclusion

This chapter has argued that New Zealand's post-war transition, from corporatist to neoliberal consensus, has greatly disadvantaged the workers in the tourist hotel sector. By highlighting the demise of both the Hotel Workers Union and the Tourist Hotel Corporation, the narrative links these losses to falling wages, worsening conditions and the disruption of training and development in the industry. By telling the story of how the hotel sector has come to be a place of low pay, high turnover and migrant dependency, the chapter concludes by exploring the possibilities of either continuing the current neoliberal approaches or seizing a different neo-corporatist future. The hope is that rather than locking readers into an 'either/or' duality, this chapter might spark debate around the possibilities of taking new initiatives in the employment relations space across the whole spectrum of political and economic approaches.

## References

Belich, J. (2001) *Paradise Reforged: A History of the New Zealanders from the 1880s to the Year 2000*. Auckland: Allen Lane.
Cropp, A. (2016) Hotel staff shortages spell trouble for tourism. *Sunday Star Times*, 31 January, p. 3.

James, C. (1986) *The Quiet Revolution: Turbulence and Transition in Contemporary New Zealand*. Wellington: Allen & Unwin.

James, C. (1992) *New Territory: The Transformation of New Zealand, 1984–92*. Wellington: Bridget Williams Books.

Kelsey, J. (1993) *Rolling Back the State: Privatisation of Power in Aotearoa/New Zealand*. Wellington: Bridget Williams Books.

Kelsey, J. (1995) *The New Zealand Experiment: A World Model for Structural Adjustment?* Auckland: Bridget Williams Books.

King, M. (2003) *The Penguin History of New Zealand*. Auckland: Penguin Books.

Lincoln University (2007) Enhancing financial and economic yield in tourism. Lincoln, NZ: Tourism Recreation Research and Education Centre, Lincoln University. See http://hdl.handle.net/10182/279.

McClure, M. (2004) *The Wonder Country: Making New Zealand Tourism*. Auckland: Auckland University Press.

Ministry of Business, Innovation & Employment (2013) *The New Zealand Sectors Report 2013 – Tourism*. See https://www.mbie.govt.nz/immigration-and-tourism/tourism-research-and-data/tourism-data-insights/tourism-sector-report/.

Molina, O. and Rhodes, M. (2002) Corporatism: The past, present, and future of a concept. *Annual Review of Political Science* 5 (1), 305–331. doi.org/10.1146/annurev.polisci.5.112701.184858

New Zealand Tourism Industry Association (2015) *People & Skills 2025*. See https://tia.org.nz/resources-and-tools/insight/people-and-skills-2025/.

Rasmussen, E. (2009) *Employment Relations in New Zealand* (2nd edn). Auckland: Pearson.

Roper, B.S. (2005) *Prosperity for All? Economic, Social and Political Change in New Zealand Since 1935*. Victoria: Thomson/Dunmore Press.

Schmitter, P.C. (1989) Corporatism is dead! Long live corporatism! *Government and Opposition* 24 (1), 54–73.

Schmitter, P.C. and Grote, J.R. (1997) The corporatist Sisyphus: Past, present and future. SPS Working Paper. See http://cadmus.eui.eu/handle/1814/284.

Stuff (2016) International tourism overtakes dairy to regain top spot as our biggest export earner. *Stuff.co.nz*, 24 December. See http://www.stuff.co.nz/business/75443924/international-tourism-overtakes-dairy-to-regain-top-spot-as-our-biggest-export-earner.

Tourism Industry Association New Zealand (2014) *Tourism 2025: Growing Value Together*. Wellington: Tourism Industry Association New Zealand. See http://www.tianz.org.nz/assets/Tourism-2025/59e63f563b/TIA_T2025+B-Doc_A4_v11.pdf

Tourism Industry Association New Zealand (2015) *Tourism 2025: People & Skills 2025*. Wellington: Tourism Industry Association New Zealand. See https://tia.org.nz/assets/86eb0cde68/People-Skills-2025-November-2015.pdf.

Trotter, C. (2007) *No Left Turn: The Distortion of New Zealand's History by Greed, Bigotry, and Right-wing Politics*. Auckland: Random House.

Walsh, P. (1991) Trade unions in New Zealand and economic restructuring. ACIRRT Working Paper No. 17, University of Sydney. See http://hdl.handle.net/2123/12352.

Wiarda, H.J. (1997) *Corporatism and Comparative Politics: The Other Great 'Ism'*. Armonk, NY: M.E. Sharpe.

Williamson, D. (2017a) In search of consensus: A history of employment relations in the New Zealand hotel sector – 1955 to 2000. PhD thesis, Auckland University of Technology. See http://aut.researchgateway.ac.nz/handle/10292/10412.

Williamson, D. (2017b) Too close to servility? Why is hospitality in New Zealand still a 'Cinderella' industry? *Hospitality & Society* 7 (2), 203–209. doi.org/10.1386/hosp.7.2.203_7

# Part 5

# Diversification into Niche Tourism

# 13 Film Tourism through the Ages: From Lumière to Virtual Reality

Peter Bolan and Mihaela Ghisoiu

## Introduction

On 28 December 1895 the Lumière brothers presented the world's first public film screening in Paris. They could not have foreseen the impact on the world, future events and indeed people's everyday lives, knowledge, values and decisions. Since the late 19th century, film, and subsequently television and digital streaming services, have increasingly influenced people's interests and decision-making processes, including their travel choices and tourism experiences. This chapter charts and examines key turning points, from the Lumière brothers' screening in the 1800s to the impact of 1940s classics such as *The Third Man* and the streetscapes of Vienna, through the 1950s and 1960s with *The Quiet Man*'s appeal to the American diaspora in visiting Ireland, the draw of *Roman Holiday* to visitation of Italian vistas, the influence of *Crocodile Dundee* on tourists to Australia in the 1980s, the *Braveheart* effect on Scottish tourism in the 1990s, The Lord of the Rings movies and their tourism impact on New Zealand, the lure of Harry Potter to a Britain associated with magic and mystery, to the enormous phenomenon of *Game of Thrones* and its impact on tourist visitation to Northern Ireland. The influences and significance of film/screen tourism through history is examined via such milestone examples, and key turning points such as *Braveheart* in 1995 and later The Lord of the Rings in 2001 in securing tourist board industry involvement at national and international levels are explored in depth. Further, the chapter looks forward to the growing influence of digital media on how viewers consume their film/television content (through digital streaming services, on mobile devices, wearable tech and evolving elements of AR, VR and holographic technology) and how this will fundamentally change the nature of such content itself and in turn the impact it will have on the tourist of tomorrow.

## Historical Turning Points

### Turning point 1: Lumière lighting the way (1800s and early 1900s)

With the creation of the *Cinématographe* in the late 1800s, Auguste and Louis Lumière opened a window to a new world – a world which has pushed the boundaries of human creativity and imagination, fuelling the human desire for escapism to new places and fantastical vistas. While origins of the moving image can arguably be traced back to magic lantern slide shows in the 17th century, it was the work of the Lumière brothers that pioneered the development of the film industry as we know it today. While the French were in many ways the instigators, they were soon joined by the British, Scandinavians, Italians, Russians, Germans, Indians, Japanese and of course the Americans.

Even the earliest film screenings of the late 1800s and early 1900s created an enticement for audiences to travel, especially since one of the very first screenings depicted a moving train. Both of these technological developments marked the beginning of mass movement of people (Karpovitch, 2010). 'Lumière's films of natural vistas extended film into the realm of travel, connected to practices of mobility and taxonomy found in the discourse of colonialism' (Friedberg, 1993, cited in Harbord, 2002: 22). By the late 1920s and early 1930s the American film industry was beginning to dominate over Europe in the film world, with the growth of Hollywood, the development of sound and an ever-increasing array of new subject matter providing escape for people into uncharted places. The concept of film tourism was well and truly under way. Beeton (2005: 3) advocates the strong influences from film having a profound effect from these earliest of times, stating: 'the link between travel and popular media, particularly in terms of imaginative literature, has long been recognised, while the pervasiveness of film in today's globalised society has strengthened this link'.

The phenomenon of film tourism has been defined as '... tourist visits to a destination or attraction as a result of the destination being featured on television, video, DVD or the cinema screen' (Hudson & Ritchie, 2006b: 256). While television, video and DVD were the stuff of science fiction to the Lumières and indeed still so in the 1920s and 1930s, nonetheless, film in its purely cinematic form was gathering pace, increasing in popularity and influencing people's desires and emotions.

### Turning point 2: From *The Third Man* to *The Quiet Man* (1940s and 1950s)

It was arguably the post-war decades of the 1940s and 1950s that saw some of the earliest truly significant examples of the medium of film influencing a strong desire to travel. It was an important turning point because this impact of some of these films still continues today (demonstrating the long-enduring impact of film tourism). Post-war films, also known as 'rubble' or 'trümmerfilme', mark a unique genre dealing with the physical and cultural devastation following WWII (Randall, 2015). This period in cinema sees some of the most influential films and their extraordinary directors, one such example being *The Third Man*. It has been listed among the 15 essential post-war films by IndieWire (2015) and as the best ever British film by the British Film Institute in 1999 (IndieWire, 2015; Zwart, 2015).

The influence of Carol Reed's 1949 film *The Third Man*, set in post-war Vienna with its atmospheric direction, iconic music and acclaimed performances from Orson Welles, Joseph Cotten and Trevor Howard, still draws tourists to the Austrian capital city (Bolan & Davidson, 2005). Interestingly, even though the streets of Vienna play a significant role in the movie, the veracity of the scenes has been called into question both by the audience of the time and later by scholars as being too unrealistic (Randall, 2015). Despite this, even today (some 70 years later), tourists can still retrace the footsteps of Welles' Harry Lime on the Third Man Sewer Tour, visit the Riesenrad (the famous ferris wheel from the film), and view props and memorabilia at the Third Man Museum (Phelan, 2017). The film was indeed so iconic that it inspired the title of the autobiography of Nobel Prize winner Maurice Wilkins, the co-discoverer of the structure of DNA (Zwart, 2015).

Moving away from the rubble of post-war city streets, films of the period also tackled the desolate rural life inhabited by many, a landscape from which one had to escape through migration (Frost & Laing, 2014). In a reverse migration, John Ford's *The Quiet Man* (1952) sees the character of Sean Thornton played by John Wayne returning to his childhood Ireland from America. In order to give the film a sense of authenticity, the director, Ford, insisted on shooting on location in Ireland, a practice that was less common in Hollywood at that time (Frost & Laing, 2014). Based on the strong ties between Irish-Americans and their homeland, especially in the 1950s, films had a great impact on tourism in Ireland (Cronin & O'Connor, 2003) and as such the country became a notable example of one of the earliest phenomena of film tourism.

Shot largely in the region of Cong in County Mayo, the film struck a particular chord with American audiences not only because of their ancestral ties with Ireland but also because of the depiction of a fairy tale village frozen in time (Frost & Laing, 2014). To this day, over 60 years later, this rural region still attracts American tourists (Bolan & Kearney, 2017). Guided tours, a Quiet Man Cottage Museum, location maps and more recently a life-size statue of John Wayne holding Maureen O'Hara in his arms are all designed to provide film tourists going to Cong with a taste of idyllic Irish village life.

### Turning point 3: The screen impact of the 1960s

The 1960s saw the emergence of an era of liberation – especially for women escaping the confines of the kitchen – with a newly found sexuality and sensuality. Nobody expressed it better than Federico Fellini's movies filmed at the famous Italian studios, Cinecittà. One of his most acclaimed creations, *La Dolce Vita*, centres around Via Veneto, a street in Rome which at the time teemed with cafés where intellectuals, journalists and famous actors could meet and converse (Kezich, 2005). In a way it echoes another famous district, Montmartre, a century older and in another city, Paris, where the bohemians of the time gathered and made art. The true protagonist of *La Dolce Vita* is Rome itself, which saw a revival following WWII and the death of Pope Pius XII in 1958 who was known for disagreeing with the city having a nightlife (Kezich, 2005). The now iconic scene when the character of Sylvia, portrayed by Anita Ekberg, plays in the Trevi Fountain must have remained stamped on the retina of the viewers, not only the ones who in the movie surround the fountain, but especially the ones beyond the screen. Rome now became synonymous with fame and glamour, and Fellini's set became a destination for tourists eager to see the actors at play (Kezich, 2005).

This echoed and built upon how Rome was seen through American eyes in the romantic comedy *Roman Holiday*, starring Audrey Hepburn and Gregory Peck. Destined primarily for American audiences, the film became an ambassador for Italy and Rome through its portrayal of a city of light, atmosphere and adventure. The audience is presented through the film with a veritable tourist map featuring the most iconic places in the city (Rodriguez *et al.*, 2011). The 1960s was therefore a golden time both for cinema and for travel. Gundle (2002, cited in Hudson & Ritchie, 2006a) discusses the impact of *La Dolce Vita* (1960) and how it transformed the image of Rome into a city of sin and pleasure, of Audrey Hepburn, Liz Taylor, Ava Gardner and Frank Sinatra, of elegance and nightclubs, of aristocrats, Latin lovers, fast cars and stylish intellectuals. This is further supported by Bolan (2010), who advocates that such imagery conjured up and displayed by Federico Fellini's film was further developed by numerous American films of the 1960s to create an image of Italian glamour and style to the world that greatly encouraged international tourist visitation.

### Turning point 4: *Crocodile Dundee* (1980s) and reinvigorating economies through film tourism

While a number of varied films through the 1970s (*Ryan's Daughter*, *Deliverance*, *Close Encounters of the Third Kind*, etc.) had an impact on tourism, it was arguably in the 1980s that the next major turning point occurred, in terms of impact and especially on the recognition from tourism authorities that here was a viable and economically significant form of generating tourism and reaching new markets. In September 1986, cinema audiences around the world were introduced to Mick 'Crocodile' Dundee, the wise-cracking Australian bush-tracker (played by Paul Hogan). Here was a character who for a time would come to define people's perceptions of the typical 'Aussie' and a film that served to showcase the country of Australia and its landscape scenery (such as the Kakadu National Park) to the world. It was a huge commercial success: from a budget of just over AU$7 million, the film earned over $300 million at the box office (Plautz, 2016).

Coinciding with a time in the late 1980s and early 1990s when the Australian dollar was low and international airfares were becoming more affordable, Australia was becoming a more viable holiday destination choice. This was then brought more prominently to the world's attention by the success of *Crocodile Dundee* and the vast reach the film had with global audiences. Riley and Van Doren (1992) were the first major paper in academic circles to credit the film's tourism impact in their destination promotion focused article in the *Tourism Management* journal.

Testament to the success of this film example from the 1980s and its impact regarding Australian tourism, a recent initiative has tried to recapture some of that film-induced tourism success and in doing so has become one of the most expensive film tourism marketing campaigns ever. A series of teaser trailers, culminating in a full trailer for a new film called *Dundee* about the son of the original Crocodile Dundee and starring a host of well-known Australian actors (such as Chris Hemsworth, Hugh Jackman, Russell Crowe and Margot Robbie) began to hit the media in late 2017. The full trailer was first broadcast during the American Superbowl in January 2018, hitting the internet very quickly afterwards. Interestingly, the trailer was for a fake movie. There was to be no new *Dundee* film. It was simply a very expensive marketing campaign (to the tune of AU$28 million) by Tourism Australia.

### Turning point 5: The *Braveheart* effect and the rise of the hero (1990s)

The 1990s saw the rise of historic freedom-fighting figures in movies, particularly William Wallace in *Braveheart* and Rob Roy MacGregor in *Rob Roy*, which echoed the realities of freedom fighters in Europe such as in Scotland, Spain or Belgium (Wallace, 1996). Although predominantly filmed in Ireland, the movie *Braveheart* (1995) marks a significant turning point in the impact of film on tourist interest and visitation to the setting of the movie, in this instance, Scotland. The direct impact on Scottish tourism from both *Braveheart* and *Rob Roy* amounted to a financial income of $30 million (Gjorgievski & Trpkova, 2012), as well as an increase in visitor numbers by 300% to the Wallace monument the year after the release of *Braveheart* (Hudson & Ritchie, 2006a).

The Scottish Tourist Board saw the potential spin-off impact from *Braveheart* in particular (a more commercially successful and Oscar-winning example than *Rob Roy*), and jumped at the chance to brand Scotland as 'Braveheart Country' (Grihault, 2003). This was a relatively early example of tourism authorities recognizing the tourism potential of a film (certainly in UK terms). Winning the Oscars for Best Picture and Best Director while also grossing over $210 million worldwide, *Braveheart* showcased a romantic story of heroism against a perceived dramatic Scottish landscape (actually Ireland) which served to inspire and encourage tourists from around the globe.

Shot entirely on location in Scotland, the film *Rob Roy* was released in the same year as *Braveheart* (1995). While not as commercially successful, the film (which portrays the story of 18th century Clan Chieftain Rob Roy MacGregor, something of a folk hero in Scotland) still picked up a number of coveted awards and added further impact to the 'Braveheart Effect' that was already occurring. These films also serve to highlight the important linkages between historically set movies, heroism/folklore and heritage tourism. While it is accepted that films can enhance the awareness of a destination (Rittichainuwat & Rattanaphinanchai, 2015), it is more that historic movies through their portrayal of castles and romantic landscapes act as a draw for tourists to visit these destinations (Croy & Walker, 2003; Frost, 2006).

## Turning point 6: The magic of Middle Earth and Harry Potter (2000s)

In the year 2001 a particular movie took the film world by storm. The movie was *The Lord of the Rings: The Fellowship of the Ring*, based on the literary works of J.R.R. Tolkien. According to Bolan (2010: 70), 'Director Peter Jackson brought to the cinema screen what many had thought to be un-filmable and produced a series of movies which received popularity and acclaim from film critics and the cinema-going public alike'. The Lord of the Rings (LOTR) films became the modern 21st century benchmark for how such a medium could significantly influence tourism. They also brought a renewed interest from academics in terms of research into the phenomenon of film tourism (Beeton, 2005, 2006; Bolan & Williams, 2008; Carl *et al.*, 2007; Jones & Smith, 2005; O'Connor & Bolan, 2008). In the wake of such research, the industry began slowly to accept that here was a form of tourism that could actually have a high-impact value and that should be encouraged, developed and leveraged far more proactively than had been the case in the past.

The New Zealand government bought into the concept from the beginning and invested some US$18.6 million on projects to promote the country in connection with LOTR (Grihault, 2003). Furthermore, according to Grihault (2007) the country also won a top 'cinema sightseeing' award after a poll by UK tour operator Thomson Holidays, and experienced growth of 50% between 2001 and 2005 from visiting UK tourists alone. The effect was profound and changed people's views of the country, the destination, its landscape and its people. 'The representation of New Zealand's landscapes in the films of The Lord of the Rings (LOTR) has led to the construction of new tourism spaces, thus, new "landscapes of pleasure" tailor-made for film tourists' consumption' (Carl *et al.*, 2007: 49).

Following a similar timeline in the early stages of its movie form, Harry Potter can be seen as one of the most eloquent examples of how literature transposed into film can influence film tourism. The first film, *Harry Potter and the Philosopher's Stone*, was released in 2001, the same year as the first LOTR movie. Based on the highly successful novels of J.K. Rowling, Harry Potter has become quite simply a phenomenon. According to Brown and Patterson (2009: 522), 'More than 100 full-length books have been published on the much-loved boy wizard'. The literary interest, developed ever more strongly by each of the subsequent films, has created a huge and highly dedicated fan base. These fans rapidly translated themselves into tourists. Alnwick Castle in Northumberland (which features as part of Hogwarts in the films) more than doubled its intake of tourist visitors following the release of the first film alone (Oxford Economics, 2007).

### Turning point 7: *Game of Thrones* (2010 onwards)

One of the greatest phenomena in the film and television world in the last 10 years has been the HBO-produced show, *Game of Thrones* (based on the literary works of George R.R. Martin). The show began development in 2007 with the first season broadcast in 2010/2011. Northern Ireland is home to the majority of the studio work (Titanic Studios in Belfast and the Linen Mill Studios in County Down). With much of the outdoor depiction of Westeros (the fictional setting of the show) also being filmed in Northern Ireland, utilizing a range of countryside and coastal locations, tourism here has benefited enormously from the impact of *Game of Thrones* (Bolan & Kearney, 2017).

Tourism Ireland ran its first official *Game of Thrones* (GOT) marketing campaign in 2014, backed by HBO (the American production company behind the show). As a result, a number of GOT location maps were created, which supports the work of authors such as Beeton (2005) and O'Connor *et al.* (2008) on the importance and relevance of movie maps and film trails in both enticing film tourists and adding value to their experience when visiting such destinations. According to Bolan and Kearney (2017: 2154), immersive experiences catering to the GOT fan base are now effectively providing the 'tangible' essence of Westeros that such visiting fans are seeking; at Castle Ward (Co. Down) which featured as 'Winterfell' in the television show, GOT medieval barbeques are now offered to tourists as part of an overall GOT experience which includes dressing in GOT-style costumes and taking part in activities such as archery (in the very courtyard that featured some famous archery scenes in Season One).

*Game of Thrones* has therefore become something of a phenomenon for both the image of Northern Ireland and its tourism industry. In much the same way as other recent examples such as LOTR and Harry Potter, GOT has provided a mechanism to coalesce the impact from the literary and filming world and tap into a highly dedicated and passionate fan base to come to Northern Ireland as visiting tourists.

## Future Turning Points

### Future turning point 1: Hyper-reality to virtual reality

Film, tourism and subsequently film tourism have come a long way since the ingenuity of the Lumière brothers in the late 1800s. Such changes and advancements have progressed through the examples and turning points already discussed. However, we are on the cusp of greater technological developments which will have a profound effect. Today's global media culture 'enables people to travel mentally and emotionally without moving in physical geography' (Jansson, 2002: 430). This means film and television allow people to experience places without the need to leave their home area.

Bolan (2010: 93) contends that 'representations of a place through film and television are traversing the landscape of the hyper-real'. Hyper-reality can be seen as a condition in which 'reality' has been replaced by simulacra. Such hyperrealism is an indicative symptom of postmodern culture. According to Iwashita (2006: 59), 'It can be argued that it is those media representations and images that people actually consume, rather than realities … through which they understand the world'.

So, the medium of film and television already provides a hyper-real version of reality, something which will inevitably grow with developments in virtual reality (VR). The concept of VR technology officially began in the 1950s, although a primitive form can be traced as far back as the 1860s (Virtual Reality Society, 2018).

> With virtual reality, information is not displayed in two dimensions via the computer monitor; instead, the user finds him/herself in the same dimension as and is immersed within the data. The experience within the realms of virtual reality is augmented with various sensory stimulations such as sight, sound and even touch, together with their respective feedback. (Cheong, 1995: 418)

Initially reserved to only a handful of domains, the potential for wide applications was quickly recognized and travel and tourism as an industry that could vastly benefit from it. The 1990s saw VR technology expanded to the wider public but the hype was quite short-lived as the technology failed to live up to its expectations (Virtual Reality Society, 2018). VR incorporated with realistic sight, sound and touch with smell and taste was still a thing of the future limited to movies such as *The Matrix* (1999). These shortcomings limited the extent of the experience and growth in its real-world applications.

In the travel and tourism industry, this technology has been incorporated on two levels: a macro level regarding tourism policy and planning, and a micro level manifested as a marketing tool for travel agencies (Cheong, 1995). Guttentag (2010) also recognizes at least six applications of VR within the tourism industry and includes preservation of heritage, management, marketing and education. Fast-forward to the second decade of the 21st century and VR technology is now increasingly connected to the visual arts, especially cinematography. AMD is an organization that provides hardware enabling VR experiences, and has its head office in Los Angeles. Strategically placed close to Hollywood, AMD can offer a platform to those studios that want to use VR experiences as means to showcase their latest releases. One such example is *Ghost in the Shell*, a sci-fi blockbuster featuring Scarlett Johansson. Premiered in Tokyo in 2017, the audience got the chance to experience being in the middle of the action via VR (BBC World Service Click, 2017). Interestingly, AMD was also listed as one of the official partners at the 2018 Baftas (Bafta, 2018), possibly indicating a future trend in the way films will be experienced.

Some of the issues being raised with VR technology and tourism refer to authenticity and motivation. Film tourists want to stand in the physical space where actors stood. It gives them the chance to feel like part of the filming experience and to re-enact certain scenes (Ghisoiu *et al.*, 2017). Although VR enables travel to places that would otherwise be difficult to reach, the question arises as to whether film tourists will be more motivated to be present physically rather than through a piece of hardware, regardless of how realistic it may render the space.

### Future turning point 2: Holographic technology

Advances in holographic technology may provide yet another avenue in creating a more immersive experience for the film and television viewer. A number of companies including the BBC have been experimenting with holographic televisions (holo TVs). Other businesses such as California-based Light Field Lab have been working on rooms with holographic video projection walls with up to hundreds of gigapixels of resolution to get close to something similar to the Holodeck from *Star Trek*.

Recently in Shanghai a landscape featuring a holographic moon above a myriad of holographic people and holographic creatures of all kinds was created in order to wow an audience of over 700,000 visitors to the city (Chen, 2018). In 2018, some 500 residents of Brussels were turned into virtual 3D statues using holographic imaging technology and put on public exhibition for a major tourism campaign to help welcome visitors back to the city after the continuing image impact of the 2016 terrorist attacks on the destination (Gianatasio, 2018). HoloLens technology may be another refinement that allows viewers (through a holographic mixed-reality headset) to watch their movies and television shows while feeling they are in the midst of the action. While such technology, already being experimented with, has not been effective enough to persuade most viewers to buy expensive headsets instead of their large flat screen television sets, as things improve this may indeed become the norm for many.

In the very near future, hologram tour guides in augmented reality are likely to be the user interface through which visitors gain a lot of their information. Such immersive digital technology has also begun to seep into the world of film and television. When we can soon inhabit the movies and television shows ourselves, surrounded by lifelike holographic characters and scenery, then the potential for influencing people to travel to the places depicted will be stronger than ever and may in time be so lifelike and immersive that they replace the 'real' holiday experience for some people. As VR and holographic technology develop, we will see in the next decade or two a level of realism that will fundamentally change how we consume film and television content, and its impact on tourism could be profound.

## Conclusion

Film tourism has come a long way since the first turning point of Auguste and Louis Lumière's public screening enticed people to travel in the late 1800s. Although we see movies as means of escapism and immersion in another reality we cannot ignore the fact that movies and television shows follow the social, economic and political trends of their particular time. This translates back to the film audience as well. Motivation in film tourism is a highly researched and contested topic but studies on motivation should never stop because they constantly change in accordance with the changes in society on all levels. There is a never-ending synergy between movies and moviegoers and their reciprocal influence.

With growing technological advancements through turning points such as VR and holographics, how we consume film and television output will be radically altered, to the point where we may actually physically inhabit the landscape of such filmic places, in a realistic virtual sense at least – no longer just watching our favourite movie or TV show, but actually feeling as if we are in it ourselves. With such a blurring of realities, how will this affect film tourism? For some it may fuel the desire to visit the real places even more, while for others a form of virtual tourism may supersede their need to actually travel in reality.

According to authors such as Huang *et al.* (2016: 116), 'With multimedia communication channels that are not constrained by geographic boundaries, virtual worlds open markets to promote in-world commodities and off-world services, providing a medium for innovative use of technology to engage customers with particular brands and eventually affecting intentions to purchase'. The world of film and television is already changing and will experience much more radical changes which in turn will fundamentally alter its impact upon tomorrow's tourist, creating whole new turning points in the world of film tourism in the future. A future with endless possibilities.

## References

Bafta (2018) *AMD Official Partner Los Angeles*. See http://www.bafta.org/amd-official-partner-los-angeles (accessed 21 March 2018).
BBC World Service Click (2017) *Virtual Reality Movie Trailer*. See http://www.bbc.co.uk/programmes/p050g7bv (accessed 21 March 2018).
Beeton, S. (2005) *Film-Induced Tourism* (1st edn). Clevedon: Channel View Publications.
Beeton, S. (2006) Understanding film-induced tourism. *Tourism Analysis* 11 (3), 181–188.
Bolan, P. (2010) Film-induced tourism: Motivation, authenticity and displacement. PhD thesis, Ulster University.
Bolan, P. and Davidson, K. (2005) Film induced tourism in Ireland: Exploring the potential. *Tourism & Hospitality in Ireland Conference (THRIC)*, University of Ulster, June.
Bolan, P. and Kearney, M. (2017) Exploring film tourism potential in Ireland: From Game of Thrones to Star Wars. *Journal of Tourism and Development* 27/28 (1), 2149–2156.
Bolan, P. and Williams, L. (2008) The role of image in service promotion: Focusing on the influence of film on consumer choice within tourism. *International Journal of Consumer Studies* 32 (4), 382–390.
Brown, S. and Patterson, A. (2009) Harry Potter and the service-dominant logic of marketing: A cautionary tale. *Journal of Marketing Management* 25 (5–6), 519–533.
Carl, D., Kindon, S. and Smith, K. (2007) Tourists' experiences of film locations: New Zealand as Middle-Earth. *Tourism Geographies* 9 (1), 49–63.
Chen, L.Y. (2018) Holograms in China captivate tourists' eyes, and wallets. *Bloomberg*, 12 February. See https://www.bloomberg.com/news/articles/2018-02-12/holographic-models-virtual-jellyfish-are-the-new-draw-in-china (accessed 21 March 2018).
Cheong, R. (1995) The virtual threat to travel and tourism. *Tourism Management* 16 (6), 417–422.
Cronin, M. and O'Connor, B. (2003) *Irish Tourism: Image, Culture and Identity*. Clevedon: Channel View Publications.
Croy, W.G. and Walker, R.D. (2003) Fictional media, film and tourism. In D. Hall, L. Roberts and M. Mitchell (eds) *New Directions in Rural Tourism* (pp. 115–133). Aldershot: Ashgate.
Frost, W. (2006) Braveheart Ned Kelly: Historic films, heritage tourism and destination image. *Tourism Management* 27, 247–254.

Frost, W. and Laing, J. (2014) Fictional media and imagining escape to rural villages. *Tourism Geographies* 16 (2), 207–220.
Ghisoiu, M., Bolan, P., Gilmore, A. and Carruthers, C. (2017) Conservation and co-creation through film tourism at heritage sites: An initial focus on Northern Ireland. *Journal of Tourism and Development* 1 (27, 28), 2125–2135.
Gianatasio, D. (2018) Holographic statues of 500 locals help boost both tourism and morale in Brussels. *AdWeek*, 24 April. See https://www.adweek.com/creativity/holographic-statues-of-500-locals-help-boost-both-tourism-and-morale-in-brussels/ (21 March 2018).
Gjorgievski, M. and Trpkova, S. (2012) Movie induced tourism: A new phenomenon. *UTMS Journal of Economics* 3 (1), 97–104.
Grihault, N. (2003) Film tourism – the global picture. *Travel and Tourism Analyst* October, 1–21.
Grihault, N. (2007) Set-jetting tourism – International. *Mintel Reports* March, 1–45.
Guttentag, D.A. (2010) Virtual reality: Applications and implications for tourism. *Tourism Management* 31, 637–651.
Harbord, J. (2002) *Film Cultures*. London: Sage.
Huang, Y.C., Backman, K.F., Backman, S.J. and Chang, L.L. (2016) Exploring the implications of virtual reality technology in tourism marketing: An integrated research framework. *International Journal of Tourism Research* 18 (2), 116–128.
Hudson, S. and Ritchie, J.R.B. (2006a) Promoting destinations via film tourism: An empirical identification of supporting marketing initiatives. *Journal of Travel Research* 44, 387–396.
Hudson, S. and Ritchie, J.R.B. (2006b) Film tourism and destination marketing: The case of Captain Corelli's Mandolin. *Journal of Vacation Marketing* 12 (3), 256–268.
IndieWire (2015) *Guilt And Rubble: 15 Essential Post-War Films*. See https://www.indiewire.com/2015/07/guilt-and-rubble-15-essential-post-war-films-261612/ (accessed 3 September 2019).
Iwashita, C. (2006) Media representation of the UK as a destination for Japanese tourists: Popular culture and tourism. *Tourist Studies* 6 (1), 59–77.
Jansson, A. (2002) Spatial phantasmagoria: The mediatization of tourism experience. *European Journal of Communication* 17 (4), 429–443.
Jones, D. and Smith, K. (2005) Middle-earth meets New Zealand: Authenticity and location in the making of The Lord of the Rings. *Journal of Management Studies* 42 (5), 923–945.
Karpovich, A. (2010) Theoretical approaches to film-motivated tourism. *Tourism and Hospitality Planning & Development* 7 (1), 7–20.
Kezich, T. (2005) Federico Fellini and the making of La Dolce Vita. *Cineaste* 31 (1), 8–14.
O'Connor, N. and Bolan, P. (2008) Creating a sustainable brand for Northern Ireland through film induced tourism. *Tourism, Culture & Communication* 8 (3), 147–158.
O'Connor, N., Flanagan, S. and Gilbert, D. (2008) The integration of film-induced tourism and destination branding in Yorkshire, UK. *International Journal of Tourism Research* 10, 423–437.
Oxford Economics (2007) *The Economic Impact of the UK Film Industry*. Oxford: Oxford Economics (supported by the UK Film Council and Pinewood Shepperton plc).
Phelan, L. (2017) *A Spotter's Guide to Film (and TV) Locations*. Franklin, TN: Lonely Planet.
Plautz, J. (2016) 'Crocodile Dundee' turns 30: How Paul Hogan changed tourism in Australia. *Travel & Leisure*, 26 September.
Randall, A.Z. (2015) Austrian Trümmerfilm?: What a genre's absence reveals about national postwar cinema and film studies. *German Studies Review* 38 (3), 573–595.

Riley, R.W. and Van Doren, C.S. (1992) Movies as tourism promotion – a pull factor in a push location. *Tourism Management* 13 (3), 267–274.

Rittichainuwat, B. and Rattanaphinanchai, S. (2015) Applying a mixed method of quantitative and qualitative design in explaining the travel motivation of film tourists in visiting a film-shooting destination. *Tourism Management* 46, 136–147.

Rodriguez, C.L., Fraiz, B.J.A. and Rodriguez-Toubes, M.D. (2011) Tourist destination image formed by the cinema: Barcelona positioning through the feature film Vicky Cristina Barcelona. *European Journal of Tourism, Hospitality and Recreation* 2 (1), 137–154.

Virtual Reality Society (2018) How did virtual reality begin. See https://www.vrs.org.uk/virtual-reality/beginning.html (accessed 21 March 2018).

Wallace, B. (1996) A nation in search of a modern Braveheart. *Maclean's* 109 (44), 28–30.

Zwart, H. (2015) The Third Man: Comparative analysis of a science autobiography and a cinema classic as windows into post-war life sciences research. *History and Philosophy of the Life Sciences* 37 (4), 382–412.

# 14 The Evolution of the Grand Tour in the Digital Society

Sabrina Seeler

## Introduction

Tourism has the flexibility to sustain continuous growth, adapt to changes and at the same time shape national identities and societies. Berghoff and Korte (2002: 1) summarized that tourism embodies the 'displacement, restlessness and image orientation' of Western life. With the acknowledgement of tourism as an important driver of economic and societal wellbeing, academic interest has grown significantly. To encourage future growth and sustain tourism as an economic driver, several studies have aimed to gain a more holistic understanding of the motivation behind leisure travel and the travel behaviour of contemporary tourists. While product and destination lifecycles have been proposed, the individual tourists' lifecycles over time remain undervalued. When looking at the history of tourism and reflecting on important historical turning points, such as the Grand Tour of Europe or the establishment of organized group travel, and at the same time considering current developments and phenomena in tourism, such as the growing demand for knowledge enhancement and digitization, several parallels become evident.

This chapter addresses some historical turning points that have helped shape national identity and that have been instrumental in the development of tourism. The chapter also reflects on recent changes in the tourism landscape, establishes links between the past and future of tourism that previous research has largely overlooked and proposes future turning points. Throughout the chapter, emphasis is given to the development of tourism in Europe and Germany. The conceptual review of the literature is underpinned by some recent findings from a study that explored 'the experienced tourist' as a promising market segment.

## Historical Turning Points

### Turning point 1: The Grand Tour of Europe – the foundation of leisure travel

The 'Grand Tour' was the first extensive movement of tourists documented by primary research material, and has gained much attention from researchers. Zuelow (2016: 3) proposed that '[t]he story of tourism as we currently understand it [...] was born of the Grand Tour'. The term Grand Tour was coined by the travel writer Richard Lassels in his book *Voyage of Italy*, published in 1670 (Chaney, 1985). The adventurous journeys of young explorers around Western Europe are evidenced by their diaries, journals and letters, sharing information about tour itineraries, impressions of countries visited, and their accommodation and modes of transportation (Towner, 1984). The Grand Tour has attracted researchers from different disciplines with non-unifying conceptions, as the Grand Tour can be defined in multiple ways, for example by the social class of the tourist or by the route undertaken.

*Early explorers and their travel motives*

The Grand Tour of Europe is mostly associated with the young elites and male aristocrats of Britain in the 17th and 18th centuries. However, the period of the Grand Tour is thought to extend further than that. Travel documentation indicates that it developed from the mid-16th century and continued into the early 19th century. Alongside the English elite, the wealthy classes of other mostly Protestant northern European countries also undertook so-called *Bildungsreisen* ('educational travel') and identified themselves as Grand Tourists (Zuelow, 2016). As upper-class men dominated early travel writings, most studies refer to the Grand Tour as a male practice and a masculine establishment. Several young females participated in these early forms of travelling, but their voices and experiences are mostly neglected in the history of tourism (Gleadhill, 2018).

Overall, the Grand Tour is described as a distinct circuit through Europe for educational purposes in order to visit places of cultural significance, particularly sights around Italy, France, Germany and Switzerland. However, after tracing over 900 published manuscript diaries and journals, Towner (1984, 1985) concluded that the underlying and true reasons for undertaking a Grand Tour remain unclear. Travelling for hedonic reasons and pure enjoyment was not tolerated in the early years of the Grand Tour period, and educational reasons, such as career or literary, were believed to be the dominant motivations. In discussing the travel experience of female Grand Tourists, Gleadhill (2018) provided the following interpretation of a Grand Tour:

> Travel promised to empower these women for the same reasons that it empowered their male counterparts – as a social marker, as a way to

familiarise themselves with Britain's Classical cultural heritage, as a means of gaining knowledge through experience and as an opportunity to network. (Gleadhill, 2018: 3)

Travelling and moving through foreign spaces while taking risks and being adventurous was seen as a metaphor for personal change and thought to be indispensable in developing 'civility of mind and manner' (Quadflieg, 2002: 23). Zuelow (2016: 21–22) further argued that the 'Grand Tour was a rite of passage from childhood to adulthood' and a 'finishing school for the rich'. The Grand Tour of Europe epitomized a form of cultural exchange – educational experiences and enlightenment that could only be obtained by crossing borders and becoming immersed in other cultures and countries.

### Spatial and temporal travel developments

Alongside the changing motives for undertaking a Grand Tour, the routes, itineraries and interpretations of tours changed over time. Although spatial patterns remained similar, the central foci of tourists changed from a *Classical Grand Tour* to a *Romantic Grand Tour* (Towner, 1985). Cities such as Paris, Rome and Venice and the sites of Renaissance antiquities were central to any Grand Tourist, yet the importance of other spatial areas fluctuated with changes in tourists' motives and desired outcomes. The emerging romanticism of tourism was the key development that led to a new appreciation of natural landscapes and awareness of the tourist destination (Berghoff & Korte, 2002). It also contributed to a more self-centred and self-realizing tourist experience (Chard, 2002).

The duration of the tour steadily shortened and the decline of the Grand Tour is attributed to various societal developments and changes. The French Revolutionary War in the 1790s was associated with a decline in the status of aristocrats and, at the same time, a growing middle class, which changed the face of travel in Europe (Duché, 2017). Technological advances during the Industrial Revolution, and particularly the introduction of steam-driven transportation in the early 19th century, further contributed to a change in travel modes and the Grand Tour lost its significance. The motivation to travel changed from pursuing knowledge and personal growth to pure enjoyment, recreation and relaxation, and tourism became a form of commercialized consumption.

## Turning point 2: The era of Thomas Cook – the development of mass tourism

With the facilitation of travel through the development of rail and steamships in the 19th century, the neo-classical enthusiasm for the Grand Tour faded and travelling became not only faster but also classless and more accessible to the masses (Zuelow, 2016). The emerging bourgeois class and growing capitalism in the 19th century resulted in an evolving leisure regime and a constantly growing leisured society (Berghoff & Korte, 2002). Tourism in the time of Grand Tourists was mainly non-institutionalized and services developed exclusively for Grand Tourists were rare. However, the dominance of some centres along the route fuelled the emergence of tourist services and the development of an organized tourism industry and eventually laid the foundation for the first travel agency initiated by the British businessperson and explorer Thomas Cook in the 1860s. Thomas Cook advocated free trade and he claimed that travel was for everyone and needed to be enjoyed beyond social elites (Berghoff, 2002).

*Technological advances and the acceleration of travel*

The new efficiencies and speed of travel of steam-driven transportation resulted in an 'annihilation of space and time' (Koshar, 2000: 3). The changing modes of transportation made cultural and natural landscapes more accessible and contributed to a democratization of travel. At the same time, these accelerated transportation styles no longer allowed tourists to experience the natural and cultural environment directly. As Koshar (2000) noted:

> They [travellers of the past] experienced space and time as unfolding and sequential. Modern railways removed travelers from their immediate surroundings, creating a 'panoramic' mode of perception […]. Space and time were to be conquered, not felt or experienced, and the process of travel itself was now diminished by the hegemony of the tourist destination. (Koshar, 2000: 3)

Despite the democratisation and facilitation of tourism for the masses, a group of travellers remained who were interested in a more authentic experience, away from the major tourist routes that had developed. These tourists aimed at safeguarding their prestige through individualized adventures and travel. Boorstin (1992) summarized that travel transformed from being an active athletic exercise to a passive spectator sport and further explained that the importance of *travellers* declined while the significance of *tourists* rose.

*Changes in travel linguistics*

Closely related to the overall changes in tourism, major discussions emerged to differentiate the often synonymously used terms, *tourist* and *traveller* (Buzard, 1993). The origin of the English word *travel* is in the French *travail*, which in turn has its roots in the Latin *trepalium*. The noun *travail* expresses a painful exertion, bodily or mental hard work, suffering or an excessive labour, while *trepalium* can be translated as an instrument of torture (Oxford English Dictionary, 2018). Consequently, *travelling* expresses endurance and perseverance and is most appropriate to describe tourists before the invention and geographical extension of rail networks. From the late 18th century, the notion of *tourist* was used synonymously with *traveller* and the appropriateness of the interchangeability of the terms was only questioned in the later 20th century. The term *tourist* has its linguistic roots in the Latin word *tornus*, which derives from the Greek and stands for a tool creating a circle (Oxford English Dictionary, 2018). With the democratization of travel and the growth in mass tourism, these definitional discrepancies became more apparent and created awareness of a changing image of tourism.

## Turning point 3: The changing image of tourism – from education to political power and consumerism

Travelling in the past went beyond self-identity; it also incorporated a search for national identity. The search for national identity has been a dominating influence on Germany's history and still shapes Germans today. WWI and WWII not only left scars on European societies, but they brought significant changes to European and German travel culture.

### German tourism in the time of the wars and beyond

While the Nazi regime claimed to have created a classless society in which everyone was eligible and able to travel, the Nazi regime soon recognized that travel among Germans was comparatively low prior to WWII (Kopper, 2009). To support the Nazi propaganda image of a *community of German people* (*Volksgemeinschaft*), a state-operated leisure organization *Kraft durch Freude* (KdF; translated as 'Strength through Joy') was initiated by Nazi Germans (Spode, 2007). Establishments such as the KdF seaside resort on the island of Rügen in the German Baltic Sea offered accommodation for 20,000 vacationers. It epitomized a leisure fabric that contributed to the immense growth of organized mass tourism in the 1930s. While civil and human rights were being dismantled, the Nazi regime tried to offer affordable leisure experiences as compensation and they encouraged social tourism.

The world wars and post-war austerity clearly influenced the development of tourism in Europe. Political regimes used travel not only as a leisurely and pleasurable distraction, but tourism became a national asset and was used as a mobilization of political ideology and thus 'carried a heavy symbolic load' (Koshar, 2000: 4). Despite the growth of pre-war travel under the Nazi regime, travelling was not possible for everyone and those who were entitled to travel had the power to re-engage in commerce, trade and education. Previous claims that travelling was a fundamental human right were challenged and tourism became seen as a form of superficial consumption encouraged by capitalist regimes. The economic and social growth, increases in paid holidays for workers, and reconstructions after WWII contributed to the revitalization of tourism in Europe, encouraged mass tourism and also supported the emergence of a new German leisure travel culture with hedonic motives (Koshar, 2000; Mandel, 2011).

*Travelling as an expression of conspicuous consumption*

Strongly associated with the institutionalization of the tourism product, tourism flourished after the wars and was no longer limited to being a luxury product for the higher social classes. Criticism began as capitalist societies contributed to a commodification of tourism and a move away from authentic experiences. Tourism was interpreted as a symbol and motor of consumerism. The German writer Gerhard Nebel described tourists as 'swarms of giant bacteria' (Nebel, cited in Frank, 2016: 200). Boorstin (1992) echoed this view and called tourists deluded, superficial, instant travellers looking for pseudo-events to satisfy extravagant illusions. Other scholars shared this pessimistic view of tourism and criticized the commodification and systematic destruction of the real meanings of places and the emergence of staged authenticity and superficial experiences (Cohen, 1988; MacCannell, 1973). A major aim of tourists in the 1970s and 1980s was to escape from their everyday life routines and working environments, and hedonistic reasons dominated travel decisions.

Alongside the vigorous discussion of tourism as capitalist consumption that destroys authenticity and culture, scholars developed tourist typologies and identified tourists who distanced themselves from organized mass tourism and continued to pursue authentic experiences and personal enrichment (Cohen, 1972; Plog, 1974). These tourists were characterized as educated, lone travellers seeking novel and off-the-beaten-track experiences. Although these early typologies demonstrated a duality of tourists, value-driven and managerial perspectives emphasized providing structured itineraries and experiences that satisfied the hedonic needs of the masses and overshadowed the long-underestimated individual travellers and their desire for knowledge enhancement. Berghoff (2002) argued that, despite fully organized group itineraries, individuals participating in those packaged tours would familiarize themselves with their trips and accumulate knowledge through trip preparation. Thus, knowledge enhancement remained central to all tourists.

## Turning point 4: Travelling to be educated – drowning in information and starved for knowledge

Although travel motives and modes changed drastically in the 20th century, education and knowledge enhancement remained an important, although often underestimated and subconsciously perceived, reason for travelling. Tourism has been identified as an open education system and scholars have proposed a general trend away from pure relaxation to more educational and communicative travel forms (Mandel, 2011).

Computerization has positively contributed to the desired education and communication outcomes in the different stages of a trip and created increasingly digitized and knowledgeable travellers.

*The digitized traveller*

Just as the development of steam-driven transportation and later air connectivity changed the modes of transportation and contributed to the growth of international tourism, technological advancements and the development of digital information have changed the information and travel behaviour of contemporary tourists. Whereas the former resulted in the growth of mass tourism, the latter contributed to increasingly aware and informed tourists and the expansion of individual travel. This correlation between digitization and independent travel behaviour is echoed in the literature. Packaged group holidays dominated the pre-internet era as travel research was time-consuming and a logistical challenge. Today's internet age facilitates communication, accelerates the information search, contributes to higher degrees of independence for tourists and spreads knowledge more broadly (Pendergast, 2010).

While the instant availability of information encourages knowledge transfer across borders, a digital overload has been recognized. Contemporary tourists are aiming for inter-human relationships, which are constantly declining in the digital context. As a response, a trend towards analogue and face-to-face interaction has emerged and is reflected in a growing demand for immersive and interactive experiences and intra-personal relationships that assist in broadening the horizon and the development of the self in a real-world context.

*Studienreisen ('study tours') – a new phenomenon or an old idea?*

German tourism operators' products for educational travel have changed and have moved from a classical *Bildungsreise/Studienreise* ('educational travel'/'study tour') towards more multisensory and experience-oriented tours. Krohm (2007) described a modern *Studienreise* as an intelligent way of travelling that encourages immersion and cultural exchange with host communities.

Similar to the tutors who were travel guides for Grand Tourists, the travel guide of a *Studienreise* is the deciding factor as to whether a satisfactory tourism experience is delivered. Weiler and Black (2015) identified today's tour guides as experience-brokers who provide access to physical spaces, encourage host–guest encounters, support cross-cultural and bilingual understanding and facilitate empathic communication. Therewith, today's tour guides go beyond the previously dominant role of entertainer and enable interactive and insightful experiences for tourists and hosts. With the growing awareness of tourists and their desire for personal interaction, staged and commodified experiences are progressively being replaced by authentic and real experiences. Conradson and

Latham (2005) proposed that language barriers, sociocultural differences and felt power relations were the prevalent reasons why desired immersion and interaction between guests and hosts was impeded. Taking the changing roles of travel guides and the constantly changing nature of *Studienreisen* into account, an evolution of the Grand Tour can be evidenced.

Another analogy between the Grand Tour of history and contemporary *Studienreisen* can be drawn with regard to the current dominant market segments. In the case of both travel phenomena, affluent and more highly educated tourists paved the way and were identified as travel pioneers. However, Grand Tourists as well as participants of *Studienreisen* became more diverse and market segments expanded (Lohmann & Mundt, 2002). There is a growing trend to travel for self-fulfilment, self-development and personal growth among today's tourists and hedonic reasons are often overshadowed by eudaemonic motives for travel (Csikszentmihalyi & Coffey, 2017). Cultural tours in the form of *Studienreisen* can emotionally and aesthetically contribute to the multisensory experiences sought by increasingly experienced tourists and support individuals' education and transformation.

A distinguishing factor between Grand Tourists and contemporary tourists is the element of time. While the former devoted several months or years to educational travel, today's tourists are often pressured for time (Mandel, 2011). At the same time, there is a growing volume of long-term travelling, which is particularly found among younger generations. Although today's younger travelling generation is not necessarily as strongly associated with higher income classes as was the case in the time of the Grand Tour, the motives of knowledge enhancement, particularly language enrichment and personal growth, are still similar.

## Turning point 5: The evolution of the Grand Tour – daring to predict tomorrow's tourists

Tourism has evolved over the past few centuries and has expanded to become accessible not only to society's higher classes, but also to the wider population. Several decisive markers have been addressed in the previous sections and a close relationship to the broader political environment has been demonstrated. Despite the constantly and dynamically changing behaviour and motives of tourists, a review of the past clearly demonstrates some parallels with contemporary tourists. Berghoff and Korte (2002: 5) noted that 'our modern notion of tourism, with its elements of temporary escape from the everyday, has its most immediate roots in romantic travel' and Stilz (2002: 85) shared that 'the tourist industry of our day tends to follow the sentimental ways of travelling through the past'. Despite the thirst for new adventures, today's tourists continue to follow classic tourist routes as they offer infrastructure and provide safety.

They also possess rich cultural memories that today's knowledge-driven tourists are looking for. Stilz (2002: 85) argued that these classic tourist spots provide 'shibboleths of cosmopolitan education which make them attractive to the provincial mind ever suffering from territorial fixations'. Thus, the sophistication of today's tourists does not necessarily inhibit their desire to broaden their historical knowledge by visiting well-known tourist destinations. This has been confirmed in a study conducted to investigate the dimensions of German experienced tourists (Seeler, 2018). Based on the results, several future turning points are proposed that show a resemblance to the historical Grand Tour and at the same time take into consideration the changes associated with the digitized and accelerated society.

## Future Turning Points

### Future turning point 1: German experienced tourists – tomorrow's Grand Tourists

With a projected increase in tourism of 3.3% per annum between 2010 and 2030, leading to 1.8 billion tourists in 2030 (UNWTO, 2017), it can be expected that tourists will continue to grow in maturity and tomorrow's tourists will be increasingly experienced. However, the fast pace of modern digitized society brings disruptions and unpredictability that challenge tourism operators' competitiveness (Yeoman *et al.*, 2012). Needs, demands and expectations are constantly changing and tourism operators are required to adjust their products and services to meet changing demands.

Today's experienced tourists are defined both by their eudaemonic motivation towards knowledge enhancement and self-development and the exploration of new places and, at the same time, by their interest in travelling to well-known destinations and must-see attractions (Seeler, 2018). Although German experienced tourists are assumed to travel independently, and in many regards match the typologies of individual travellers, their knowledgeability has changed their perspectives on organized group travel. German experienced tourists value certain forms of group travel where knowledge enhancement can be achieved. Thus, they will seek types of travel equating to the concept of *Studienreisen* and will be identified as tomorrow's Grand Tourists (Seeler, 2018). Generally, experienced tourists are considered to travel in a more socially responsible and ethical way and appreciate a slower pace of travel, which suits the concept of *Studienreisen* and cultural immersion. Catering to their needs will benefit the sustainable development of any tourist destination and understanding experienced tourists while acknowledging their desire for knowledge enhancement and cultural education will bring new opportunities for tourism providers and destination marketers. However, as tomorrow's Grand Tourists will seek educational enlightenment through meaningful and co-created experiences that offer opportunities for self-exploration, self-actualization and personal transformation, the concept of *Studienreisen* will need to be adjusted.

## Future turning point 2: Co-creating educational travel – the future of *Studienreisen*

The increasingly connected, engaged and knowledgeable tourists aim to actively participate in the creation of their subjective meaningful experiences (Campos *et al.*, 2018). These experiences will contribute to the desired transformation of self-identity sought by tomorrow's Grand Tourists. At the same time, experienced tourists pursue personal enrichment, want to be challenged in new places, and want to define boundaries and test personal limits (Seeler, 2018). Although these travel motives resemble those of history's Grand Tourists, co-creation was previously an unknown driver behind experience accumulation and knowledge enhancement. Advancements in online technologies have radically changed the accessibility of information and are benefiting contemporary tourists who aim to co-create their personal journeys.

Although *Studienreisen* have changed into more experience-oriented tours, they are mostly organized group travel with predefined itineraries and globalized tour outcomes. Today's educational tourists pursue co-created experiences that are personally enriching and contribute to an individual's self-development. Tourism operators of *Studienreisen* will need to facilitate more individualized and personalized tours which will contribute to positive memorable experiences and satisfy the demand of tomorrow's Grand Tourists. Bearing in mind the influence of history's Grand Tour on the development of tourism, it should be the aim of any tour operator to create educational experiences that encourage memorability and have the potential to become decisive in shaping the future of tourism. The memorability of experiences has been associated with autobiographical representations of travel, such as travel writings and photos (Alù & Hill, 2018). Digital technologies will lead to new forms of travel narratives as they expand the opportunities for contemporary tourists to share and store their memories and provide advanced digital personal archives.

## Future turning point 3: Digital narratives – the travel writings of tomorrow's Grand Tourists

Travel writings have existed throughout the centuries, but their content and distribution channels have changed. Technological advances and digitization have contributed to the emergence of travel blogs and other user-generated content published online. This means that highly subjective and individual transformative experiences are more readily available to the public, which results in faster and wider dissemination of knowledge. Travel writings were of great importance to Grand Tourists, yet the narratives of early explorers were mostly for personal purposes and functioned as reflective tools. Today's travel narratives go beyond personal diaries, incorporate different forms of media and utilize microblogging and online social networking sites. Although the travel writings of Grand Tourists became very decisive in the shaping of tourism, their influence was only fully comprehended much later. Through digital storytelling today, travel writers and bloggers immediately become social media influencers and their stories become decisive in travel inspiration, information searches and eventual decision making (Pera, 2017).

Social media influencer marketing has recently been recognized by tourism practitioners, yet its actual implementation in tourism marketing strategies is still limited. Considering the close link between educational travel, transformative experiences and autobiographical representations and at the same time the continuing importance of word-of-mouth marketing, operators of *Studienreisen* need to develop tools to encourage tourists to share their memories online. Guiding tourists in sharing their transformative experiences will not only contribute to the tourists' satisfaction, but will also facilitate new marketing tools with personalized voices. Bearing in mind the general trend of experience co-creation, the desire for knowledge enhancement and, at the same time, the accessibility of knowledge, tourists will increasingly accumulate secondary ('reflective') experiences through other people's narratives (Seeler, 2018). The actual impact of secondary experiences is still unknown, but there is agreement that secondary experiences through digital narratives will not replace primary experiences. They are believed to function as an addition to or expansion of the actual lived experience.

### Future turning point 4: Wearable, virtual and augmented reality – new forms of knowledge creation and experience consumption

Further expansions of the actual lived experiences are generated through new technological developments related to wearables and virtual and augmented reality (Tussyadiah *et al.*, 2018). The opportunities these new technologies bring to support on-trip experiences and knowledge creation will be relevant for historical sites: where witnesses of the past have disappeared, stories can be retold with a leveraged support such as 360-degree views of the landscapes of the past, and value can be added as educational components (Seeler, 2018). Thus, tomorrow's Grand Tourists will be provided with additional tools to achieve the desired knowledge enhancement. At the same time they are seeking inter-human connections and personal interaction. This will bring challenges for tourism operators, as they will need to find the balance between expansion of on-site experience through the utilization of new technologies and ensuring personal connections through inter-human exchange.

## Conclusion

In retrospect, it is evident that the motivations to travel for leisure purposes, like other consumer trends, recur through history. Although the particulars change and travelling in general has become more accessible and feasible, the underlying mechanisms and reasons to travel today are not so different from those in the time of the Grand Tour. Grand Tour destinations are still must-see tourist sites, and cultural immersion and knowledge enhancement are still among the most important reasons for people to travel. While early explorers mostly travelled on their own or were chaperoned by a tutor, they followed a predefined route. With technological advancement and the emergence of organized package tours, group travel grew and structured itineraries followed. Although today's tourists travel more independently and often aim for off-the-beaten-track adventure, some degree of structure and predefined itinerary remain for the modern majority.

This chapter has demonstrated the resemblance of today's experienced tourists to history's Grand Tourists, particularly with regard to their shared desire for knowledge enhancement and the way that modern digital narratives and storytelling mirror the handwritten diaries, journals and letters of former times. This review of the history of tourism in comparison to contemporary tourism has revealed its iterative nature, which is exemplified in the changes in dominant travel forms and the importance of technological developments for the advancements of tourism. Early explorer and Grand Tourist travel was independent but, with the foundation of the first travel agency, packaged group tours became increasingly preferred; mass tourism further grew during the wars. However, the development and expansion of online information and the growing sophistication of tourists has contributed to a return to more independent travel forms. At the same time, experienced tourists understand the value of organized group tours and appreciate the services provided by tourism specialists. This clearly indicates the iterative nature of travel trends and hints that the next phase of tourism may again become more organized.

This cyclical development of tourism is also strongly related to the source market of tourists. In this chapter, emphasis was given to the Grand Tourists of Europe, particularly German (experienced) tourists. Other and newer travel markets are in different stages of travel maturity, yet follow similar patterns. Germans were not among the early pioneers and followed English explorers, yet they are one of the most important international travel markets today – in terms of both travel intensity and travel trends. To predict the behaviour of travellers in the future, one needs to look at the development of other more mature tourism markets, but one also needs to be aware of tourism behaviour in the past.

## References

Alù, G. and Hill, S.P. (2018) The travelling eye: Reading the visual in travel narratives. *Studies in Travel Writing* 22 (1), 1–15.

Berghoff, H. (2002) From privilege to commodity? Modern tourism and the rise of the consumer society. In H. Berghoff, B. Korte, R. Schneider and C. Harvie (eds) *The Making of Modern Tourism: The Cultural History of the British Experience, 1600–2000* (pp. 159–180). New York: Palgrave.

Berghoff, H. and Korte, B. (2002) Britain and the making of modern tourism: An interdisciplinary approach. In H. Berghoff, B. Korte, R. Schneider and C. Harvie (eds) *The Making of Modern Tourism: The Cultural History of the British Experience, 1600–2000* (pp. 1–20). New York: Palgrave.

Boorstin, D.J. (1992) *The Image: A Guide to Pseudo-Events in America*. New York: Vintage Books.

Buzard, J. (1993) *The Beaten Track: European Tourism, Literature, and the Ways to 'Culture', 1800–1918*. Oxford: Oxford University Press.

Campos, A.C., Mendes, J., do Valle, P.O. and Scott, N. (2018) Co-creation of tourist experiences: A literature review. *Current Issues in Tourism* 21 (4), 369–400.

Chaney, E. (1985) *The Grand Tour and the Great Rebellion: Richard Lassels and 'The Voyage of Italy' in the Seventeenth Century*. Genève: Slatkine.

Chard, C. (2002) From the sublime to the ridiculous: The anxieties of sightseeing. In H. Berghoff, B. Korte, R. Schneider and C. Harvie (eds) *The Making of Modern Tourism: The Cultural History of the British Experience, 1600–2000* (pp. 47–68). New York: Palgrave.

Cohen, E. (1972) Toward a sociology of international tourism. *Social Research* 39 (1), 164–182.

Cohen, E. (1988) Authenticity and commoditization in tourism. *Annals of Tourism Research* 15 (3), 371–386.

Conradson, D. and Latham, A. (2005) Friendship, networks and transnationality in a world city: Antipodean transmigrants in London. *Journal of Ethnic & Migration Studies* 31 (2), 287–305.

Csikszentmihalyi, M. and Coffey, J. (2017) Why do we travel? A positive psychological model for travel motivation. In S. Filep, J. Laing and M. Csikszentmihalyi (eds) *Positive Tourism* (pp. 122–132). New York: Routledge.

Duché, E. (2017) Revolutionary ruins: The reimagination of French touristic sites during the Peace of Amiens. In R. Sweet, G. Verhoeven and S. Goldsmith (eds) *Beyond the Grand Tour: Northern Metropolises and Early Modern Travel Behaviour* (pp. 203–221). London: Taylor & Francis.

Frank, S. (2016) *Wall Memorials and Heritage: The Heritage Industry of Berlin's Checkpoint Charlie*. New York: Routledge.

Gleadhill, E. (2018) Improving upon birth, marriage and divorce: The cultural capital of three late eighteenth-century female Grand Tourists. *Journal of Tourism History* 10 (1), 21–36.

Kopper, C.M. (2009) The breakthrough of the package tour in Germany after 1945. *Journal of Tourism History* 1 (1), 67–92.

Koshar, R. (2000) *German Travel Cultures*. Oxford: Berg.

Krohm, C. (2007) Was gutes Reisen besser macht: Die Qualität von Studiosus Studienreisen. In A. Grünewald Steiger and J. Brunotte (eds) *Forum Kulturtourismus: Qualitäten des kultivierten Reisens*, Vol. 32 (pp. 95–106). Wolfenbüttel: Wolfenbütteler Akademie-Texte.

Lohmann, M. and Mundt, J.W. (2002) Maturing markets for cultural tourism: Germany and the demand for the 'cultural destination'. In R. Voase (ed.) *Tourism in Western Europe: A Collection of Case Histories* (pp. 213–225). Wallingford: CABI.

MacCannell, D. (1973) Staged authenticity: Arrangements of social space in tourist settings. *American Journal of Sociology* 79 (3), 589.

Mandel, B. (2011) Kulturelle Lernorte im (Massen-)Tourismus? Potentiale und Strategien kultureller Bildung von Musentempel bis Disneyland. In A. Hausmann and L. Murzik (eds) *Neue Impulse im Kulturtourismus* (pp. 175–198). Wiesbaden: VS Verlag für Sozialwissenschaften.

Oxford English Dictionary (2018) See http://www.oed.com/ (accessed 20 January 2018).

Pendergast, D. (2010) Getting to know the Y generation. In P. Benckendorff, G. Moscardo and D. Pendergast (eds) *Tourism and Generation Y* (pp. 1–15). Wallingford: CABI.

Pera, R. (2017) Empowering the new traveller: Storytelling as a co-creative behaviour in tourism. *Current Issues in Tourism* 20 (4), 331–338.

Plog, S.C. (1974) Why destination areas rise and fall in popularity. *Cornell Hotel and Restaurant Administration Quarterly* 14 (4), 55–58.

Quadflieg, H. (2002) Approved civilities and the fruits of peregrination: Elizabethan and Jacobean travellers and the making of Englishness. In H. Berghoff, B. Korte, R. Schneider and C. Harvie (eds) *The Making of Modern Tourism: The Cultural History of the British Experience, 1600–2000* (pp. 21–46). New York: Palgrave.

Seeler, S. (2018) Continuum of an experienced tourist's multidimensionality – explorations of the experience levels of German and New Zealand tourists. PhD thesis, Auckland University of Technology.

Spode, H. (2007) Some quantitative aspects of Kraft durch Freude tourism, 1934–1939. In M. Dritas (ed.) *European Tourism and Culture* (pp. 123–133). Athens: Livanis.

Stilz, G. (2002) Heroic travelers – romantic landscapes: The colonial sublime in Indian, Australian, and American art and literature. In H. Berghoff, B. Korte, R. Schneider and C. Harvie (eds) *The Making of Modern Tourism: The Cultural History of the British Experience, 1600–2000* (pp. 85–108). New York: Palgrave.

Towner, J. (1984) The Grand Tour: Sources and a methodology for an historical study of tourism. *Tourism Management* 5 (3), 215–222.

Towner, J. (1985) The Grand Tour: A key phase in the history of tourism. *Annals of Tourism Research* 12 (3), 297–333.

Tussyadiah, I.P., Wang, D., Jung, T.H. and tom Dieck, M.C. (2018) Virtual reality, presence, and attitude change: Empirical evidence from tourism. *Tourism Management* 66, 140–154.

UNWTO (2017) *UNWTO Tourism Highlights: 2017 Edition*. Madrid: World Tourism Organization. See http://www.e-unwto.org/doi/pdf/10.18111/9789284419029.

Weiler, B. and Black, R. (2015) The changing face of the tour guide: One-way communicator to choreographer to co-creator of the tourist experience. *Tourism Recreation Research* 40 (3), 364–378.

Yeoman, I., Yu, R.L., Mars, M. and Wouters, M. (2012) *2050 – Tomorrow's Tourism*. Bristol: Channel View Publications.

Zuelow, E.G.E. (2016) *The History of Modern Tourism*. New York: Palgrave.

# 15 Shopping on the Edge: Identifying Factors Contributing to Tourist Retail Development in Heritage Villages

Gianna Moscardo, Laurie Murphy, Karen Hughes and Pierre Benckendorff

**Introduction**

Tourist shopping villages (TSVs) are small towns that base their visitor appeal on heritage and retailing (Getz, 2000). They are spaces that bring together visitor experiences, retail settings and regional community development. As researchers have begun to focus attention on this phenomenon, some of the following issues have been identified: negative environmental and social impacts; authenticity and commodification; a failure to provide expected benefits to regional communities; and undesirable changes in the nature of the village and retail experiences offered to residents and visitors (Murphy *et al.*, 2011). One option to address these challenges is to understand the different evolutionary pathways that have led to positive or negative outcomes for both visitors and the local communities in which they are situated.

This chapter reports on a detailed historical case study analysis of three TSVs: Hahndorf in Australia, St Jacobs in Canada and Cheddar in England. These three villages are well established tourist destinations with diverse tourism development histories. The chapter will argue that an analysis of the historical development of these three villages can identify important turning points where decisions were made that led down particular evolutionary pathways. The aim of this historical case study analysis is to identify the key factors connected to this evolution. A better understanding of that process could inform decisions made about alternative futures.

The chapter will briefly identify challenges in managing and understanding TSV development. The three case studies will be introduced and then used to identify and describe key common turning points in the evolutionary pathways that typically emerge for this type of tourism. The chapter will then critically examine current forces on TSV development before concluding with some suggested alternative future pathways.

## Challenges in TSV Development

The relationship between shopping and tourism is complex and contested. It is clear that retail shopping is a major tourist activity (Jin *et al.*, 2017), a common travel motivation and destination choice factor (Moscardo, 2004), a contributor to destination image and loyalty (Suhartanto, 2018) and an important source of revenue for destination communities (Timothy, 2018). Despite this importance in practice, tourist shopping has been given only scant academic attention (Choi *et al.*, 2016; Jin *et al.*, 2017) and is rarely overtly acknowledged in destination marketing (Murphy *et al.*, 2011). Clearly there is some ambivalence in the way shopping is viewed within the tourism context and this may be partly due to some of the negative impacts associated with the development of tourist shopping.

There have been two main approaches to examining the consequences of tourist shopping developments in village communities. First, cross-sectional analyses have explored the impacts and processes associated with different villages at one point in time (Murphy *et al.*, 2008). Secondly, longitudinal case studies have examined the processes of development and change for a particular village over time. While longitudinal case studies offer the opportunity to gain a better understanding of these processes, they are rare in this area. One exception is a series of single village case studies conducted in Canada and China by Mitchell and colleagues (Fan *et al.*, 2008; Mitchell, 2003; Mitchell & Coghill, 2000; Mitchell & de Waal, 2009). These studies identified a common development pathway linked to 'creative destruction'. In these cases, the development of tourist shopping enhanced the touristic appeal of villages offering heritage buildings and a sense of the rural idyll. The growing numbers of visitors attracted external investors, sparking a process of creative destruction whereby new business replaced old, with a resulting shift in the size of the village and the nature of its shopping and negative changes to the rural idyll. This work suggested a development lifecycle with five main stages (Fan *et al.*, 2008; Mitchell, 2003; Mitchell & Coghill, 2000; Mitchell & de Waal, 2009).

- Early commodification sees the emergence of tourism alongside continuing commercial activity aimed at local residents.
- Advanced commodification occurs as tourism grows and tourist-focused activities displace local-focused businesses.

- Pre-destruction is characterized by considerable investment in the renewal of buildings, landscaping and tourist facilities with locals no longer catered for in the village precinct.
- Advanced destruction results in an outflow of local residents, crowding, congestion, and a change in merchandise from local to mass produced.
- Post-destruction is associated with the arrival of franchises, factory outlets and chain stores and is seen as providing an inauthentic and unsatisfactory tourist experience linked to the decline of the TSV.

Getz (2000) and Murphy *et al.* (2011) identified a number of the negative impacts associated with the TSV growth reported by Mitchell and colleagues, including problems with traffic congestion and crowding, changes to the physical landscape, particularly to heritage buildings, and loss of authenticity in the experience. This loss of authenticity and shift from local to mass-produced merchandise is a common issue identified in cross-sectional analyses. Interviews with stakeholders in six TSVs in the New England region of the United States, for example, provided evidence that the retention of specialist local shops can be difficult in the face of outlet shopping developments, especially in TSVs located within easy driving distance of major urban centres, with cheap land and less stringent local development regulations (Murphy *et al.*, 2011).

A study evaluating 29 TSVs in Australia, New Zealand and Canada concluded that local or regionally distinctive craft and food, in addition to heritage conservation and the presence of a major anchor tourist attraction, were important for a successful tourist experience (Murphy *et al.*, 2008). Another cross-sectional study of two TSVs in Australia also found that opportunities to purchase and see the production of locally made products resulted in higher tourist satisfaction and greater expenditure, highlighting the importance for TSV communities of retaining specialist regional shopping (Murphy *et al.*, 2013).

While there is some consistency across the available research about the importance of authenticity, heritage preservation and tourist access to regionally distinctive or locally produced or crafted merchandise, it is not clear that the process of destruction suggested by Mitchell's research group is inevitable or widespread. Murphy *et al.* (2011) analyzed 65 TSVs across eight different countries and found that the stages outlined by Mitchell's analyses of creative destruction could be found in many cases, but that they did not apply to all TSVs and that other factors were critical, including the development and/or loss of a major tourist attraction, the geographical features of the village which in some cases meant that tourist and local precincts were separate, and the extent to which the village had already lost local-focused businesses to larger nearby centres. Overall, the existing research offers some insights into the nature of the consequences of tourism and shopping development on host villages but there is still only limited information on the detail of the processes that have contributed to these consequences. The detailed examination of specific case studies focusing on their evolution offers an alternative way to access this type of detail and thus provide guidance for improving the management of this type of tourism.

## Introducing the Cases

Hahndorf, located in South Australia between Adelaide and the winegrowing region of the Barossa Valley, has the shortest tourist development history, with recognition and celebration of the region's German heritage beginning in the 1970s. Initial tourism in the 1970s was built around German food and wine festivals, and the establishment of a museum and art gallery in the house of one of Australia's eminent landscape artists Hans Heysen (Adelaide Hills Online, 2012). The whole village was declared a State Heritage Area with 22 separate buildings also listed in 1988 (Hahndorf Business & Tourism, 2018). Currently the town offers a heritage and food and wine based experience with an extensive events programme and many retail shopping opportunities (Hahndorf Business & Tourism, 2018).

Cheddar village, located in Somerset, England, is close to Cheddar Gorge which is a National Nature Reserve popular for rock climbing and caving. The caves in the gorge are also the location of important archaeological discoveries, including Cheddar Man, the oldest complete skeleton found in Britain and dated at 9000 years old. Cheddar is also a dairy farming region, and the famous Cheddar cheese was first produced here and matured/stored in the Cheddar caves. Popular tourism began with the opening of the Cheddar Valley Railway in 1869/1870, making it the oldest developed TSV of the three cases (Cheddar Parish Council, 2007). Over time, caves in the Gorge were opened to visitors and the nearby village was developed as a service hub for the caves with additional tourist attractions built including a lookout tower and prehistory museum. Shopping developed around these tourist attractions and currently there are many specialty shops mostly selling cheese, local products and souvenirs; it now includes an outlet centre (Visit Somerset, 2018).

St Jacobs village in Canada has been a centre for Mennonite settlement since the 18th century and the continued traditional rural lifestyle of the group continues to be a major attraction for tourists. Development of the village for tourism began in the 1970s with the opening of restaurants with a Mennonite cuisine theme, followed by the conversion of existing heritage buildings into speciality arts and crafts stores. Up to the 1990s the TSV developed in a pattern similar to that of Hahndorf. In the 1990s an outlet mall was built close to the village, causing considerable debate among stakeholders about the appropriateness of this kind of development. A large Farmers' Market is also an important part of the TSV (St. Jacobs Country Tourism, 2018).

### Evolutionary Pathways for TSVs

The current situation and development history of each of the TSV cases was examined through a range of sources including websites, tourist travel blogs, government reports, published academic research, visitor survey information and site visits and audits. On completion of the analysis of each case, the results were compared and contrasted both with each other and with information available from the Murphy *et al.* (2011) analyses to highlight key points in the evolution of this type of tourism. These analyses suggest key turning points where the nature of the decision made by planners, residents and both internal and external entrepreneurs contributed to a major change in the nature of the tourist experience provided: the discovery of the village by tourists or tourism entrepreneurs and the establishment of a key attraction; local retail shifting attention from locals to tourists; the arrival of external entrepreneurs and departure of residents; tourism reaching a critical mass; and creative destruction, replacement, re-invigoration or disruption.

### Historical Turning Points

#### Turning point 1: The discovery and establishment of a tourist attraction

Murphy and colleagues (2011) identified three main catalysts for the establishment of a rural village as a TSV – location, anchor attractions and entrepreneurs. The three case studies for the present analysis both provide an example of each of these catalysts and demonstrate their interaction. Tourism to Cheddar, for example, was and still is based primarily on its location as an access point to Cheddar Gorge and its caves. The development of tourism in Cheddar was slow, extending over more than a century (Irwin, 1986). It could be argued, however, that the opening of different caves over time provided an increasing stock of attractions which was supplemented by museums focused on the area's prehistory and most recently cheese making. Hahndorf offers an example of a village where tourism was promoted by the opening of an attraction, the house of Hans Heysen, a prominent Australian landscape artist. Its development as a TSV was, however, a little more complex than that of Cheddar. The opening of the Heysen house as an art gallery and museum after the artist's death in 1968 certainly drew some attention to the area, but it also coincided with considerable media attention given to the government's response to conflict over a proposed retail and residential development in the late 1970s (Hutchings, 1998). Hutchings (1998) offers Hahndorf's subsequent heritage listing and protection as an early example of history and heritage being considered as important aspects of local amenity. The widespread public debate over the nature and extent of the heritage listing highlighted the town's historical significance and German heritage. These two factors, combined with the increasing appeal of the rural idyll as an escape for city dwellers (Winchester & Rofe, 2005), supported the growth of tourism to this TSV. While similar themes of escape to a rural idyll and the importance of and interest in heritage buildings can also be linked to the development of St Jacobs as a TSV (McClinchey & Carmichael, 2010), this case provides an example of the power of an entrepreneur to generate tourism interest and development. According to Mitchell and de Waal (2009) the investment by one entrepreneur of approximately $8m in the building or renovation of retail space in the late 1980s prompted a significant increase in tourism. In both the Hahndorf and St Jacobs cases, ethnic heritage, linked to German settlers and Mennonites, respectively, was also a major attraction for visitors.

### Turning point 2: Local retail shifts attention from locals to tourists

In all three cases, the next stage in their evolution was very similar. As the number of tourists increases, local businesses, governments and specific interest groups such as historical societies combine to formally advertise the village attractions and provide more amenities for tourists such as additional walking trails, access to historic sites and programmes of festivals and events. This continued growth supports the development of restaurants and accommodation enhancing longer stays in the village. This combination of increased visitors and longer stays encourages tourists to shop, primarily for souvenirs. In the cases of Hahndorf and St Jacobs, these souvenirs included and often focused on the arts, crafts and food linked to the German and Mennonite heritage of these two locations. In Cheddar, the main focus was on souvenirs linked to the caves with the emergence of an interest in the local cheese appearing much later in its history. The Cheddar Gorge Cheese Company shop and attraction, for example, did not open until 2003 (Cheddar Gorge Cheese Company, 2018). While this interest in tourist shopping can create or revive an interest in local production of arts and food, it is also associated with changes in convenience for local residents as shops that serve mostly locals move away and are replaced by tourist shops. It is important to note, however, that this shift of shops and services from smaller rural villages to regional centres is not unique to shopping villages and is linked to larger patterns of rural–urban change (Amcoff et al., 2011).

### Turning point 3: Arrival of external entrepreneurs and departure of local residents

As tourist numbers increase and the businesses serving them grow and profit, TSVs often begin to attract the interest of external investors and entrepreneurs who see an opportunity to make money. Some build or open new attractions including food and accommodation, but others invest in property, raising rents for shop owners and local businesses. In St Jacobs these new investors focused on additional retail space and outlet shopping malls (Mitchell & de Waal, 2009). For some TSVs located close to urban centres this process of gentrification and escalating property values also changes the residential population, with urban dwellers moving in, attracted by the new amenities provided for tourists. In others that are more distant from urban centres, there may be a rise in second home ownership, although this was not apparent in any of the three cases studied here. These trends often lead to an exodus of the local population who are either escaping the changes or taking advantage of the increased property values to move to a better lifestyle elsewhere. Sometimes these departing residents are replaced by newcomers attracted by the changes, but sometimes there is just a decline in local residential populations, as was the case for Hahndorf prior to the start of the 21st century (ABS, 2003).

### Turning point 4: Tourism reaches a critical mass

In St Jacobs, the new investor focus on additional retail space and outlet shopping malls changed the nature of the village experience, attracting different tourist markets including those seeking only generic budget shopping opportunities (Mitchell & de Waal, 2009), reflecting a pattern identified in several TSVs across the United States and Canada (Murphy *et al.*, 2011). This shift in St Jacobs also resulted in the appearance of outlets for national and international franchises (McClinchey & Carmichael, 2010). In Hahndorf, increasing commercial rentals resulted in a change from shops offering local products to shops offering more expensive nationally and internationally branded homewares and luxury goods. Both trends result in a decline of locally based businesses and local production of arts and food. At the same time, increasing numbers of tourists can create issues of crowding and congestion, changing the balance of tourism costs and benefits significantly for each TSV (Getz, 2000).

### Turning point 5: Disruption, creative enhancement and/or rejuvenation

Up to this point, the pattern described in the present study can be linked to Mitchell's first four stages of creative destruction. At this point, however, the two analyses diverge with Mitchell suggesting an inevitable decline in the nature of the tourist experience and the village overall. Such a decline was not evident in the three cases examined for this paper, although the changes in the nature of the shopping offered and the corresponding change in tourist markets precipitated a major change in the TSVs' evolution. Decline has been noted in other TSVs but it is linked to a variety of external factors such as increasing competition from other locations, natural disasters or the closure of major attractions and not just to internal changes in the experience offered. The three cases analyzed suggested that this disruption point can lead to some combination of creative enhancement and rejuvenation. While in the case of St Jacobs, Mitchell's (1998) original examination of the village suggested that creative destruction inevitably led to decline in activities other than tourism, her more recent analyses (Mitchell, 2013) suggest that this process of creative destruction was more complex than she originally thought. In more recent analyses, Mitchell (2013) notes the existence of a parallel process of creative enhancement where the changes made can improve innovation and village functionality, supporting population growth and the retention of many services for both locals and tourists. In Cheddar, this disruption point has been associated with extensive rejuvenation of tourist facilities and an extension into local food and craft production. In a similar fashion, Hahndorf is undergoing a planning and public consultation process to guide a rejuvenation of the town and its tourist offerings (Mt. Barker District Council, 2017), with the area also experiencing population growth (Mt. Barker District Council, 2018).

### Re-imagining TSV Futures

Many trends are likely to impact on the future of TSVs and although many of these are likely to be localized, several more global megatrends highlight opportunities and challenges for TSVs seeking to reinvent themselves. These include the growing influence of Asian consumers, generational changes, the counterculture rebellion against the globalization of consumer culture, the impact of technological advances and the emergence of sustainability as a lens for examining development.

In emerging economies such as China and India, increasingly affluent middle-class travellers are driving retail spending. Sociodemographic trends in these Asian countries also point towards increasing levels of urbanization and education, with migration from relatively poor rural areas to wealthier urban areas largely driven by younger (aged 15–39 years) workers and consumers (Jones, 2016). These increasing levels of urbanization, education and affluence are likely to drive ongoing demand for short-stay trips to rural locations to allow travellers to escape the daily grind of city life. Villages located within an hour or two of major cities in Asia are likely to benefit from these trends. In addition, the tremendous growth of outbound tourism from China, India and the Middle East to developed Western nations also suggests new potential markets for TSVs located close to major international gateways. The growth potential from domestic tourism in developed Western countries seems more subdued. However, despite also displaying a population shift towards major urban conurbations, Western countries such as the United States, Canada,

Australia, the UK and parts of Western Europe have also shown evidence of 'amenity migration' from cities to regions with high-quality landscapes and lifestyle amenities (Gurran et al., 2016).

From a generational perspective, in addition to the Baby Boomers and Generation Xers, the vast majority of Generation Ys have now graduated from the education system and are a major source of employment and consumer spending. Research on the tourist shopping preferences of these three generations is limited, but related literature does provide some indication of the demand for TSV experiences. For example, a recent study indicated that Gen Y university students in the United States had a strong interest in some forms of authenticity (Chhabra, 2010). Moscardo et al. (2011) also reported that Generation Y had more positive attitudes towards diversity and social issues and were seeking local culture and social experiences. There is also a much stronger tendency for Generation Ys to seek employment opportunities that provide a better work–life balance. These observations suggest that TSVs may be well positioned, in terms of demand for products and experiences as well as offering a lifestyle that may be attractive to the next generation of TSV entrepreneurs.

Aside from generational changes, the subcultures within generations are of relevance to the discussion about the future of TSVs. Of particular note are counter-cultures such as 'hipsters', whose defining trait appears to be to assert individuality by rejecting the uniformity created by a globalized, neoliberal consumer culture. Lanham et al.'s (2003) *The Hipster Handbook* explains that hipsters prefer authenticity, small artisan shops, unbranded products and independent or organic food outlets, which would appear to be a good match for the types of products and shopping experiences offered by many TSVs. It has been argued that hipsters play an important role in modern society, because they expose mainstream consumers to a wider variety of underground trends (Nordby, 2013).

The major technological trends influencing the wider retail space cannot be overlooked. While shopping may not be the primary motivator for visiting a TSV, many villages depend on the economic benefits that result from turning visitors into shoppers. In other retail settings, online stores have become a major threat to traditional 'bricks and mortar' stores. However, there is considerable evidence that rather than purchasing arts and craft on the internet, visitors want to fully experience and evaluate products in a store before they buy (Stoddard et al., 2012). The experiential aspects of meeting artisans in person, seeing how products are made or co-creating products during the visit also contribute to visitor satisfaction (Murphy et al., 2011). The internet plays a key role in the 'inspiration' phase of travel decision making, but very little has changed since Murphy et al. (2011) observed almost a decade ago that the vast majority of TSV websites fail to present villages as part of a wider customized itinerary to help visitors plan aspects of their visit (e.g. activities, food and parking). Likewise, few villages effectively use social media to connect with visitors to build relationships which encourage return visits. There is an opportunity for the more strategic use of tools such as snapchat filters, Instagram frames, Instawalks, selfie-spots, 'retailer/café of the week', product profiles and contests that encourage visitors to share content. The ubiquity of smart devices creates new opportunities for 'smart TSVs' to learn more about visitor behaviour while providing visitors with more personalized and customized information.

Finally, the emergence of the term 'overtourism' (Alexis, 2017) and a range of recent critical reviews suggesting that little, if any progress, has been made to improve the sustainability of tourism practices (Moscardo & Murphy, 2014) suggest that some major shifts are needed in the way tourism researchers and practitioners think about the nature and impacts of tourism. Arguably, academics in this area have been concerned with the sustainability dimensions of TSVs, with many offering critical examinations of the negative impacts that tourist shopping and the changes in local businesses, landscapes and populations associated with this type of tourism can have on village residents. Unfortunately, as with many other areas of tourism, there has been little connection between these analyses and any set of recommendations provided to TSV planners, policy makers or developers. This is partly because the nature of tourism diffuses responsibility for its management across a diverse range of often disconnected stakeholders (Alexis, 2017). An alternative and stronger force has been the dominance of neoliberal discourses about business and market forces in tourism development practice, making academic suggestions about regulating TSV aspects such as what shops are allowed to sell or controlling shop rental costs for local producers very unlikely to be considered. The dominance of neoliberalism is, however, now being challenged (Centeno & Cohen, 2012) with new models emerging, especially in Western countries, often focusing on greater community empowerment in local development and planning (Giddens, 2013). This may change the way TSVs in some places change in the future.

Combining all these factors and linking them to the previously evolutionary turning points in the development of TSVs, we can suggest two future turning points: the emergence of Asian TSVs and the return to a rural idyll. These are not alternatives to one another and reflect a likely divergence in pathways for two entirely different types of TSV.

**Future Turning Points**

### Future turning point 1: The emergence of new Asian TSVs

The growing affluence and experience with international travel in numerous Asian countries is likely to support the adoption of tourist products and services in the domestic sector to match those seen overseas. A number of TSVs have already been identified and examined in China (Mitchell, 2013). These analyses suggest that while there are similarities between these Asian TSVs and existing Western TSVs, such as a focus on heritage architecture and access to special locations with the development of shopping as a major ancillary experience, there are also significant differences. The greater degree of government control over tourism development in China, for example, is connected to very different development trajectories and possibly fewer negative impacts for local residents (Qun et al., 2012). One possibility is a continuing growth of this type of tourist experience in Asia with a different evolutionary trajectory influenced by the different cultural, socio-economic and environmental contexts.

### Future turning point 2: A return to the rural idyll

The changes in values associated with the ageing of different generations, the rise of alternative lifestyles and the search for better lifestyles in rural regions outlined in the previous sections may support a turning point specific to existing TSVs in Western countries. The movement of new residents and entrepreneurs seeking business opportunities in arts, crafts and artisanal food production into existing TSV regions could revive the shopping focus on locally produced goods and a return of the balance of business ownership to locals. These forces may also encourage the development of new TSVs in rural areas, dispersing tourist pressures. In combination with changes in planning approaches that more fully embrace sustainability, it is possible that many TSVs will return to the style of experiences that were associated with their early development.

Although this turning point can be predicted based on current forces and pressures, it also reflects the history of the evolution of TSVs. Decisions made by tourism governance organizations that encouraged growth without consideration of impacts, a shift in attention from locals to tourists in retail and services, and external investment to support growth all contributed to Turning Points 2 and 3 and, through them, Turning Point 4. Current tourism decision makers are hopefully now aware of the unexpected negative consequences of these earlier decisions and so it is likely that this return to the rural idyll will remain focused on improving the wellbeing of local resident communities. An examination of the history of TSV development is therefore critical in understanding the links between decisions and consequences and thus guiding the future evolution of this tourism sector.

### References

ABS (2003) *Census of Population and Housing: Selected Characteristics for Urban Centres and Localities, South Australia, 2001.* Canberra: Australian Bureau of Statistics.

Adelaide Hills Online (2012) Town history of Hahndorf. See http://www.adhills.com.au/tourism/towns/hahndorf/history.html (accessed 27 March 2019).

Alexis, P. (2017) Over-tourism and anti-tourist sentiment: An exploratory analysis and discussion. *Ovidius University Annals, Economic Science Series* 17 (2), 288–293.

Amcoff, J., Moller, P. and Westholm, E. (2011) The (un)importance of the closure of village shops to rural migration patterns. *International Review of Retail, Distribution and Consumer Research* 21 (2), 129–143.

Centeno, M.A. and Cohen, J.N. (2012) The arc of neoliberalism. *Annual Review of Sociology* 38, 317–340.

Cheddar Gorge Cheese Company (2018) About us. Our story so far … See https://www.cheddaronline.co.uk/cheesemaking (accessed 27 March 2019).

Cheddar Parish Council (2007) About Cheddar. See https://cheddarparishcouncil.org/about-cheddar/ (accessed 27 March 2019).

Chhabra, D. (2010) Back to the past: A sub-segment of Generation Y's perceptions of authenticity. *Journal of Sustainable Tourism* 18 (6), 793–809.

Choi, M.J., Heo, C.Y. and Law, R. (2016) Progress in shopping tourism. *Journal of Travel and Tourism Marketing* 33 (suppl. 1), 1–24.

Fan, C., Wall, G. and Mitchell, C.J.A. (2008) Creative destruction and the water town of Luzhi, China. *Tourism Management* 29 (4), 648–660.

Getz, D. (2000) Tourist shopping villages: Development and planning strategies. In C. Ryan and S. Page (eds) *Tourism Management: Towards the New Millennium* (pp. 211–225). Oxford: Elsevier Science.

Giddens, A. (2013) *The Third Way: The Renewal of Social Democracy*. Cambridge: Polity Press.

Gurran, N., Norman, B. and Hamin, E. (2016) Population growth and changes in non-metropolitan coastal Australia. In R. Ganser and R. Piro (eds) *Parallel Patterns of Shrinking Cities and Urban Growth: Spatial Planning for Sustainable Development of City Regions and Rural Areas* (pp. 165–184). New York: Routledge.

Hahndorf Business and Tourism (2018) Hahndorf. See http://hahndorfsa.org.au/ (accessed 27 March 2019).

Hutchings A. (1998) Planning history in practice. *Australian Planner* 35 (3), 122–126.

Irwin, D.J. (1986) Gough's Old Cave – its history. *Proceedings of University of Bristol Spelaological Society* 17 (3), 250–266.

Jin, H., Moscardo, G. and Murphy, L. (2017) Making sense of tourist shopping research: A critical review. *Tourism Management* 62, 120–134.

Jones, G.W. (2016) Migration and urbanization in China, India and Indonesia: An overview. In C.Z. Guilmoto and G.W. Jones (eds) *Contemporary Demographic Transformations in China, India and Indonesia* (pp. 271–276). Cham: Springer.

Lanham, R., Nicely, B. and Bechtel, J. (2003) *The Hipster Handbook*. New York: Anchor.

McClinchey, K.A. and Carmichael, B.A. (2010) Countryside capital, changing rural landscapes, and rural tourism implications in Mennonite country. *Journal of Rural and Community Development* 5 (1/2), 178–199.

Mitchell, C. (1998) Entrepreneurialism, commodification and creative destruction: A model of post-modern community development. *Journal of Rural Studies* 14 (3), 273–286.

Mitchell, C. (2003) The heritage shopping village: Profit, preservation and production. In G. Wall (ed.) *Tourism: People, Place and Products* (pp. 151–176). Waterloo: University of Waterloo.

Mitchell, C. (2013) Creative destruction or creative enhancement? Understanding the transformation of rural spaces. *Journal of Rural Studies* 32, 375–387.

Mitchell, C. and Coghill, C. (2000) The creation of a cultural heritage landscape: Elora, Ontario, Canada. *The Great Lakes Geographer* 7 (2), 88–105.

Mitchell, C. and de Waal, S.B. (2009) Revisiting the model of creative destruction: St. Jacobs, Ontario, a decade later. *Journal of Rural Studies* 25, 156–167.

Moscardo, G. (2004) Shopping as a destination attraction: An empirical examination of the role of shopping in tourists' destination choice process and experience. *Journal of Vacation Marketing* 10, 294–307.

Moscardo, G. and Murphy, L. (2014) There is no such thing as sustainable tourism: Re-conceptualizing tourism as a tool for sustainability. *Sustainability* 6 (5), 2538–2561.

Moscardo, G., Murphy, L. and Benckendorff, P. (2011) Generation Y and travel futures. In I. Yeoman, C.H.C. Hsu, K.A. Smith and S. Watson (eds) *Tourism and Demography* (pp. 87–100). Oxford: Goodfellow.

Mt. Barker District Council (2017) *Hahndorf Township Plan: Draft for Consultation*. See https://www.mountbarker.sa.gov.au/webdata/resources/files/A1001_V1_170809_Consolidated%20Plan.pdf (accessed 27 March 2019).

Mt. Barker District Council (2018) Community profile. See https://profile.id.com.au/mount-barker/population?WebID=120 (accessed 27 March 2019).

Murphy, L., Moscardo, G., Benckendorff, P. and Pearce, P. (2008) Tourist shopping villages: Exploring success and failure. In A. Woodside and D. Martin (eds) *Tourism Management: Analysis, Behaviour and Strategy* (pp. 405–423). Wallingford: CABI.

Murphy, L., Benckendorff, P., Moscardo, G. and Pearce, P.L. (2011) *Tourist Shopping Villages: Forms and Functions*. New York: Routledge.

Murphy, L., Moscardo, G. and Benckendorff, P. (2013) Tourist shopping experiences on the margins. In J. Cave, L. Jolliffe and T. Baum (eds) *Tourism and Souvenirs: Glocal Perspectives from the Margins* (pp. 132–146). Bristol: Channel View Publications.

Nordby, A. (2013) What is the Hipster? *Spectrum* 25, 52–64.

Qun, Q., Mitchell, C. and Wall, G. (2012) Creative destruction in China's historic towns: Daxu and Yangshuo, Guangxi. *Journal of Destination Marketing and Management* 1 (1–2), 56–66.

St. Jacobs Country Tourism (2018) St. Jacobs Country. See https://stjacobs.com/ (accessed 27 March 2019).

Stoddard, J.E., Evans, M.R. and Shao, X. (2012) Marketing arts and crafts: Exploring the connection between hedonic consumption, distribution channels, and tourism. *International Journal of Hospitality and Tourism Administration* 13 (2), 95–108.

Suhartanto, D. (2018) Tourist satisfaction with souvenir shopping: Evidence from Indonesian domestic tourists. *Current Issues in Tourism* 21 (6), 663–679.

Timothy, D. (2018) Shopping tourism. In S. Agarwal, G. Busby and R. Huang (eds) *Special Interest Tourism* (pp. 134–144). Wallingford: CABI.

Visit Somerset (2018) Cheddar. See https://www.visitsomerset.co.uk/discover-somerset/towns-villages/north-somerset/cheddar-(1) (accessed 27 March 2019).

Winchester, H.P.M. and Rofe, M.W. (2005) Christmas in the 'Valley of Praise': Intersection of the rural idyll, heritage and community in Lobethal, South Australia. *Journal of Rural Studies* 21 (3), 265–279.

# 16 Tourism and Religion: Pilgrims, Tourists and Travellers – Past, Present and Future

Richard Butler and Wantanee Suntikul

## Introduction

Of all the links between tourism and other human activities, that between tourism and religion is one of, if not the, oldest (Butler & Suntikul, 2018). People have been travelling for religious reasons for millennia, whether to visit holy people, to worship at holy sites, or to pay respects to deceased family and friends or at other death-related sites. Such travel is often termed as pilgrimage (travel made to a holy place for religious or spiritual reasons), and is the subject of a large body of literature, although relatively little of that relates to tourism and its issues (for example, Coleman & Elsner, 1995; Eade, 2015; Stoddard & Morinis, 1997). In the past, the needs of religiously motivated travellers have given rise to a number of key elements of modern tourism and hospitality, including accommodation and facilities for travellers, organized tours, financial elements, security arrangements, transport facilities and infrastructure, and political arrangements. Some of these elements are related to specific events and their effects and several of these are discussed below. In other cases, while the motivation for travel may be clear, the precise origins are unclear. Such is the case with much pilgrimage travel; when the first of such travels was made is often not recorded, although the event giving rise to the travel may be quite specific. To project into the future is always difficult, but a number of forthcoming events suggest that tourism may well be influenced by events of religious significance. While it may be fashionable to discount the importance of religion in Western contemporary life, such a viewpoint is incorrect and inappropriate in many other parts of the world and particularly in non-Christian societies (Akkok, 2015).

## Major Religions, Key Events and Turning Points

All of the major religions (those with several million or more believers) involve pilgrimage or travel for religious purposes to specific sites and events. Thus the establishment of these religions has been of major significance in terms of influencing and creating tourist travel, both short-term and short-distance and in many cases long-haul and longer term journeys. Thus Christianity, Islam, Judaism, Buddhism, Hinduism and Mormonism are considered in some detail, in terms both of their past influence and of their likely future significance for tourism. Within these major religions there are a number of divisions, such as: Sunni and Shia in particular in Islam; Catholic, Protestant and Orthodox in Christianity; and Orthodox and Sephardic in Judaism. These divisions have in turn given rise to specific events and commemorations, and also resulted in disagreements, prohibition and conflict between themselves, all affecting tourism. Other faiths, such as Shinto and Sikh, also involve travel to sacred sites, although in the case of Shinto (Nakanishi, 2018) much of the travel is short term in relation to the time and distance involved, and in the case of Sikh pilgrimage, mostly to one site. Travel by the adherents of some faiths can involve specific requirements in terms of accommodation, food and access to places of worship (Weidenfeld, 2006) and such requirements are likely to increase in terms of demand and specific characteristics and thus become a major influence on the nature of religious tourist travel.

Table 16.1 shows a timeline illustrating significant past events of a religious nature that have impacted greatly on tourism. In some cases an event has led directly to pilgrimage and tourism to the locations at which they occurred; in other cases there have been a number of developments stemming from the original event that have in turn led to tourism, not exclusively in the form of pilgrimage.

In terms of turning points, that is, events of such significance that they have radically altered and/or grown international tourism, two common elements emerge from this brief examination of the relationship between the major faiths and tourism.

**Table 16.1** Significant events of a religious nature

| Religion | Events | Dates | Individuals | Locations |
| --- | --- | --- | --- | --- |
| Judaism | Exodus from Egypt | Circa 15th–13th centuries BC | Moses | Middle East |
| Hinduism | Vedic Age | Circa 15th–7th centuries BC | Rishis (sages) | South Asia |
| Buddhism | Life of Buddha | Circa 6th–4th centuries BC | Gautama Buddha | South Asia |
| Christianity | Life of Christ | Circa 0–30 CE | Jesus Christ | Middle East |
| Islam | Life of Mohammed | 570–632 CE | Mohammed | Middle East |

### Historical turning point 1: Births of specific individuals

For most of the major faiths Turning Point 1 has been the establishment of the individual faith, generally following the birth of specific individuals. This has resulted in religiously inspired travel related to the birth and subsequent life and travels of those individuals. In subsequent years non-secular travellers also visit the same places, motivated by curiosity and seeking knowledge, inspiration and the cultural heritage of such sites.

### Historical turning point 2: Division

In general there is rarely a second specific event that might warrant being termed a turning point, but many of the major faiths have seen a common if unfortunate development that has had important impacts on international tourism. This is a schism or division within individual faiths, Shia and Sunni, or Catholic and Protestant, in the case of Islam and Christianity, for example. Such divisions have led to the identification of additional places to be visited by the faithful and others, but also, regrettably, antagonism and even violence between followers of the different forms of the faiths concerned. This has inevitably had negative effects on international tourism, even leading to terrorism and war over specific locations. Travel within Europe in the 16th and 17th centuries was disrupted by such conflicts, as is travel within the Middle East and also Burma, Kashmir and Tibet today.

*Christianity*

For Christianity, the major single turning point is clearly the birth of Jesus Christ (0000 BC), commemorated as Christmas in the Christian calendar – an event which involves a great deal of travel, much of it tourism. It ranges from short journeys to local churches and to visit family members, to pilgrimages to the Holy Land and other sacred sites, and is unlikely to change greatly in the future, although the year 3000 may experience global celebrations if humanity and civilizations have survived to that date. Whether the birth of Christ is the starting point for Christian travel can be argued, as one might go further back in time and take the Exodus from Egypt and the subsequent establishment of the Jewish settlement in Palestine as the real beginning. The birth of Christ (a Jew) was only possible in Bethlehem following the founding of Israel and its conquest by Roman forces, and has resulted in continued visitation there since 0000 BC. This travel led to the Christian settlements in Jerusalem, and then to the Crusades in order to achieve and maintain access to the holy sites of Christendom in Palestine. In part these travels led to an ongoing conflict between the Christian and Muslim faithful over access to the Holy Land and other areas, coupled with the spread of Islamic conquests from the 7th century to the 15th. From the Crusades came the establishment of the orders of Knights (O'Gorman, 2018) Hospitallers and Templars, and in the 6th century the beginning of the provision of formal hospitality and shelter for travellers. St Benedict established the basic rules of hospitality from which have emerged the foundation of inns, hostels and hotels, following his basic principles laid out initially for religious travellers and visitors to monasteries (O'Gorman, 2018).

Following the establishment of the Christian Church and its spread throughout Europe came the conversion of the Roman Empire to Christianity by Constantine in AD 312, the establishment of the Papacy and the pre-eminence of Rome as the centre of the church, itself becoming a site of pilgrimage, in part related to the relics of St Peter and partly as the home of the Popes. With the schism in the Catholic Church in the 11th century and the establishment of the Eastern Papacy in Constantinople, increased travel to that city began until its conquest by the Turks in 1453. In Europe other sites assumed Christian religious significance and attracted visitors: Avignon in France for the period during which the Papacy was based there; sites of miracles such as Lourdes (France) and Fatima (Portugal); sites of relics such as Turin (Italy), home of the Shroud of Turin, and Santiago de Compostela, location of the relics of St James and the end point of several Caminos (routes) which crossed Europe and have been used by pilgrims for several centuries. These Caminos are currently increasing in popularity and similar pilgrim-inspired routes are being established in several European countries (Butler & Suntikul, 2018). All of these events and features relate to the birth of Christ which is, therefore, one of the major turning points shaping tourism historically and thus, because of the inertia involved in tourism whereby each generation's travel patterns are influenced by the destinations chosen by the preceding one, is also continuing to influence tourism at the present time.

*Islam*

The key event and turning point in Islam is the birth of Mohammed (541 BC) and from this, the pattern of his life and travels. Of major significance is his ascension to Allah from the Dome of the Rock in Jerusalem (Islam's third most religious site after Mecca and Medina) and his tomb in Mecca, the most sacred site for Muslims. His tomb is the scene of the annual *Hajj*, a pilgrimage required at least once in the life of every Muslim able to make such a journey, with several million making the trip each year (Raj & Kessler, 2018). The advent of modern transportation methods has greatly reduced the difficulty of this journey for those living outside the Arabian Peninsula and numbers have increased accordingly, along with problems of congestion, health and security issues and political problems. In recent years the increasing hostility between branches of Islam, particularly Sunni and Shia, has reached a head with restrictions being placed on Iranians visiting Mecca by Saudi Arabia, the guardian of the Holy Sites in its country. Increasing development of tourist facilities around the Kaaba has resulted in considerable criticism from Islamic religious leaders, with accusations of the *Hajj* becoming a tourist trip and losing its religious commitment (Sardar, 2014) The *Hajj* is unlikely to reduce in size, and all indications are that it will increase in the years ahead as Muslim numbers grow around the world and air travel in particular makes access to Saudi Arabia much easier for Muslims living large distances away. The Millennium celebration of the birth and death of Mohammed has passed and thus any specific additional celebration would not occur until much later in the 21st century (AD 2600 and 2700) However, tourism to other Islamic sacred places such as Jerusalem is likely to increase as numbers of the faithful increase and become more affluent. This could pose problems of security and access in Israel at least, where the ongoing conflict between Israel and Palestine shows little sign of ending. Elsewhere, the growth of Muslim tourism has been noted (Isaac, 2018) and the provision of Muslim-friendly facilities is likely to increase, even while restrictions on Muslims are imposed in some destinations.

*Judaism*

The origins of Judaism and Israel have been noted above in the context of the foundation of Christianity, and while Judaism does not formally involve pilgrimage, as Collins-Kreiner and Luz (2018) have noted, travels by the Jewish faithful to specific sites is common, particularly to Jerusalem and the Wailing Wall and related sites. Beyond the Exodus from Egypt and the resulting establishment of Jewish settlement in Palestine noted earlier, other important turning points include the Holocaust and related events and the formal establishment of the State of Israel. Travel by the Jewish diaspora to Israel, in particular, has become linked to demonstrating support for the continued existence of the State of Israel (Mansfeld, 1994) and this is likely to continue in the future, barring catastrophic events in the Middle East. In addition, as a result of the Holocaust (1940–1945), Jewish travel to Holocaust-related sites has grown steadily, to Auschwitz in particular, from around 300,000 in 1959 to over 2 million in 2017 (Memorial and Museum Auschwitz-Birkenau, 2018). It is possible that the centenary of the opening of the museum (1947) may see a significant increase in the number of visitors but this would be primarily of significance to Poland rather than globally. A definitive and lasting peace settlement in Israel/Palestine would have major implications for the growth of tourism in the Middle East as a whole (as discussed below), as many visitors to the Holy Land might also visit neighbouring countries such as Jordan and Egypt which also have both Christian and Jewish sites of interest.

*Mormonism*

The Mormon faith is one of the smallest and newest religions, having been founded in the United States in 1830, but has been involved in tourism and travel almost since its inception. Apart from the actual establishment of the church itself, other turning points include the migration of members of the Church of Latter-day Saints to the State of Utah, which in turn resulted in the creation of Salt Lake City and its monuments and other sites of particular interest, both to the adherents of the Mormon faith and to many other secular tourists. Since the establishment of the church there has been a strong evangelical movement resulting in the global spread of the religion, and the faithful visit Salt Lake City and other sites related to the faith, and engage in re-enactments of events (Olsen & Timothy, 2018). While more visible globally than simple numbers would suggest, and with the next centennial anniversary of the founding of the church occurring in the relatively near future, this may encourage additional travelling by the faithful. However, the dimensions of such travel are not likely to affect global international tourist numbers to any degree. The Mormon sites of interest are likely to experience higher visitation in the future but would involve a relatively small number of travellers on the global scale.

*Buddhism*

The Buddhist faith has vast numbers of adherents and visiting the places to which the Buddha travelled is important to many; thus tourist travel is significant in terms of numbers and scale, although much of the travel is within the travellers' own countries and thus is not international tourism. The birth of the Buddha (2600 years ago) is the only real turning point in this faith, followed by his life and particularly his travels and the site of his death; these remain major pilgrimage sites for the faithful. Most of the faithful live in Asia, but it has become a global faith and has received increased attention following the Chinese persecution in Tibet and the Dalai Lama fleeing from there in 1959. In recent years the Chinese authorities have relaxed their prohibition on religion and Buddhism has recovered in numbers and importance in China. The Dalai Lama being awarded the Nobel Peace Prize in 1989 increased awareness of Buddhism and in the past decades the image of Buddhism as a peaceful faith has earned it many supporters; however, its image has been severely damaged by the recent actions of the Myanmar government against Muslims in that country (Mercer, 2018). The centenary of the Dalai Lama's flight from Tibet might prompt some commemorations but the Chinese authorities are unlikely to welcome large numbers of visitors to Tibet to commemorate that event. Thus, while Buddhism may witness an increase in numbers including those engaged in pilgrimage, it is unlikely to cause any major change in the pattern of religious tourism in the foreseeable future.

*Hinduism*

Of all attendees at religious events, the numbers of the followers of Hinduism are the largest, but as with Buddhism, most of those involved are travelling within their own country and are thus not international tourists. There is no clear turning point in the context of Hinduism and international tourism, reflecting the numerous gods and religious icons and the domestic focus of most travel. While religious events in India have attracted independent travellers from the West with a spiritual or counter-cultural interest, particularly since the 1960s, the numbers of these tourists are minute compared to the masses of domestic tourists travelling within India for spiritual reasons. The most significant religious events in terms of the number of travellers they attract are the *Kumbh Melas* – mass pilgrimages to bathe in sacred or holy rivers – which take place at four locations in India where tradition holds that Lord Vishnu let drops of Amrita (the elixir of immortality) fall to earth. It is uncertain when the *Kumbh Mela* began but it was inscribed on the UNESCO list of Intangible Cultural Heritage in 2017. A *Kumbh Mela* is celebrated every 12 years at a given location, so that a *Kumbh Mela* will take place somewhere in India every one to three years. *Ardh* (half) *Kumbh Melas* are celebrated every six years in Allahabad and Haridwar, and *Maha* (great) *Kumbh Melas* every 144 years, only in Allahabad. These are among the largest peaceful gatherings on earth, for example with 120 million people attending the 2013 *Maha Kumbh Mela* in Allahabad over a two-month period, with 30 million visitors on 10 February, the 'day of silence' or *Mauni Amavasya*. The next *Maha* (great) *Kumbh Mela* will not be held until the year 2157, and it too is likely to be dominated by domestic travellers.

## Discussion

It is important to consider the role of religion in global life in the future if one is to anticipate the likely impact of religious events and the emergence of future turning points on tourism. Despite evidence that participation in religion, e.g. attending services in churches, is declining in many Western European countries (Akkok, 2015), visits to religious sites and artefacts is increasing throughout the world, reflecting both the sacred role such places play in people's lives, and their attractiveness as tourist sites for travellers whose visits are not primarily spiritually motivated. This reflects their architectural appeal, their beauty, their historic interest and their 'marking' as places to visit. In London, Westminster Abbey and St Paul's Cathedral are two of the most popular sites for tourists, as are Vatican City in Rome, the Blue Mosque and Hagia Sophia in Istanbul, Manger Square in Bethlehem, and the Wailing Wall, the Church of the Holy Sepulchre and Al Aksah Mosque in Jerusalem. Eastern and southern Asia are also replete with such sites, including the Buddhist sites of Wat Arun in Bangkok, Jokhang Temple in Lhasa, the Todaiji Temple in Nara, the Shwedagon Pagoda in Yangon, and Hindu sites such as the *ghats* of Varanasi, the Golden Temple of Amritsar and the temple complex of Hampi. Other sacred sites such as the Temple of the Delphi Oracle in

Greece, Uluru in Australia, Angkor in Cambodia and the Pyramids of Egypt are global tourist attractions, only in part because of their religious connections and heritage.

In the long term, shifts in religious demographics will have an effect on the patterns of spiritually motivated travel. Christianity is expected to have the most adherents through the middle of this century, but will be surpassed by Islam, the fastest growing faith in numbers, by the century's end. Historically and currently, Europe has been the home of the greatest number of Christians of any continent, but by 2050 more Christians are expected to live in Africa than in any other continent. The number of Christians in Europe is set to decline, not just in terms of the proportion of the population (with inward migration of non-Christians and increasing prevalence of irreligiosity likely contributing to this), but also in absolute numbers (Pew Research Centre, 2014). As European Christian tourists become fewer and older, overtaken by younger, less affluent Africans, patterns of religiously inspired tourism among Christians is certain to change in character and probably in numbers.

By 2050, the Pew Report (Pew Research Centre, 2014) predicted that 30% (2.8 billion) of the population will identify themselves as Muslim compared to 31% (2.9 billion) identifying themselves as Christian. In Europe, it is suggested that by 2050, 10% of the continent will be Muslim and in the United States it will become the second-largest faith. According to the report authors, the increase in the Muslim population is because those following the faith are younger and infant mortality rates are falling. Sixteen percent of the population was unaffiliated to a religion in 2010 and Pew predicted that by 2050 this would fall to 13%, mainly because individuals in this group are older and have fewer children.

**Future Events**

The identification of future turning points among the world's major faiths that would affect global and international tourism is extremely difficult, because even predicting the secular future at the best of times is fraught with problems and generally results in inaccurate forecasts. In general secular terms, one is often able to use an evolutionary approach and anticipate current trends continuing to some degree. In the case of religious turning points, however, this is not a viable approach, as the events shown in the timeline above (Table 16.1) were all unexpected, at least in their timing and certainly in their effects upon tourism in the 21st century, and thus were impossible to predict. One might expect researchers discussing the future pattern of international tourism in general to include some discussion of the potential of religious events to influence tourism, but such is not the case. Scott and Gössling (2015), for example, identify five major factors as influencing tourism development in their paper discussing the next 40 years of tourism (social, technology, economic, environment and political), and while they list political instability and terrorism as sub-factors of the first and last major issues, there is no specific mention of religion. Dwyer (2015) examines drivers of tourism globalization and identifies five similar elements (economic, technology, demographic, social and political), but again makes no reference to religious influence, although he does note changing attitudes towards human rights, societal values and movement of people. Webster and Ivanov (2015) do mention religion, but only as one of many factors related to one of their geopolitical drivers of future tourism flows, in the context of 'increased global political instability' as one of several issues affecting the fragmentation of countries. One might examine the likely travel and personal habits of future tourists, as Richards (2015) has done, but he also says nothing about increased interest in, or influence of, religion among the young 'global nomads'. Buckley *et al.* (2015) include ideology as a possible driver of change but in the context of governments and not religion, while Moutinho (2016) focuses more on broad social and technological changes. Gallego and Li (2017: 1), in their introduction to a special issue of the *Journal of Policy Research in Tourism, Leisure and Events* (Volume 9, Issue 1) on Instability and Tourism, do mention 'the rise of Islamic State' as having adverse effects on tourism, but none of the other papers in that issue discusses religion or its influence on tourism. Thus, one has to fall back on imagining some specific events that might be of such a dimension as to qualify as turning points that would significantly affect global tourism patterns.

## Future Turning Points

### Turning point 1: Discoveries and events

The first potential turning point would be connected to the discovery of new religious artefacts, sites or events. For example, this could be a significant find relevant to Christians (such as the discovery of the Ark of the Covenant or genuine relics of saints and other early sacred persons), as this would attract large numbers of both religious and secular visitors. Another example would be if new sites confirmed as having been visited by the Buddha were discovered and verified, which would have a similar effect. Additionally, another example would be the discovery of new evidence of the activities of Mohammed or his close relations. Such an event might also have impacts on the divisions within Islam and a reduction in mutual antagonism.

All of such events could result in an increase in numbers of international tourists as well as a possible relocation of religiously inspired travel. Even where sites with supposedly religious connections are not genuine, they are often visited by large numbers of tourists (Page, 2010), because of the power of social media and marketing making any new significant curiosity likely to be exploited to attract tourists. Sites of 'miracles', such as Lourdes and Fatima, have attracted the faithful in search of redemption or cures for illnesses or misfortune since the earliest times, and in the modern era they attract large numbers of secular tourists as well. The most likely locations for future significant religious-related discoveries are in the Middle East, and in the Holy Land in particular, given the significance of this region to three of the major religions, or in India and surrounding countries for sites relevant to Hinduism or Buddhism. Whether any such discovery would result in significant tourism growth or changes in current patterns of visitation depends heavily on the political situation in the particular region at the time.

### Turning point 2: Peace

Perhaps the most likely religious turning point that would have the greatest effect on tourism is one that would not be primarily religious at all, but would be political, namely the establishment of a lasting and universally accepted peace between Israel and all of its Arab neighbours. Such an event would allow greater tourism visitation without any security concerns, and might even change the occurrence of terrorist events globally, which would justify its being regarded as a turning point. It would allow a major rise in Islamic tourism to sites in Israel from beyond the current borders and remove other security problems currently deterring such tourists (Isaac, 2018). The cessation of any Islamic fundamentalist inspired terrorism would almost certainly see the return to growth of tourism to Egypt and Tunisia, if not also to Syria, Lebanon, Iran and Iraq. While the lack of a permanent peace between Israel and its Arab neighbours is not the only reason behind Islamic fundamentalist terrorism, it is a major factor in influencing the shape and dimension of tourism to much of the eastern and southern Mediterranean and the Middle East.

In the same vein, agreements between the opposing main branches of Islam – Sunni and Shia – would also be a major factor in reducing concerns over violence and its resulting effects on tourism, as well as religious travel, including pilgrim travel such as the *Hajj*, in many areas in the Middle East. While such inter-sectoral conflicts have religious interpretation at the base of the problems, the situations are made more complex and irreconcilable by political intervention by a number of states and their governments. The religious issues are unlikely to be resolved in religious terms alone, just as many of the political disagreements will require religious solutions as well (Timothy, 2013). Resolution of the Kashmir problems between India and Pakistan could result in significant increases in tourism to that region, both by Sikhs and other religious tourists, and by secular tourists drawn to the many attractions of the border region.

Religious sites are important destinations for adherents of Hinduism, Buddhism and Taoism, and are also among the most popular and distinctive sites for non-religious tourists, many of whom come from abroad. With religious tensions beginning to flare up with greater frequency, there are also implications for tourism. Violent Islamist movements in the Philippines and Thailand sometimes intentionally target tourism hotspots, frightening tourists away from some destinations (Nicolas, 2017). In contrast, the extreme persecution of the Muslim Rohingya minority in Myanmar continues, yet the violent confrontations are far from the popular tourist areas and few tourists seem to be avoiding the country out of fear or protest, with tourist arrivals to Myanmar rising 18% in 2017, the peak year of the atrocities against this minority (Thu & Koutsoukis, 2018).

Other negative influences of religious disagreement and conflict can be seen in the recent (2016–2017) avoidance by tourists of established destinations in the eastern and southern Mediterranean which had either experienced terrorism (for example, Tunisia and Egypt) or were experiencing issues with refugees from the Middle East and Africa (Turkey and Greece, in particular). In the case of 'The Troubles' in Ireland, while the issue of unification of Northern Ireland and the Republic of Ireland and the ending of British governance in Northern Ireland were primarily politically motivated, there were clearly longstanding religious differences underlying much of the political struggle. The Good Friday Peace Agreement (1998) has demonstrated that the ending of conflict and the establishment of good relations between communities can result in not only a resumption of tourism, but also significant investment and an increase in numbers of tourists to both communities (Boyd, 2017).

## Conclusion

Thus the relationship between religion and tourism is two-sided. Religious heritage (buildings, artefacts, myths and lifestyles) can be extremely attractive to tourists and can draw many visitors from throughout the world to specific locations, a large portion of whom are secular rather than spiritual travellers. On the other hand, religious differences and conflicts can lead to violence, war, terrorism and persecution, all having catastrophic impacts upon tourism and deterring potential tourists from visiting areas so affected, or perceived to be affected. There is not a great likelihood of any significant events of a religious nature that are likely to affect tourism numbers or patterns in the foreseeable future but, equally, the religiously related attractions are unlikely to diminish in appeal in the future either.

Continued or increased religious conflicts, on the other hand, are quite possibly going to continue and even increase in number and severity in the future, in which case they could have major negative impacts on tourism globally and regionally. Major terrorist events could end, at least temporarily, tourism to specific destinations, even to countries and regions, and reduce global tourism, as happened briefly after the Twin Towers attacks in New York in 2001. Such events can shake general tourist confidence in the security of travel as well as the safety of specific locations and features. Heightened security arrangements for travel, particularly air travel, have restored confidence generally in the movement of people, but specific locations are still suffering from the after-effects of particular events. While Paris and London suffered only very short-term minor declines in tourism as a result of atrocities there, in the case of Tunisia it has taken two years for governments to lift a travel advisory on the country and for tourists to return. Unfortunately, the predictability of any future religiously motivated terrorist attacks is low, and to tourists with 'the whole world as their oyster' because of the wide choice of locations and the ease and lower cost of global travel, any threat to security generally means a decision not to include such destinations in their range of potential destinations when selecting the location of their next holiday.

It is impossible to separate religion and religious events and their influences on tourism from other world issues and problems. Many of the problems associated with religion in the context of tourism are also closely related to political issues in specific parts of the world, particularly the Middle East. The likelihood of major religious events occurring in the future that would significantly affect global tourism is low. The birth of Christ, although forecast in old texts, was a surprise, and the births of Mohammed, Buddha and other religious and sacred figures were generally unnoticed at the time and it was only after their activities during their lifetimes (miracles, preaching and travelling) that they became famous, worshipped and had impact, not just on tourism but on the world in general. Some events are predictable – anniversaries and commemorations in particular – but none is anticipated in the next half-century that might be significant enough to change tourism patterns or numbers. Political changes, as noted above, are much more likely to take place and to have major influence on religious matters, disputes, celebrations, gatherings and particularly pilgrimages. Much of what is often termed religious tourism is actually secular tourism to religious sites and artefacts, and this is likely to continue growing; some political events may, if positive in terms of mitigating or stopping conflict, cause major rises in numbers of visitors to specific sites. Similarly, pilgrim numbers would also increase under such scenarios. On the negative side, an increase in religious hostility, in religious-inspired terrorism or in political conflict resulting in war would cause tourist numbers in general to decline, and in specific areas to diminish to a tiny percentage of the level they had achieved during peaceful times, as witnessed in Egypt and Tunisia in recent years. Thus religion is likely to remain an influence on global tourism for the foreseeable future, both positively and negatively.

## References

Akkok, R. (2015) Mapped: What the world's religious landscape will look like in 2015. *The Telegraph*, 8 April. See https://www.telegraph.co.uk/news/worldnews/11518702/Mapped-What-the-worlds-religious-landscape-will-look-like-in-2050.html (accessed 15 March 2015).
Boyd, S.W. (2017) Tourism and political change in Ireland, North and South. In R.W. Butler and W. Suntikul (eds) *Tourism and Political Change* (pp. 153–168). Oxford: Goodfellow.
Buckley, R., Gretzel, U., Scott, D. and Becken, S. (2015) Tourism megatrends. *Tourism Recreation Research* 40 (1), 59–70.
Butler, R.W. and Suntikul, W. (2017) *Tourism and Political Change*. Oxford: Goodfellow.
Butler, R.W. and Suntikul, W. (2018) *Tourism and Religion: Issues and Implications*. Bristol: Channel View Publications.
Coleman, S. and Elsner, J. (1995) *Pilgrimage Past and Present: Sacred Travel and Sacred Place in the World Religions*. London: British Museum Press.
Collins-Kreiner, N. and Luz, N. (2018) Judaism and tourism over the ages: The impacts of technology, geopolitics and the changing political landscape. In R.W. Butler and W. Suntikul (eds) *Tourism and Religion: Issues and Implications* (pp. 51–57). Bristol: Channel View Publications.
Dwyer, L. (2015) Globalization of tourism: Drivers and outcomes. *Tourism Recreation Research* 40 (3), 326–339.
Eade, J. (2015) *International Perspectives on Pilgrimage Studies*. London: Routledge.
Gallego, M.S. and Li, S.N. (2017) Special issue on tourism and instability. *Journal of Policy Research in Tourism, Leisure and Events* 9 (1), 1–2.
Isaac, R. (2018) Religious tourism in Palestine: Challenges and opportunities. In R.W. Butler and W. Suntikul (eds) *Tourism and Religion: Issues and Implications* (pp. 143–160). Bristol: Channel View Publications.
Mansfield, Y. (1994) The Middle-East conflict and tourism to Israel 1967–90. *Middle Eastern Studies* 30, 644–667.
Memorial and Museum Auschwitz-Birkenau (2018) 2,1 million visitors at the Memorial in 2017. See http://auschwitz.org/en/museum/news/2-1-million-visitors-at-the-memorial-in-2017,1292.html (accessed 1 March 2018).
Mercer, D. (2018) Marketing Myanmar: The religion/tourism nexus in a fragile polity. In R.W. Butler and W. Suntikul (eds) *Tourism and Religion: Issues and Implications* (pp. 161–181). Bristol: Channel View Publications.
Moutinho, L. (2016) What will future bring for tourism and travel? *Advances in Tourism Hospitality and Research* 4 (2), 137–139.
Nakanishi, Y. (2018) Shintoism and travel in Japan In R.W. Butler and W. Suntikul (eds) *Tourism and Religion: Issues and Implications* (pp. 68–82). Bristol: Channel View Publications.
Nicolas, B.D. (2017) Terror threats make it harder for PH to become key tourist destination. *BusinessInquirer*, 29 May. See https://business.inquirer.net/230330/terror-threats-make-harder-ph-become-key-tourist-destination (accessed 3 March 2018).
O'Gorman, K. (2018) Origins of hospitality in monastic and Christian orders. In R.W. Butler and W. Suntikul (eds) *Tourism and Religion: Issues and Implications* (pp. 18–32). Bristol: Channel View Publications.
Olsen, D.H. and Timothy, D.J. (2018) Tourism, Salt Lake City and the cultural heritage of Mormonism. In R.W. Butler and W. Suntikul (eds) *Tourism and Religion: Issues and Implications* (pp. 250–269). Bristol: Channel View Publications.
Page, J. (2010) Tomb tourists told he's not the Messiah. *The Times*, 19 July, p. 11.
Pew Research Centre (2014) Religious landscape study. See https://www.pewforum.org/religious-landscape-study/ (accessed 3 March 2018).

Raj, R. and Kessler, K. (2018) Inspiration for Muslims to visit mosques. In R.W. Butler and W. Suntikul (eds) *Tourism and Religion: Issues and Implications* (pp. 33–50). Bristol: Channel View Publications.

Richards, G. (2015) The new global nomads: Youth travel in a globalizing world. *Tourism Recreation Research* 40 (3), 340–352.

Sardar, Z. (2014) *Mecca: The Sacred City*. London: Bloomsbury.

Scott, D. and Gössling, S. (2015) What could the next 40 years hold for global tourism? *Tourism Recreation Research* 40 (3), 269–285.

Stoddard, R.H. and Morinis, A. (1997) *Sacred Places, Sacred Spaces: The Geography of Pilgrimages*. Baton Rouge, LA: Geoscience Publications.

The Times (2016) Israel halts Holy Sepulchre church tax. *The Times*, 28 February, p. 40.

Thu, K. and Koutsoukis, J. (2018) The Rohingya crisis hasn't hurt Myanmar's tourism industry. Bloomberg, 21 February. See https://www.bloomberg.com/news/articles/2018-02-21/myanmar-sees-rise-in-tourist-numbers-despite-rohingya-crisis (accessed 3 March 2018).

Timothy, D.J. (2013) Tourism, war, and political instability: Territorial and religious perspectives. In R.W. Butler and W. Suntikul (eds) *Tourism and War* (pp. 13–25). London: Routledge.

UNWTO (2018) On the Camino to a better world. See http://media.unwto.org/press-release/2018-03-15/walking-talk-value-human-rights-camino-de-santiago (accessed 16 March 2018).

Vukonic, B. (2002) Sacred places and tourism in Roman Catholic tradition. In D.J. Timothy and D.H. Olsen (eds) *Tourism, Religion and Spiritual Journeys* (pp. 237–253). London: Routledge.

Webster, C. and Ivanov, S. (2015) Geopolitical drivers of future tourist flows. *Journal of Tourism Futures* 1 (1), 58–68.

Weidenfeld, A. (2006) Religious needs in the hospitality industry. *Tourism and Hospitality Research* 6 (2), 143–159.

# 17 The History and Future of Mountaineering Tourism

Ghazali Musa and Md Moniruzzaman Sarker

## Introduction

Tourism requires physical and mental immersion when exploring a new frontier. To some tourists, this endeavor is a lifestyle and not merely a recreational activity. People's interest in tourism has shifted increasingly from traditional mass tourism to special interest adventure tourism activities. In adventure tourism, the levels of risk and uncertainty outcomes are higher than for traditional tourism (Weber, 2001). Studies have revealed that adventure tourism is a growing phenomenon among international travelers, with Beedie (2015) reporting that around 42% of travelers engaged in adventure tourism activities during their previous holidays. UNWTO defines adventure tourism as the presence of two of these three elements in tourism activities: physical activities, natural environment and cultural immersion. Additionally, risk perception, uncertainty of outcome, requirement of skill and competency and requirement of equipment and/or transport are also elements involved in adventure tourism. These elements, however, differ in their intensity depending on the type of activity. For example, overland tourism is the pursuit of exploring or enjoying the environment in a natural setting where common activities encompass bird-watching, enjoying natural beauty, exploring new cultures or remote environments, etc. These activities are considered as less risky or challenging than being involved in mountaineering, climbing, abseiling, kayaking, skiing, etc. (Weber, 2001). Scholars term these more risky and challenging activities as adventure tourism. Within this tourism domain, mountaineering is appraised as one of the most popular adventure tourism activities, entailing high risk with uncertain outcomes, a high level of self-motivation and personal competencies, and access to remote locations.

In many cultures and religions of the world, mountains are acknowledged as sacred destinations where devotees perform religious rituals and become closer to God. Apart from religious and spiritual significance, travelers visit mountains for recreational purposes such as trekking, skiing, photography, mountaineering, etc. (Musa *et al.*, 2015a). A study reported that mountain areas are the second most popular tourist destinations, generating 15–20% (US$70–90bn) of the annual global tourism income (Charters & Saxon, 2007). This firmly indicates the popularity of mountaineering tourism among tourists in the 21st century. However, not all of these mountaineering activities are equal with regard to the level of involvement, risk and uncertainty. Mountaineering tourism may be categorized as hard (e.g. high-altitude climbing, rock climbing) and soft adventure (e.g. mountain trekking), based on Ewert and Jamieson's (2003) categorization. In this chapter we adopt Musa *et al.*'s (2015a: xxi) rationale to define mountaineering tourism as '… the activities of mountaineering tourists, their interplay with members of the climbing community and all associated stakeholders, together with associated impacts and management at the environmental and local community level'.

Until now, researchers have devoted their efforts to conceptualizing mountaineering tourism by studying various aspects, including a conceptual framework (Nepal & Chipeniuk, 2005), mountaineering from the perspectives of activity, people and place (Musa *et al.*, 2015a), mountain tourism experience, environments, communities and sustainability (Richins & Hull, 2016), among others. Previous studies have presented mountaineering tourism in relation to geography, social and historical development, issues related to the people involved in mountaineering, and environmental management. With increasing demand for mountaineering tourism, impacts on the ecosystem have been well documented, including some proposed managerial solutions (Musa *et al.*, 2015b; Richins & Hull, 2016). Although general concerns about environmental and socioeconomic issues have been observed in previous studies, knowledge about the historical aspects and future development of mountaineering tourism with regard to activity, people and place is scarce. This chapter aims to provide brief ideas about the historical development of how mountain activity and tourism have evolved, along with the future of mountaineering activity.

## The History of Mountaineering

### Historical turning point 1: Mountain climbing in the early era (before 1854)

Mountaineering tourism is a recent development. The earliest recorded history of mountain ascent was by a Japanese monk named En no Ozunu on Mount Fuji (3776 m) in AD 633. Even though it is acknowledged as the first ascent to any mountain peak, its purpose was to explore minerals and hunt for crystals (Beedie, 2015). The first ascent driven by aesthetics and/or mystical quests was recorded when in 1336 Petrarch, an Italian alpinist, reached the summit of Mt Ventoux (1912 m), which is located in the southern province of France. Later, in 1492, the French alpinist Antoine de Ville with a small team ascended Mt Aiguille (2085 m). This climb was significant for two reasons. First, alpinists for the first time used ropes and ladders to climb to the summit, and secondly it is acknowledged as the true beginning of the age of mountaineering (Sedghi, 2016). However, until the 18th century, there were no other such mountaineering activities recorded in which the purpose of the ascent was only for pleasure or sport. Then, from the 1700s, visiting mountains emerged as a recreational activity. William Blake (1757–1827) famously stated that extraordinary things happen when men and mountains meet. Before this time, people's intentions were not to embrace the adventurous spirit of the mountainous areas, but rather to explore economic opportunities or fulfil spiritual needs. Therefore, although the climb on Mt Aiguille marked the beginning of mountaineering, it is not regarded as the start of what we have conceptualized as modern mountaineering tourism in this chapter. Modern mountaineering is accredited as the hard adventure activities in which mountaineers embrace risk and uncertain outcomes through sport, an adventurous spirit and the desire to know the unknown. Macfarlane (2003) opined that the instinct to get involved in dangerous activities through climbing mountains has only been observed over the last three centuries. Hence, we categorize the early era of mountaineering as between pre-early era (before 1700) and post-early era (1701–1853). In the pre-early era, mountain climbing was dominated by spiritual or economic quests, whereas the post-early era is considered to be the beginning of the modern mountaineering age.

During the post-early era, mountain climbing was seen as a new approach to enjoying the mountains' spirit. Alpinists began to ascend mountain peaks by deliberately accepting the challenge of climbing or in order to be thrilled. In 1741 there was a historic visit to Chamonix Valley by two Englishmen, Richard Pococke and William Windham, who told of the amazing mountain discoveries there. Chamonix is a community of the Haute-Savoie department in southeastern France which consisted of 16 villages. This historic visit had an impact on other adventurous minds at the time. As a result, a Genevan scientist, Horace-Bénédict de Saussure, attempted to scale Mt Blanc in 1757, which is the highest peak among the Chamonix commune in France and also the tallest peak in Europe. During his visit to Chamonix, he struggled to scale the peak of Mt Blanc (4808 m) and could not reach the summit. Despite his unsuccessful effort to reach the peak, he offered prize money to anyone who could conquer it. Following his announcement, in 1786 two alpinists from the Chamonix commune, named Jacques Balmat and Michel G. Paccard, scaled the peak and claimed the prize. Just a year later in 1758 de Saussure himself climbed the mountain successfully. In mountaineering history, this event left a footprint among alpinists to embrace the challenge of climbing mountains. In fact, the adventurous spirit of enjoying mountaineering flourished from this event. In that spirit, Marie Paradis became the first woman to summit Mt Blanc in 1808.

During this era, a few successful climbs were also recorded in North America. The first ascent on the Rocky Mountains was one example. Although these mountains were discovered in 1806, Pikes Peak (4394 m) in the Rockies was first summited in 1820 by three climbers including Edwin James. Among the Rockies, Fremont Peak (4189 m) was the most glaciated mountain which was conquered by John C. Fremont and two others in 1842. In 1848 some US soldiers climbed the highest peak in Mexico, Pico de Orizaba (5636 m). Although there were a few remarkable events in mountaineering history before 1853, events that denoted mountaineering as an adventurous type of leisure activity were absent.

### Historical turning point 2: Mountaineering during the golden era (1854–1865)

During the golden era (1854–1865), mountaineering emerged as an adventurous activity alongside a few significant incidents. Whereas early mountain scaling was motivated by religious, economic and/or social reasons, during the golden era it became fashionable to climb mountains. During this era, climbers scaled all the high mountains in the European Alps. At the beginning of the era in 1854, an English alpinist, Sir Alfred Wills, ascended a peak of the Swiss Alps called the Wetterhorn (3692 m), and popularized mountaineering among alpinists with the establishment of the first mountaineering club, the 'Alpine Club', in 1857 in England. This adventurous spirit did not stop there, as in 1863 alpinists launched the *Alpine Journal*, publishing their experiences and activities. A prestigious group of guides was established who shared their experiences and invented new techniques to climb rock surfaces or ice walls. They also frequently organized mountain climbing competitions which displayed a spirit of pure sportsmanship during this age. However, a turning point occurred in 1865 when tragedy struck the famous Whymper Expedition on the Matterhorn (4478 m). Despite a successful attempt to reach the peak, four of the climbers died on the way down. This accident inspired a public response demanding that mountaineering should be made safer through the hire of experienced local guides (Musa *et al.*, 2015). Following this, a formal mountaineering training and education platform was established by people who were familiar with mountains and mountaineering activities. Soon after that, mountain guiding became popular and was transformed into the commercialization and commodification of the mountaineering experience. With this incident, the golden age of mountaineering came to an end.

During the golden age of alpinism, there were a few significant landmarks that built the maiden spirit among adventurous alpinists. First, mountaineering became part of the range of sports and entertainment through which an elite group of people engaged in and popularized the activity as adventure. Secondly, a new and different group of tourists emerged during this period who deliberately participated in this challenging activity. They scaled many hazardous mountains which would psychologically allow them to express their desire for adventure. Thirdly, they invented techniques, equipment and technical know-how for the scaling of hazardous mountains. The formation of the Alpine Club and the *Alpine Journal* were efforts to introduce mountaineering as a professional pursuit. Fourthly, the incidence of the Whymper Expedition evolved the philosophy of mountaineering sportsmanship into professional mountaineering and within this the concept of trained, professional mountain guides emerged.

### Historical turning point 3: Mountaineering after the golden era (post-1865 to the early 20th century)

Following the golden age of mountaineering, the phenomenon expanded throughout the world. It achieved an international flavor from 1865 to the beginning of the 20th century. This spirit stirred alpinist minds globally, fueling many successful ascents of mountains in different parts of the world such as the North American, South American, and the African and Oceanian mountains. In 1897 an Italian mountaineer and member of the Royal House of Savoy named Duke of the Abruzzi and a few others summited Mt St Elias (5498 m), situated on the border of Alaska and Yukon. He also summited Margherita Peak (5119 m), which is the tallest peak in the Ruwenzori range area in East Africa. However, the tallest peak in Africa is Mt Kilimanjaro (5895 m). This mountain was summited by German geologist Hans Meyer and Austrian alpinist Ludwig Purtscheller in 1889. North America's tallest peak, Mt McKinley (6168 m), was climbed by Hudson Stuck in 1913. Later, in 1925, Mt Logan (5959 m) – the tallest peak in Canada – was conquered. Another North American mountain named Grand Teton in the Grand Teton National Park was also climbed during this period. There was a disagreement about the timing of the first ascent on Grand Teton (4197 m), the highest peak in the Teton range. There were two claims of the first ascent on this peak. The first was in 1872 by Nathaniel P. Langford and James Stevenson, and the other in 1889 by William O. Owen. This confusion was settled in 1965 when an American mountaineer and author named Leigh Ortenburger concluded that Langford and Stevenson may have ascended the mountain, but probably did not reach the top, whereas Owen certainly reached the actual peak of the Grand Teton.

During this period climbers ventured into almost all other remarkable mountain areas as well. For instance, English alpinist Whymper began his exploration of the highest peaks in the Andes in South America. He scaled Chimborazo (6268 m) in 1880, yet it was not the highest peak in that area. Three years later, in 1883, Paul Gussfeldt climbed a volcano named Maipo (5260 m) and also attempted to climb Mt Aconcagua (6961 m) in the same year, but the venture was unsuccessful. Mt Aconcagua is the tallest elevation in the Andes, and a Swiss alpinist,

Matthias Zurbriggen, was the first to scale it successfully in 1897. This expedition began in 1896 and was led by Edward FitzGerald. The mountaineering spirit was also expanding in the Oceanic region during this period. In 1882 an Irish climber, William Spotswood Green, scaled the Southern Alps of New Zealand. Although a little later, mountaineering in Asia was observed at the beginning of the 20th century. Mt Gongga, known as Minya Konka, is the tallest mountain in the Sichuan state of China with a height of 7556 m. Two American mountaineers, Terris Moore and Richard Burdsall, were the first to reach its peak in 1932 with the purpose of accurately measuring its altitude.

### The last frontier of mountaineering

The Himalayan expedition is regarded as the last frontier of mountaineering. Before the expedition into the world's tallest mountain region, the Himalayas, alpinists from different countries had ascended the other unexplored mountains all over the world. Even when those mountains had been climbed before, mountaineers tried to summit these peaks in different fashions using innovative techniques, more difficult routes, a different set up, etc. Although mountaineering history had been led by the British, people from Germany, Australia, Italy, China, France, Japan, Russia and India also embraced a similar adventurous spirit during this era (Kiesinger & Smith, 2017). For example, from 1933 to 1936, Communism Peak (which was later renamed Imeni Ismail Samani Peak of Pamirs), Siniolchu and Nanda Devi were climbed by Russian, German and English teams, respectively. During this era, mountain expeditions became specific target-oriented expeditions. The Himalayas and neighboring sites were eventually targeted by adventurous mountaineers, and the alpinists started an expedition to the world's tallest mountain area in the Himalayas as early as 1892. However, it became more popular after the 1950s when the highest peak on earth, Mt Everest (8850 m) was conquered in 1953. For this reason, the Himalayas is acknowledged as the last frontier of mountaineering tourism (Beedie, 2015). All the highest elevations on earth are situated in the Himalayas, of which 14 are at a height of over 8000 m (Johnston & Edwards, 1994; Messner, 1988). In contrast to the Himalayas, the tallest mountain outside Asia is situated in the Andes, Mt Aconcagua (6961 m). This comparison provides an indication of the height and probable difficulties associated with climbing mountains in the Himalayas.

Historically, mountaineering in central Asia was carried out for purposes of spiritual, religious and economic activity. Specifically, Tibet and the Himalayan regions are well known for their sacred and spiritual mountain places. The Karakoram mountains began to be explored in 1892 by Sir William Martin Conway. He managed to scale to a height of 7000 m, although the highest elevation of the Karakoram range is K2 (8611 m). Later, in 1954, an Italian team directed by Ardito Desio successfully summited K2. Before that, English mountaineers Eckenstein and Crowley attempted to ascend K2 but only managed to ascend to 6700 m due to the challenging weather. In 1895, a four-member team (J. Norman Collie, C.G. Bruce, Geoffrey Hastings and Albert Mummery) attempted to climb another mountain in the Karakoram named Nanga Parbat (8126 m), but were unsuccessful. It is one of the 8000 m mountains and is acknowledged as tough to climb. That four-member group stopped scaling this mountain when Mummery disappeared during the expedition. There were also seven unsuccessful attempts to climb Nanga Parbat before the successful ascent in 1953 by Herman Buhl. Among the unsuccessful attempts, many climbers died on the mountain due to difficulties and unfavorable weather conditions. Although many unpleasant incidents occurred during that time, there were also a series of success stories that took place after the 1950s. Along with the successful summit of Nanga Parbat, Maurice Herzog and Louis Lachenal were the first to ascend Annapurna (8091 m) – another 8000-m peak in the Himalayas, in 1950.

The high point of mountaineering history was recorded in 1953 when Sir Edmund Hillary and the Tibetan Sherpa Tenzing Norgay successfully summited the tallest mountain on Earth, Mt Everest (8850 m). The expedition was led by a British army officer, John Hunt, and it was the eighth attempt in 30 years to scale the mighty mountain. Among the unsuccessful attempts to climb Everest, the expedition in 1922 was painfully tragic. Seven porters died during that expedition although they reached a height of 8320 m. Two mountaineers also disappeared in 1924 after they had reached a higher elevation compared to the effort in 1922. Then there was another historic event in mountaineering when Kanchenjunga (8586 m) was scaled in 1955 by Charles Evans. He was the leader of an expedition team from England. Kanchenjunga is the third tallest mountain on earth and is acknowledged as the most difficult to climb compared to other mountains. There were also several unsuccessful attempts recorded and few fatal incidents occurred before the successful ascent took place. Another summit in the Himalayas named Lhotse I (8516 m) was climbed by a Swiss team in 1965. Also in 1964, a Chinese expedition team led by Xŭ Jìng first scaled Shishapangma (8027 m), which was the lowest among all the 8000-m peaks in the Himalayas and the only one in China. With that expedition, it was claimed that all the 10 tallest mountains in the Himalayas had been conquered by mountaineers. Up to 2017, more than 7600 mountaineers have summited Mt Everest and around 300 people have died during the expeditions (Encyclopedia Britannica, 2018). After conquering all the tallest mountains on earth, alpinism has reached a new point with the age of tourism.

*The age of tourism*

There has been huge concern about guided mountaineering which takes inexperienced climbers to high mountains including the Himalayas.

In fact, after conquering Mt Everest in 1953, mountaineering began to be accepted as a tourism product and this attitude permeated into adventurous minds around the world. Many adventure tourists started to explore previously scaled mountains in the Himalayas with the help of trained Sherpas. Even though the mountains had already been conquered, a new style of climbing evolved to set new records. The idea is to scale the same elevations but to use more difficult routes, different techniques, in different weather, etc. (Johnston & Edwards, 1994). For instance, among the different faces by which to climb Everest, the southwest face is more predictable in weather patterns, whereas the west face is regarded as the most difficult route to scale. In 1963, an American team of William F. Unsoeld and Thomas F. Hornbein scaled Mount Everest using the west face. Later, the use of artificial aids and advanced scaling methods were applied as new scaling techniques to climb the rock faces of mountains. These new techniques were applied to climbing the difficult southeast face of El Capitan (1100 m) in 1970 by American mountaineers. Situated in the Yosemite National Park in North America, this has the smoothest vertical faces of granite monoliths anywhere in the world. Table 17.1 reports some of the remarkable events in the history of mountaineering.

**Table 17.1** Remarkable events in the history of modern mountaineering

| Year | Name of mountaineer(s) | Name of mountain | Significance |
| --- | --- | --- | --- |
| 1799 | Miss Parminter | Le Buet (3096 m) | First recorded climb by a woman |
| 1808 | Marie Paradis | Mt Blanc (4807 m) | First female ascent |
| 1975 | Junko Tabei | Mt Everest (8850 m) | |
| 1978 | Reinhold Messner and Peter Habeler | Mt Everest (8850 m) | First climb without supplementary oxygen |
| 1980 | Leszek Cichy and Krzysztof Wielicki | Mt Everest (8850 m) | First winter ascent |
| 1980 | Reinhold Messner | Mt Everest (8850 m) | First solo ascent |
| 1988 | Vern Tejas | Mt McKinley (6190 m) | First solo ascent in the winter season |
| 2003 | Ming Kipa Sherpa | Mt Everest (8850 m) | First youngest summit by a 15-year-old Nepalese girl |
| 2011 | Apa Sherpa | Mt Everest (8850 m) | Ascended 21 times |
| 2013 | Sherpa Phurba Tashi | | |
| 2014 | Karl Egloff | Mt Aconcagua (6959 m) | Fastest ascent to Mt Aconcagua (11 h 52 min) |
| 2015 | Anne-Marie Flammersfeld | Mt Kilimanjaro (5895 m) | Fastest female ascent (8 h 32 min) |

Source: Encyclopedia Britannica (2018).

During this era, guided mountaineering and the use of equipment were vastly commercialized because of the high demand among novice adventure tourists. Commercial operators tried to provide a comparatively safer experience through properly planned expeditions, including managing equipment, planning routes and providing training before the expedition, etc., which were not available during the golden era or earlier. As a result, novice mountaineers became interested in taking this opportunity and enjoying the adventure of climbing. Mountaineering involves high costs in tourism expenses and is hardly affordable at an individual level. Thus, a guided mountaineering concept often grouped mountaineers together, allowing them to share the guiding expertise and logistical arrangements required for a successful mountaineering expedition. Mountaineering activity has evolved in adventure tourism, which was initiated in the golden era and popularized after the 1950s.

## The Future of Mountaineering

Adventure tourism found its peak of acceptance at the beginning of the 21st century. Parallel to what Charters and Saxon (2007) mentioned earlier, Sandler Research (2016) reported that the adventure tourism industry earned US$7.88 trillion in 2015 and is expected to grow by 46% during the period 2016–2020. This signals the potential of mountaineering tourism among the next generation of tourists. Although mountaineering is acknowledged as parallel to lunacy (Macfarlane, 2003) and involves high risk and uncertainty (Ewert & Jamieson, 2003), the demand for it among adventurous spirits is unceasing. It has gained massive acceptance among inexperienced climbers through the commodification of mountaineering (Johnston & Edwards, 1994). Specifically, commercial mountain tour operators have exerted a significant effort in marketing mountaineering activities among adventure tourists. Nonetheless, the provider side of mountaineering tourism is not the only fuel expediting its acceptance; the mentality of adventure tourists is also a crucial element behind this growth. Mountaineering activities involve specialized equipment, technical skills and physical strength, and at the core of it all is an adventurous mind (tourist lifestyle) that triggers the experience of uncertainty (Pomfret, 2006). Musa and Thirumoorthi (2015) added that for mountaineers to be successful they need medical clearance, physical fitness, technical skills and the presence of a strong mind to handle adverse situations.

In the early age of modern mountaineering, expeditions were led by an elite group of alpinists, specifically the British, Europeans and Americans, and their adventurous spirit was financially supported either individually or as a group. The establishment of the Alpine Club in 1858 and the *Alpine Journal* in 1863 are classic examples in this regard. These expeditions were usually managed by government funds, whereas the current alpine expedition is run by either self-funding or managing sponsorship from private companies. Hence, it is quite challenging for novice mountaineers to climb any mountain without the support of experts in mountain guiding. To a great extent, individual expeditions would be impossible as they require many technical types of equipment that are impossible to manage individually because of cost and accessibility (i.e. oxygen kits, food and safety equipment). Commercial tour operators provide a more inexpensive opportunity by supplying all the necessary materials including trained mountain guides. As a result, there are a huge number of guided mountaineering tours that take inexperienced climbers to high mountains including Mt Everest.

## Future Turning Points

### Future turning point 1: Climate change

Regardless of prior arrangements and planning for mountaineering, history has recorded many tragic fatal incidents from the Matterhorn to expeditions in the Himalayas. Mountains are unpredictable and dangerous environments. This unpredictability and danger can be reduced by the application of improved technology in terms of ropes, boots, clothing, axes and gloves, etc. However, the main challenges of mountaineering now and in the future are the inevitable consequences of climate change (Hall, 2016) and the influx of inexperienced climbers who inflict detrimental impacts on the economy, social culture and environment of mountain regions (Musa *et al.*, 2015). Global warming is one of the environmental challenges causing melting ice and unstable rock surfaces all over the world, including in mountain areas such as the Himalayas and the Andes. Due to the high-altitude nature of mountains, speedy winds and snow storms are commonly observed in almost every mountain area. As a result, the majority of accidents have occurred due to avalanches in the mountains along with rockfall, landslides, slashing stones from rock slopes and glacial deposits (Ritter *et al.*, 2012). Inexperienced climbers are also responsible for polluting the mountain environment through littering and improper sewage disposal. Hence, the future sustainability of mountaineering largely depends on maintaining a healthy mountain ecosystem on the routes, valleys and peaks (Johnston & Edwards, 1994). Future mountaineering requires the further enforcement of codes of conduct for climbers, guides and operators during expeditions.

### Future turning point 2: Entrepreneurial adaptations and technology

Despite the climate change issue, the future of mountaineering tourism remains good, with anticipated entrepreneurial adaptations and advances in technology in terms of equipment, communication and weather forecasts (Johnston & Edwards, 1994). Specifically, technological advancement will dominate future mountaineering tourism activities during climbing or on an expedition overall. The use of computers, mobile apps, sensors, GPS and personal locator kits, etc., will be readily available to the mountaineers (Musa *et al.*, 2015b). These devices will inform climbers about different types of routes, weather updates and forecasts and the nature of potential difficulties.

*Equipment*

During the tourism age of mountaineering, almost all the equipment has been modernized with technological advancements. When Mt Everest was conquered in 1953 by Edmund Hillary and Tenzing Norgay, they individually carried loads weighing around 44 pounds (Regenold, 2012). Nowadays, individual climbers carry on average half of this weight. This has become possible because modern equipment (ice axes, ropes, oxygen cylinders, communication gear, food, tents, backpacks, clothing, helmets, etc.) has been improved with regard to weight and strength. For example, mountaineers can now load almost all their equipment inside or outside the chambers of their backpacks, which are now stronger and more durable than in earlier times. Future improvements in mountaineering backpacks will be based on the evolution of other equipment. Oxygen cylinders are another essential piece of equipment for alpinists that need to be carried during climbing. On the mountain expedition by Hillary and Norgay, oxygen tanks weighed 20 pounds; with the advancement of technology they now weigh only six pounds. The future development of oxygen tanks would include a more efficient flow which lasts longer. Likewise, ladders and crampons will be stronger and lighter so they can be comfortably carried with other apparatus. Furthermore, the innovation of synthetic materials has brought major advancement in ropes, tents, trekking suits, sleeping bags, etc. These are now stronger and more durable, even in icy, rainy or windy conditions on mountain areas, and will become even stronger, lighter and more breathable. For instance, trekking suits, sleeping bags, face masks and tents are used to cover up the whole body including the face; hence these items must be designed in a way that helps easy breathing while moving or resting. Overall, existing equipment will become stronger and more adaptable to face the adverse conditions in mountain areas. As an example, Scarpa is a newly designed mountaineering boot commercialized by Ribelle Tech which claims to be lighter to allow easier running, climbing and staying dry during mountain ascents (see https://m.epictv.co.uk/media/podcast/scarpa-ribelle-tech-od-the-future-of-mountaineering-boots/606920). Stronger yet lighter equipment will be the next major advancement in mountaineering equipment which the manufacturing companies are relentlessly trying to devise.

*Safety and security*

Another major development in mountaineering safety is the use of automatic identification systems (AIS) which will be able to detect the exact location of mountaineers. Google Maps technology, in fact, unveiled the way to use AIS devices to analyze routes in advance. The design of different types of mobile apps related to mountaineering will also emerge to monitor, track and guide climbers (i.e. monitoring blood pressure, GPS location finders, communicating with other expedition members, potential hazard updates, etc.) during trekking.

*Food technology*

Food is another important element that mountaineers must carry. Hillary and Norgay carried regular packaged food such as chicken noodle soup, canned fish, apricots, tea, etc., to meet their daily intake needs. Along with the advancement of climbing equipment, there has been an improvement in food and cooking. Currently, mountaineers take specially packaged foods that are high in calories and nutrition. For example, Clif Bar and Company produces many types of food bars and gels for mountaineers which contain all the necessary nutrition and energy (http://www.clifbar.com/products/clif/shot-energy-gel). In the future, an expected improvement in food technology will be food pills that can fulfil the energy and nutrition intake of climbers.

*Medical technology*

Portable hyperbaric chambers are a completely new addition to medical technology in mountaineering which will give climbers access to emergency treatment if anyone experiences high-altitude problems such as acute mountain sickness (UIAA, 2018). Further advancements will be in the improvement of medicine and treatment methods in mountaineering medicine. The emergency base camp will be equipped with all the necessary apparatus to survive in adverse conditions. Finally, the number of permanent structures on the mountains will increase to allow greater access for climbers and greater security.

*Virtual experience and entertainment*

The demand for an 'armchair' experience of mountain climbing is also inevitable. It will attract more novice climbers to experience mountain adventure. Mountaineering as an entertainment activity will be common for those who do not want to physically visit but would like to get thrilled without climbing. As a result, an increasing demand for a 'virtual reality' mountaineering experience along with video games and movies is expected to increase in the future. This trend currently prevails among the general public as there have been more than 40 movies and documentaries filmed to date. Among the mountaineering movies, *Touching the Void* (2003), *North Face* (2008), *The Summit* (2012), *Everest* (2015), *Beyond the Edge*

(2013), *127 Hours* (2010) and *K2* (1991) are very popular with mass audiences (IMDb, 2015; Peter, 2018).

*Novel adventure spirit*

One last new development will take place as the very experienced climbers find climbing the same mountain monotonous. Adventurous minds usually nurture their spirit through enjoying uncertainty, which vanishes once a mountain has been scaled. Although the world's tallest mountains have already been conquered, the number of climbers involved in mountaineering is steadily increasing. A future focus will move on to scaling comparatively smaller yet more difficult and sensational-looking mountains. For instance, Adam Ondra scaled the 45 m long Silence (situated in Flatanger, Norway) in September 2017 with no equipment, and described it as the hardest climb ever. No-one has so far been able to repeat this ascent on Silence. Future mountaineering adventures of this nature are expected to increase. When humans colonize Mars in the future, mountaineers will shift their gaze to conquering Olympus Mon (25,000 m), which is almost three times taller than Mt Everest on Earth.

## Conclusion

The human instinct for experiencing the uncertain is ancient and the emergence of adventure tourism can be seen as one of the greatest endeavors to embrace that desire for uncertainty. Throughout history mountaineering has provided a plethora of examples of daring individuals and inspirational expeditions and it has presented a wealth of knowledge for adventurous minds about how to achieve their next, great mountain challenge. Despite the unpredictable challenges and adverse difficulties associated with it, mountaineering tourism has already reached landmark popularity and is expected to grow in the near future. The spirit of uncertainty not only satisfies curious minds when they experience adventure but also expedites the economic well-being of associated stakeholders, now and in the future.

## References

Beedie, P. (2015) A history of mountaineering tourism. In G. Musa, J. Higham and A. Thompson-Carr (eds) *Mountaineering Tourism* (pp. 40–54). New York: Routledge.
Charters, T. and Saxon, E. (2007) *Tourism and Mountains: A Practical Guide to Managing the Environmental and Social Impacts of Mountain Tours*. Paris: UNEP Division of Technology, Industry and Economics (DTIE). See https://www.unenvironment.org/resources/report/tourism-and-mountains-practical-guide-managing-environmental-and-social-impacts (accessed 21 February 2018).
Encyclopaedia Britannica (2018) Mount Everest: Mountain, Asia. See https://www.britannica.com/place/Mount-Everest (accessed 18 March 2019).

Ewert, A. and Jamieson, L. (2003) Current status and future directions in the adventure tourism industry. In J. Wilks and S.J. Page (eds) *Managing Tourist Health and Safety in the New Millennium* (pp. 67–83). Oxford: Elsevier Science.

IMDb (2015) Climbing movies, mountain high. See http://www.imdb.com/list/ls072524193/ (accessed 19 March 2019).

Johnston, B.R. and Edwards, T. (1994) The commodification of mountaineering. *Annals of Tourism Research* 21 (3), 459–478.

Kiesinger, C.D. and Smith, G.A. (2017) Sport: Mountaineering. *Encyclopaedia Britannica*. See https://www.britannica.com/sports/mountaineering (accessed 19 March 2019).

Macfarlane, R. (2003) *Mountains of the Mind: A History of a Fascination*. London: Granta.

Messner, R. (1988) *All 14 Eight-Thousanders*. Seattle, WA: Cloudcap Press.

Musa, G. and Thirumoorthi, T. (2015) Health and safety issues in mountaineering tourism. In G. Musa, J. Higham and A. Thompson-Carr (eds) *Mountaineering Tourism* (pp. 294–312). New York: Routledge.

Musa, G., Higham, J. and Thompson-Carr, A. (eds) (2015a) *Mountaineering Tourism*. New York: Routledge.

Musa, G., Thompson-Carr, A. and Higham, J. (2015b) Mountaineering tourism: Looking to the horizon. In G. Musa, J. Higham and A. Thompson-Carr (eds) *Mountaineering Tourism* (pp. 328–348). New York: Routledge.

Nepal, S.K. and Chipeniuk, R. (2005) Mountain tourism: Toward a conceptual framework. *Tourism Geographies* 7 (3), 313–333.

Peter (2018) 25 best mountaineering movies ever made. *Atlas & Boots Outdoor Travel Blog*, 24 February. See https://www.atlasandboots.com/best-mountaineering-movies-ever-made/ (accessed 21 March 2019).

Pomfret, G. (2006) Mountaineering adventure tourists: A conceptual framework for research. *Tourism Management* 27 (1), 113–123.

Regenold, S. (2012) Everest climbing gear – then and now. *National Geographic*, 5 March. See https://www.nationalgeographic.com/adventure/features/everest/gear-edmund-hillary-hilaree-oneill/ (accessed 21 Marcy 2019).

Richins, H. and Hull, J. (2016) *Mountain Tourism: Experiences, Communities, Environments and Sustainable Futures*. Wallingford: CABI.

Ritter, F., Fiebig, M. and Muhar, A. (2012) Impacts of global warming on mountaineering: A classification of phenomena affecting the Alpine trail network. *Mountain Research and Development* 32 (1), 4–15.

Sandler Research (2016) Adventure tourism market growing at nearly 46% CAGR to 2020. See https://www.prnewswire.com/news-releases/adventure-tourism-market-growing-at-nearly-46-cagr-to-2020-597059331.html (accessed 17 October 2017).

Sedghi, S. (2016) Time travel: The history of mountaineering, *Paste Media Group*, 12 April. See https://www.pastemagazine.com/articles/2016/04/time-travel-the-history-of-mountaineering.html (accessed 22 March 2019).

UIAA (2018) A guide on when and how to use portable hyperbaric chambers. See https://www.theuiaa.org/mountaineering/a-guide-on-when-and-how-to-use-portable-hyperbaric-chambers/ (accessed 21 March 2019).

Weber, K. (2001) Outdoor adventure tourism: A review of research approaches. *Annals of Tourism Research* 28 (2), 360–377.

# 18 Sustainability, Ecotourism and Scotland: Concerns, Complaints, Conflicts and Conservation

Alastair Durie, Ian Yeoman and
Una McMahon-Beattie

## Introduction

It is impossible now not to be aware of ecotourism and the issues related to how and where we travel for pleasure. There is certainly a growing recognition that travel and tourism may have a negative impact on the environment of the host country or region, or indeed in global terms (Yeoman, 2012; Yeoman & McMahon-Beattie, 2006). Sustainability and the challenges associated with the ethical use of natural resources have indeed become a significant issue in the environment–tourism relationship. Twenty years ago Romeril (1989: 204) aptly noted that 'the finite nature of natural resources, which also serve as tourism resources, makes it imperative that their enduring and sustainable use is reconciled with the continuing pursuit of social and economic goals'. This imperative remains today and we continue to debate the environment–tourism relationship. Is it harmful or symbiotic? Is there accord or discord?

Tourism in Scotland is an industry of major importance, outliving and outperforming many of the traditional pillars of the Scottish economy such as shipbuilding or coal. Historically, by the end of the 19th century, no part of Scotland was untouched by tourism. As Durie (2017) notes:

> Some places were destinations to which tourists went, others places from which holidaymakers came, and some both source and destination with a crossover: as the visitors arrived so the locals departed, as was true of Edinburgh. Early visitors in Scotland had favoured such destinations as Callander and the Trossachs, Dunkeld and Iona, and these retained their appeal and popularity, but others, thanks to better transport, were later

additions to the tourist map. Even the remotest parts saw visitors. Trips to St Kilda at the turn of the twentieth century were advertised with the slogan 'come and see Britain's modern primitives'. Locals would turn a profit by selling eggs and tweeds, as would the local post office from the sale of postcards, letters and stamps. Scots were partakers in tourism as visitors and holidaymakers and some were providers of tourist facilities. (Durie, 2017: 3)

The debate about sustainability is not new. As Hobsbawm (1995) reminds us, deliberations from the past recur and will continue to do so in the future; the only things that fundamentally change are the actors and stakeholders. We can trace the debates about sustainable tourism in Scotland back to the 1970s (Yeoman *et al.*, 2007) and issues of mass tourism. Then, Scotland was one of the earliest countries to experience tourism as a large-scale activity, with a substantial part of its appeal lying in its natural resources of land and scenery, mountain and sea, air and water, the seaside and its moors for health and for sport. It was, and is, well known for nature and for natural history. It might be considered, therefore, a place where environmental tourism, even if not under that label, could and should have made an early appearance in some form or another, perhaps if only in ingredients or strands which now have come together (Smout, 2000).

## A Definition

We are concerned with the interaction of tourism with cultural, economic and ecological resources, which together have been labelled the environment. Two views, philosophies or ideologies of use of the environment have been identified. The first is what has been called the *dominant* view (Behrensmeyer, 2006; Schaltegger & Hörisch, 2017), which espouses that the environment exists to be exploited by mankind. This view was not necessarily built on the assumption that resources are plentiful and change beneficial, but on the experience that livelihoods had to be wrung from the environment. Man might have been given dominion, as the biblical text put it, over creation, but that control was hard wrung and uphill. This classical view was more concerned about the environment's effect on man rather than vice versa: Hippocrates' *Epidemics: Air, Waters and Places* (Adams, 1891) looked at the influence on human health of nature and natural factors, climate, wind and water. In a world where life was hard and uncertain and famine was a constant threat, as in pre-modern Scotland before it began to industrialize, the problem was how to wring the maximum advantage from limited resources. Hence human interest came first. For example, the wolf was wiped out in Highland Scotland without any concern, by a certain MacQueen in 1743 (Nilsen *et al.*, 2007), because it was a threat to a subsistence pastoral economy. There was, of course, a degree of respect for the land and nature born out of communal self-interest, the need to allow a period of fallow, the restriction of hunting during the breeding season, and so on. But what mattered was what best suited man and his interests; nature was subordinate.

The alternative, *eco-centric* (Markley, 2015) view posits a rather different approach. Resources are not infinite and humanity does not have the right to take what it wants, nor to alter without restraint or consideration. The environment has 'rights' and commands respect. Tourism enters the equation in that it is a dynamic factor, bringing with it dangers of cultural and environmental corrosion, and commercialization is seen as bringing bad as well as good (Yeoman & Lederer, 2005; Yeoman *et al.*, 2007). This view, in the form articulated here, is built round a new set of questions and is one which would have made little sense to the Scots or any other poor nation in the world before tourism had emerged, or indeed during the rise of tourism as a major industry, when only the income from tourism was what mattered (Durie, 2012, 2013).

Yet there had been an alternative to the dominant ideology which was and is rooted in an equally biblical emphasis and of which the Scots became increasingly conscious. This was the notion of stewardship (Paterson, 2002) – that what people held was not just theirs to dispose of as they liked but was held in trust for future generations. This principle, whether applied to reserves of timber or stock or seed, might well come under pressure from an immediate need for cash or for food in times of need, but it was there. As and when society became able to take a longer term view, so stewardship became a more significant issue (Yeoman & Lederer, 2005).

In the pre-modern world, there was concern over the availability of particular resources, such as timber, gold and strategic materials (Yeoman, 2012). To what extent was there an awareness in pre-modern Scotland that resources were limited? Peat, once cut, was gone but there was always more. With fish, the challenge was to catch even a fraction of what was thought to be a boundless resource (Paterson, 2002). The Victorians added some debate, for example, over the extent of coal reserves (Marwick, 1973). But not until the 20th century was there any idea that resources were being depleted in a global sense. There was some sense, however, that land was a resource that, unlike the American West (Milner *et al.*, 1994), was fixed and over which there would be competing demands, with the newly appeared tourist interest only one of several. Moreover, tourism was itself divided; the interests of the stalker and of the walker conflicted (Durie, 2012). Hydro-electric schemes and forestry plantings had their claims and costs: the Falls of Foyer was but one scenic casualty (Griffiths & Morton, 2010). There was the political question of who owned the land, and overlapping with that was the question of how the countryside was to be used and in whose interest (Naomi & Margaret, 2018).

## Historical Turning Points

The remainder of this chapter will examine some of the forms of tourism in Scotland to see what concerns, if any, there were as to the longer term use of the environment and whether tourism might abuse or degrade the environment. It will reflect on concerns in the past about the impact of tourism on the physical, the natural (i.e. flora and fauna) and the cultural environment. It will identify key historical and future turning points for Scottish tourism.

### Turning point 1: Vandals, communities and tourism

At the time when tourism was still an elite experience, early concerns in relation to the environment included vandalism. Tourists, even those who were supposedly respectable, were apt to be rather unrestrained in their behaviour when on someone else's property. It might be simply the picking of flowers as a remembrance: the guides at the Duke of Atholl's estate at Dunkeld were expressly instructed in 1814 not to leave tourists unsupervised in the pleasure gardens lest they do damage (Durie, 2013). Another example would be the widespread practice of carving their names or inscribing doggerel verse on any accessible surface. At least Robert Burns left some sharp lines to mark where he had travelled or stayed (Whatley, 2018). More serious was the damage done to the attractions by, for example, lighting tar barrels to illuminate the interior of caves, which blackened them. Above all there were the souvenir hunters, a bad tradition that had long been part of European tourism, from which neither classical ruin nor Egyptian pyramid nor geological feature was safe (Zuelow, 2016). The English barrister, W.C. Dendy, complained in 1859 about a kind of visitor whom he dubbed the 'desecrating vagrant', whose 'highest pride was to carve his initials on one of the holy tombs on Iona, chip off a morsel of moulding from the prentice pillar in Roslyn and filch one of Macalister's stalactites in the spar cave of Strathaird' (Dendy, 1859). The schoolmaster on Iona, who because he had English doubled as the island's guide to visitors, slept with a stone lion from an early tomb under his bed, which he had confiscated from a souvenir-seeking tourist (MacArthur, 1995).

### Turning point 2: Tourism as a hobby and amateur science

The Victorians' passion for amateur science, for investigating and collecting everything from orchids to birds' eggs or butterflies could have been an ecological disaster. Travis (1993) has described what the craze for ferns meant in the southwest of England, with professional hunters moving in to supply demand and stripping Devon bare of the rarer species and other botanical specimens. Yet there was no parallel in Scotland.

Heather sprigs could be spared, as could Cairngorms gemstones. It is recognized that amateur botanists did good work in recording what was to be found and this was a positive contribution to science. This was, however, offset by returning home with specimens as trophies, regardless of how rare they were. There appears to have been no sense of despoliation. This is shown in an account given by a small party of Lancashire collectors to the December meeting of the Manchester Botanists Association in December 1876. Theirs had been a summer expedition in the Grampians to find rare Scottish alpine plants, and their success was not just in locating them but in bringing home specimens of most of the rarities, e.g. *Lychnis alpine* (Rogers, 1877). Nor were they alone: in Glen Clova they met a Dr Thompson who was just leaving with a gig full of botanical papers and plants. Their work was applauded. There was no sense of what this taking away might do to the environment, any more than there was concern about what tourists might be bringing in unwittingly. After all, gardens and grounds were being deliberately planted with seeds and seedlings carried from abroad by travellers (Durie, 2012).

## Turning point 3: The beginnings of organized tourism

Tourists took a carefree approach to flowers, fish and fowl; they tended to do what they liked, unless restrained by keepers and land managers. In 1861 Thomas Cook (Brendon, 1991: 76), in what has been called an early instance of the 'ecological blight that tourism so often brings in its wake', became embroiled in the alleged shooting by one of his tourists of an eagle on Iona. This was a slightly surprising furore, given that gamekeepers and their masters on the big Highland estates were no friend to this bird or any raptor. It was bad publicity for him and his tourists, which could not be deflected by the spin that the eagle was intended for a museum of natural history on the island. Nor was it offset by Cook's collection of funds from his visitors for fishing boats to help the locals, one of which was called *The Thomas Cook*.

## Turning point 4: Mass tourism

The seaside in Scotland was a magnet for increasing numbers of visitors, of all social classes (Durie, 2012). But while there was the collection of shells and fossils, the raiding of rock pools for crabs and sea life, the cheerful use of the sands, there was little worry about the impact on the beaches. There was, however, concern over the impact of mass tourism at the seaside, in terms of the behaviour of the day-tripper and the excursionist. There were tensions over mixed bathing, over the use or non-use of the Sunday, over dress and language. There was occasional damage to property, and more regularly to public order. But there seems to have been no concern in Victorian times over the use of the sea or the condition of the beaches, although access to the seaside itself could provoke objections from local landowners. Bigger numbers did lead to concerns over amenities at the seaside resorts: there were real questions of water supply, sewage and sanitation for the swollen summer populations. The overloading of systems could lead to outbreaks of epidemic disease, e.g. the typhoid epidemic at Bournemouth in 1936, which was due to contaminated ice cream and milk (Hassan, 2003). But resort enteritis, or beach tummy, was a small price to pay for the pleasure of a summer break from the harshness of the urban environment.

## Turning point 5: Mobility

Transport issues made an appearance. There was the distaste of some for the impact of the railways on the landscape and in terms of whom they carried, the so-called common order. 'What beasts the English are, the middle orders when they go touring. The railroads are the great curse of this country', moaned one snob in 1864 (Babbington, 1864: 1). Indeed, some landowners could and did obstruct whether or where lines were built. John Stirling of Kippendavie near Dunblane insisted on a tunnel to hide the line from sight as it passed through his pleasure grounds. But while there were concerns about the soot and the noise, there was nothing in terms of their effect on the environment, other than an occasional complaint about the effect on fishing (Marshall, 1998). There was to be much more concern about the motor car, cycle and bus, when they made their appearance in the years before WWI, over the dust, mud, noise, smell and dangers brought. Livestock, horse-drawn traffic and bystanders were all at risk from the reckless and untrained driver (there was no driving test until 1936), and there was much resentment. Sunday travel was a particularly vexed issue in Scotland. 'You cannot hear the kirk bells for the hooting of horns', insisted one writer to *Motor World* in 1910 (Devine & Wormald, 2012). The bona fide legislation, which meant that to get a drink at a hotel on a Sunday you had to have travelled, brought another hazard to the day of rest; the over-refreshed motorist.

## Turning point 6: Cultural gain – Scotland as a cultural oasis

The cultural dimension is a significant one. While scenery, sport, nature and of course the cult of Ossian were major draws for early visitors to the Highlands, part of what drew Johnson north to the Highlands was the search for a 'primitive society' (Durie, 2012). This was a society untouched and therefore unspoilt by industrialization, but emerging from feudalism, in which old and untainted values such as hospitality to the visitor still stood as part of manners. A restatement of this urge for the unspoilt remote was to re-emerge in the form of what Burchardt (2002) has called 'rural nostalgia' in the 1920s, a desire to preserve the countryside as a reservoir of values in an increasingly urban world. Preservation was not in the mind of early visitors to the Highlands, who observed and described, but a few did begin to question what the influx of tourists was doing. Nineteenth century writers were concerned by the impact on Highland values, the growth of begging and a dependency culture in places such as Iona and St Kilda. 'Now' said one Glaswegian visitor of the St Kildans in 1888, they appear 'to be not a great deal above the level of professional beggars' (Robins & Meek, 2008: 76). Later commentators from a socialist perspective, such as Edwin Muir (1935), criticized what they saw as the debasement, which tourism had induced, of Highland society into a culture of servility, and the deference to sporting interests, 'performing whatever menial services, these people may require'.

### Turning point 7: Cultural loss – Gaelic identity

In terms of the distinctive identity of Highland Scotland, there was a marked decline in Gaelic. While many factors were at work, this was hastened by the value of English when dealing with the tourists. In 1807 a Quaker lady from Hull, Mrs Robinson, was told by the girl ferrying her on Loch Lomond that while her mother spoke good Gaelic, it was little in use among them now, 'especially young people' (Robinson, 2006: 62). Tourists did not need Gaelic, and guidebooks to the Highlands, such as Andersons or Blacks, unlike those for France, gave no guidance sentences or phrases for the visitor; the best they did was to list some Gaelic topographical words. There was one peculiarity: visitors might not learn the language of the Highlands, but they did 'go native' in terms of adopting its dress – it became a standing joke that anyone seen in full Highland gear in the North could be assumed to be an English tourist. It was part of the cult, or kilt, of Balmoralism.

### Turning point 8: Hospitality and manners

There was a general awareness that the influx of tourists, not only the wealthy, was part of what was changing society, in re-orientating it from feudalism to commercialism. There was an impact on manners, in practice towards strangers, of which hospitality is an example (Durie, 2012). Accommodation, food and drink had in the past been given free, but that was to change as the numbers grew and social composition widened. In 1788 Mrs Diggle, thanks to letters of introduction, could stay free at Murthly Castle (Griffiths & Morton, 2010: 263) 'in the stile of ancient hospitality before inns were invented'. But a century later, only the select few were houseguests, with the remainder being charged, and charged highly which was a complaint often to be heard. After Culloden chieftains had become landlords, wealth was measured in income rather than in muscle, and many of the new owners were incomers who saw their estates, or so critics alleged, only in terms of their amusement, not their responsibilities.

### Turning point 9: Land use

Probably the greatest tension that tourism induced was over land use away from the coast, especially in terms of questions of access to inland and Highland areas. Issues of access to the countryside, particularly in the Highlands, were always problematic. There were questions about privacy and of private rights: with the support of the law, landowners could restrict or deny access. There were points of increasing friction over rights of way, over alleged interference with grouse moors, deer forests and game preservation. A Scottish Rights of Way Society was established to contest the ground, and in 1847 it fought the Duke of Atholl in a very high-profile case. Botanists, naturalists, ramblers and mountaineers all laid claim to access, which was largely resisted by landowners pre-1914, and then very reluctantly accommodated (Lambert, 2001).

Deer forests were a source of contention, with observers worried about the effect on local populations of the conversion into sporting preserves. It was a debate in which economics and ethics, more than the environment, featured large. In 1926 the writer J.M. MacDiarmid issued a powerful pamphlet that criticized deer forests as a waste of land, misused to provide sport for a few rich people. He held out a vision of the land as a vast area which might be turned from deer forests to provide a livelihood for tens of thousands of landholders, foresters, hotelkeepers and others. Others argued that the economic gain of this kind of sporting tourism substantially outweighed the losses; a 'golden stream of tourism' was a term much used, part of progress for the receiving society, essential for estates, hotels and the railway companies. It brought a lot of money into the Highlands, or so the Duchess of Atholl argued, because it brought holidaymakers and sportsmen. The case for the defence of sporting rights was not helped by actions of people like the American millionaire W.L. Winans (Orr, 1982), who attempted to clear his ground of crofters and even went to law against a crofter whose pet lamb had strayed onto Winan's ground. He had also sent a servant down to chase a naturalist away who was studying flies! The ironic thing is that for all his litigiousness, and alleged enthusiasm for deer, he left his Highland properties unvisited year after year.

### Turning point 10: Game

Hunting is often seen as a culture of killing, the antithesis of an environmental or conservationist force. Yet the management of stock, when a longer term perspective was induced by commercialization, did require restraint and control. Preservation of game was of course something of a misnomer in that preservation meant the extermination of predators (crows, foxes, hawks and polecats) in favour of game (Smout, 2000). The objective of a Highland moor was not diversity but a monoculture of grouse. What added a challenge to the management of the grouse was that, unlike partridge or pheasant, it could not be bred, and there was a total reliance on wild stock, in which predation and weather, rather than shooting, were key variables. Among some landowners there did develop the notion of stewardship. King George V summed this up in his statement that 'the wildlife of today is not ours to dispose of as we see fit. We hold it in trust for future generations' (Smout, 2000: 134). But this perspective appears to have been narrowly focused on game.

### Turning point 11: Societies and trusts

Tourism did feed into the preservation of the historic environment, of battlefields such as Culloden, of cultural sites (e.g. Burns' Cottage, Abbotsford), of the ruins at Iona, the cathedrals and castles, the heritage of Scotland. Before the state became involved, this was undertaken partly for pride and partly for profit. It could be the work of individuals. The ruins of Elgin Cathedral were lovingly tended in the early 19th century by a 'worthy garrulous' guide (Cockburn, 1889: 9), John Shanks, kept going in whisky and enthusiasm by visitors' coins. It could be landowners: for example, in the 1830s the Duke of Argyll fenced off the ruins on Iona. Alternatively it could be groups: the Society of Antiquaries undertook major preservation work here in the 1850s. Interest groups for the preservation of this and that were to proliferate in the later 19th century, for scenery, buildings and wildlife, e.g. the Society for the Protection of Birds in 1889. That protection and preservation might require purchase, especially as land ownership and land use were in such flux, was recognized, and led to the foundation of the National Trust for Scotland in 1931. Its objective was not just to preserve 'beautiful country, good architecture and ancient monuments, battlefields like Bannockburn and Culloden' (Calder, 1990: xv), but to safeguard natural historical features, including slums. Its membership then was small and elite; the 1939 Council was knee deep in the great and the titled good, but it was to enjoy great expansion post WWII, as did other bodies such as Historic Scotland.

### The Future: Concluding Thoughts

There have long been concerns over the impact of tourism on Scotland, as to cultural, economic and other effects. Against the adverse effects which some saw and the serious tensions over land use which were a result of growing numbers of tourists of all classes and differing interests within the tourist ranks, there was the reality that tourism brought welcome income. It brought it into cities such as Edinburgh, to seaside and summer resorts and to areas where scenery and sport combined to provide pleasure. The cost to the environment was not yet a campaigning issue, although aspects of the environmental agenda had begun to emerge.

From its humble beginnings, the tourism industry in Scotland has stayed the course and grown exponentially compared to the heritage industries of shipbuilding, coal and oil; it is an industry that is everlasting. Its historical development highlights many of the issues impacting tourism today and in the future from a sustainable perspective, such as community engagement, overtourism, conservation and access. Thus, the debate about the future of tourism can be found in its past.

### Future turning point 1: Climate change

The natural and distinctive beauty of Scotland and its Highlands (Yeoman & McMahon-Beattie, 2006), as described by Scott in the *Lady of the Lake* (Scott, 1810, 1896), will potentially be impacted by climate change. Changes to the ecosystem would mean that the Bass Rock Seabird Centre would no longer exist as Scotland's unique coastal seabirds disappear. Grouse shooting on the Perthshire estates would be considerably curtailed and osprey tours in the Highlands would no longer be available. Although these events would not happen overnight, a series of incremental, small-scale changes would significantly affect the overall health of the tourism sector. The visual effect on vegetation and the landscape generally would become a serious issue for Scotland's scenery-based tourism industry as climate change would reformulate mountain landscapes. While the loss of the arctic alpine ecosystem on the Cairngorm plateau would probably only affect specialist visitors, losing iconic emblems would be detrimental even to the average visitor. As such, Scotland would lose its romantic appeal as alluded to by Scott (1810, 1896).

### Future turning point 2: A sustainable approach to the future

The continued success of global tourism (UNWTO, 2016) and Scottish tourism (Bhandari, 2016; Durie, 2017; Ryan *et al.*, 2018; Yeoman *et al.*, 2009) will lead to increased pressure on tourism hotspots and local communities. Popular rural destinations such as the Spey-Aviemore corridor already experience inadequate water supply during the summer months. The pressure of increased visitor numbers combined with drier springs and summers could stretch the capacity of these areas to cater for visitors and residents in a sustainable manner (Creaney & Niewiadomski, 2016). The major golfing areas could also negatively impact on water supply should visitor numbers rise dramatically. The increased extraction of water for fairways would have knock-on impacts on local human populations and also on marine and river wildlife (Jönsson, 2016).

## References

Adams, F. (1891) *The Genuine Works of Hippocrates.* New York: William Wood.
Babbington, M. (1864) *M. Babbington to Mrs Archibald Smith.* Glasgow: Glasgow City Archives.
Behrensmeyer, A. (2006) Climate change and human evolution. *Science (Washington)* 311, 476–478.
Bhandari, K. (2016) Imagining the Scottish nation: Tourism and homeland nationalism in Scotland. *Current Issues in Tourism* 19 (6), 913–929.
Brendon, P. (1991) *Thomas Cook: 150 Years of Popular Tourism.* London: Warburg.
Burchardt, J. (2002) *Paradise Lost: Rural Idyll and Social Change since 1800.* London: I.B. Tauris.
Calder, J. (1990) *Scotland in Trust: The National Trust for Scotland.* Glasgow: Drew.
Cockburn, H.C. (1889) *Circuit Journeys* (2nd edn). Edinburgh: D. Douglas.
Creaney, R. and Niewiadomski, P. (2016) Tourism and sustainable development on the Isle of Eigg, Scotland. *Scottish Geographical Journal* 132 (3–4), 210–233.
Dendy, W. (1859) *The Wild Hebrides.* London: British Library.
Devine, T.M. and Wormald, J. (2012) *The Oxford Handbook of Modern Scottish History.* Oxford: Oxford University Press.
Durie, A.J. (2012) *Travels in Scotland, 1788–1881: A Selection from Contemporary Tourist Journals.* Woodbridge: Boydell Press for the Scottish History Society.
Durie, A.J. (2013) Sporting tourism flowers – the development from c. 1780 of grouse and golf as visitor attractions in Scotland and Ireland. *Journal of Tourism History* 5 (2), 131–145.
Durie, A.J. (2017) *Scotland and Tourism: The Long View, 1700–2015.* London: Routledge.
Griffiths, T. and Morton, G. (2010) *History of Everyday Life in Scotland, 1800 to 1900.* Edinburgh: Edinburgh University Press.
Hassan, J. (2003) *The Seaside, Health and the Environment in England and Wales since 1800.* Aldershot: Routledge.
Hobsbawm, E.J. (1995) *Age of Extremes: The Short Twentieth Century, 1914–1991.* London: Abacus.
Jönsson, E. (2016) Trump in Scotland: A study of power-topologies and golf topographies. *International Journal of Urban and Regional Resear*ch 40 (3), 559–577.
Lambert, R.A. (2001) *Contested Mountains: Nature, Development and Environment in the Cairngorms Region of Scotland, 1880–1980.* Cambridge: White Horse Press.
MacArthur, E.M. (1995) *Columba's Island: Iona from Past to Present.* Edinburgh: Edinburgh University Press.
MacDiarmid, J. (1926) *The Deer Forests and How They Are Bleeding Scotland White.* Glasgow: Scottish Home Rule Association.
Markley, O. (2015) Aspirational guidance for wiser futures: Toward open-sourced ascension from ego-centric to eco-centric human communities. *Foresight* 17 (1), 1–34.
Marshall, P. (1998) *The Scottish Central Railway: Perth to Stirling.* Usk: Oakwood Press.
Marwick, W.H. (1973) *Economic Developments in Victorian Scotland.* Clifton, NJ: Augustus M. Kelley.
Milner, C.A., Connor, C.A. and Sandweiss, M.A. (1994) *The Oxford History of the American West.* New York: Oxford University Press.
Muir, E. (1935) *Scottish Journey.* London: W. Heinemann in association with V. Gollancz.
Naomi, L.-J. and Margaret, M.S. (2018) *Four Nations Approaches to Modern British History: A (Dis)United Kingdom?* London: Palgrave Macmillan.
Nilsen, E.B., Milner-Gulland, E.J., Schofield, L., Mysterud, A., Stenseth, N.C. and Coulson, T. (2007) Wolf reintroduction to Scotland: Public attitudes and consequences for red deer management. *Proceedings of the Royal Society B: Biological Sciences* 274, 995.

Orr, W. (1982) *Deer Forests, Landlords and Crofters: The Western Highlands in Victorian and Edwardian Times*. Edinburgh: John Donald.
Paterson, A. (2002) *Scotland's Landscapes: Endangered Icon*. Edinburgh: Polygon.
Robins, N. and Meek, D. (2008) *The Kingdom of MacBrayne*. Edinburgh: Birlinn.
Robinson, A. (2006) *Seeking the Scots: An English Woman's Journey in 1807*. York: Arthur Robinson.
Rogers, T. (1877) *A Fortnight on the Grampians*. Oldham: Manchester Botanists' Association.
Romeril, T. (1989) Tourism and the environment – accord or discord? *Tourism Management* 10 (3), 204–208.
Ryan, C., Bolin, V., Shirra, L., Garrard, P., Putsey, J., Vines, J. and Hartny-Mills, L. (2018) The development and value of whale-watch tourism in the west of Scotland. *Tourism in Marine Environments* 13 (1), 17–24.
Schaltegger, S. and Hörisch, J. (2017) In search of the dominant rationale in sustainability management: Legitimacy or profit-seeking? *Journal of Business Ethics* 145 (2), 259–276.
Scott, W. (1810) The Lady of the Lake: A poem. London: John Ballantyne and Longman, Hurst, Rees, Orme, and Brown, and W. Miller, Edinburgh.
Scott, W. (1896) *Scott's Lady of the lake: With Introduction, Notes, and Appendices*. London: George Bell & Sons.
Smout, T.C. (2000) *Nature Contested: Environmental History in Scotland and Northern England Since 1600*. Edinburgh: Edinburgh University Press.
Travis, J. (1993) *Rise of Devon Seaside Resorts, 1750–1900*. Liverpool: Liverpool University Press.
UNWTO (2016) *UNWTO Tourism Highlights, 2016 Edition*. Madrid: UNWTO.
Whatley, C.A. (2018) *Contested Commemoration: Robert Burns, Urban Scotland and Scottish Nationality in the Nineteenth Century*. Oxford University Press.
Yeoman, I. (2012) *2050: Tomorrow's Tourism*. Bristol: Channel View Publications.
Yeoman, I. and Lederer, P. (2005) Scottish tourism: Scenarios and vision. *Journal of Vacation Marketing* 11 (1), 71–87.
Yeoman, I. and McMahon-Beattie, U. (2006) Understanding the impact of climate change on Scottish tourism. *Journal of Vacation Marketing* 12 (4), 371–379.
Yeoman, I., Brass, D. and McMahon-Beattie, U. (2007) Current issue in tourism: The authentic tourist. *Tourism Management* 28 (4), 1128–1138.
Yeoman, I., Greenwood, C. and McMahon-Beattie, U. (2009) The future of Scotland's international tourism markets. *Futures* 41 (6), 387–395.
Zuelow, E.G.E. (2016) *A History of Modern Tourism*. London: Palgrave.

# Part 6

# Evolution

# 19 Does the Past Shape the Future of Tourism? A Cognitive Map(s) Perspective

Ian Yeoman and Una McMahon-Beattie

## Introduction

Forecasting, by its very nature, is a hazardous exercise. Challenges such as predicting future international tourism flows has become more difficult over time, given disruptive events such as 9/11 or the economic disorder caused by the Global Financial Crisis (GFC). Econometric forecasting is based upon the variables of GDP, exchange rates, price elasticity and transport connectivity. These variables are used to predict in the short term. Looking beyond the short term means dealing with increased uncertainty, which is not about forecasting an exact future but working with plural futures. Long-term futures consider shifts in society, climate change, new technologies, demography, wealth and a range of other drivers of change (Yeoman, 2012). Given the range of variables, future planning is about understanding the range of possibilities, mitigating for risk, creating visions of the future we want or monitoring trends. However, some would argue that the future is just a repetition of the past (Elliott, 2010; Hobsbawm, 1995). Tourism has always been about such things as nature, beaches and romance: the only thing that is different is that there are more tourists given the growth of wealth, population and the middle classes (Yeoman, 2012).

According to Zuelow (2016: 181), the famous historian A.J.P Taylor said:

> Dear boy, you should never ask an historian to predict the future – frankly we have a hard time predicting the past.

However, we do believe in the past and we have a belief in Hobsbawm's (1995) philosophy. Thus, this is why this book adopts Mannermaa's (1991) evolutionary paradigm from futures studies in order to identify key turning points in the evolution of tourism (McMahon-Beattie et al., 2016). Indeed, one of the roles of futures research is to model the development of society, looking for signals, social movements, technological advancements and signs of change at the point of evolution, or what Gladwell (2002) calls tipping points. Therefore, this book takes a historical perspective and identifies future turning points to demonstrate future evolution, an approach that has been used successfully elsewhere (Blass, 2003; Weatherford, 2016; Yeoman & McMahon-Beattie, 2016).

Overall, the purpose of this chapter is to help those interested in the future of tourism by looking at the past and identifying historical turning points which set the course for the future: this has always been seen as a challenge in tourism research (Pearce & Butler, 2010). The authors' contributions have been analyzed through a discussion of the various layers, links and concepts presented and each chapter has been interpreted through a cognitive map from which an aggregate map has been developed. The key historical turning points, which determine the future past of tourism, have been identified as: *mindfulness; mobility; step changes determining mass tourism*; and *the leisure class of consumption*. These turning points have been used to develop a conceptual model of the future of tourism which focuses on four future turning points. These future turning points have then become the basis for two scenarios, namely: *Degradation – we didn't learn from the past* and *A balanced future – learning from the past*.

## Cognitive Mapping Perspective

Cognitive maps (also known as mental maps, mind maps, cognitive models or mental models) are a type of mental processing composed of a series of psychological transformations by which an individual can acquire, code, store, recall and decode information about the relative locations and attributes of phenomena in their everyday or metaphorical spatial environment (Eden & Ackerman, 1998). Applied as a research methodology, they are used to represent cognition of the researched thoughts through a series of links as a map or conceptual framework. Jones (1993) states that a cognitive map:

> ... is a collection of ideas (concepts) and relationships in the form of a map. Ideas are expressed by short phrases which encapsulate a single notion and, where appropriate, its opposite. The relationships between ideas are described by linking them together in either a causal or connotative manner.

The method used by the authors is drawn from Eden and Ackermann's (1998) use of cognitive mapping in strategic management and management science which used personal construct theory or personal construct psychology (PCP; Kelly, 1977). PCP was developed by the psychologist

George Kelly (1955) in the 1950s and helped patients to uncover their own personality 'constructs' with minimal intervention or interpretation by the therapist. The repertory grid was later adapted for various uses within organizations including decision making and interpretation of other people's worldviews. Eden and Ackermann's approach to cognitive mapping centres on the idea of concepts. These are short phrases or words which represent a verb in which ideas are linked through as cause/effect, means/end or how/why, meaning a cognitive map is a representation of a person's perceptions about a situation in terms of bipolar constructs where the terms are a contrast with each other. For example, 'the invention of the steam train' may lead to 'holidays by the seaside ... congestion and pollution'. The result is not unlike an influence diagram or casual loop diagram, although it is explicitly subjective and uses constructs rather than variables (Mingers, 2014). Eden and Ackermann (1998) suggest that cognitive mapping can also be used to record transcripts of interviews in a way that promotes analysis, questioning and understanding. However, the literature on the application of cognitive mapping (Yeoman, 2004) is bastardized as researchers adapt the theory based upon their own skills and research philosophies.

## Decision Explorer

According to Silver and Lewins (2014), a computer assisted qualitative data analysis (CAQDAS) approach assists in the automation of processing data and the speeding and capturing of concepts. It helps the modeller view relationships of phenomena and data through the ability to trace and track data and provides a formal structure for notes and memos to develop an analysis platform, which is consistent with grounded theory (Corbin, 2015). Decision Explorer (DE) is an interactive tool for assisting and clarifying problems (Huff & Jenkins, 2002), using the principles of cognitive mapping (Ackermann, 2011; Eden & Ackerman, 1998) within the realm of CAQDAS. DE allows a visual display and analysis of cognitive maps in such a manner that it permits 'multiple viewpoints', 'holding of concepts', 'tracing of concepts' and 'causal relationship management'. It is a rich interactive tool that allows for the movement of concepts and connections in order for the modeller to be able to identify turning points. This allows the modeller to draw conclusions and construct a meaningful future, piecing together the research to produce a close set of practices and interpretations that present a series of findings that 'make sense' (Weick, 1989; Yeoman, 2004). The most important feature of DE is the ability to categorize concepts, values and emergent themes (Eden & Ackerman, 1998). It allows the modeller to elicit data and code concepts, for example, using 'set management' commands. Overall, DE is a process of allowing the modeller to emerge or stand back from the data. This approach to modelling and map building is well documented by authors in tourism research.

Examples include exploring emergent themes in family tourism (Schänzel & Yeoman, 2014), the future of events (Yeoman *et al.*, 2014), a conceptualization of food tourism and futures (Ellis *et al.*, 2018; Yeoman *et al.*, 2015) and demography trends (Yeoman & Watson, 2011).

## The Contribution of Each Chapter

In this section, we identify the contribution each chapter makes in terms of a cognitive map, underlying key concepts, summative meanings and contribution.

## Chapter 2: History of the Future of Tourism

In this chapter, van der Tuuk notes that the emergence of tourism in the Middle Ages was religiously motivated by the hope of meeting God. From the Renaissance period, humans focused more on the materialistic world. The new travellers were scientists, merchants and explorers seeking knowledge and economical and political expansion. In the 17th and 18th centuries, tourism became associated with leisure. It was seen as a way to gain knowledge and to develop oneself. The journey brought people into contact with other cultures and ways of life, other ideas and habits. In the 18th century, it was common in the higher social circles of England that young male adults made a Grand Tour of the European continent. During this 'rite of passage' they learned not only more about the classical arts, politics and culture but many indulged in their first sexual encounters. Tourism in the modern sense emerged in the late 18th century when travel was the goal itself. The traveller wanted to visit authentic places not affected by civil and industrial society. The dark forest, the coast, the vast sea, the quiet lake, the deep valley and the high mountain formed the new 'holy' places for the tourist.

From the 1960s, the custom of travelling for pleasure accelerated. Travelling became a way of showing who one is and where one belongs. Cultural norms, learned tastes, the norms and values from their environments started to determine their holiday plans. Where the Romantic

**Table 19.1** Chapter 2: Historical and future turning points

| Historical turning points | Future turning points |
| --- | --- |
| Grand Tours of Europe | The future of the global economy of tourism |
| Mass tourism | Future cultural capital |
| Modern tourism | Family structures |
| Nature | |
| Pilgrimage | |

Does the Past Shape the Future of Tourism? 247

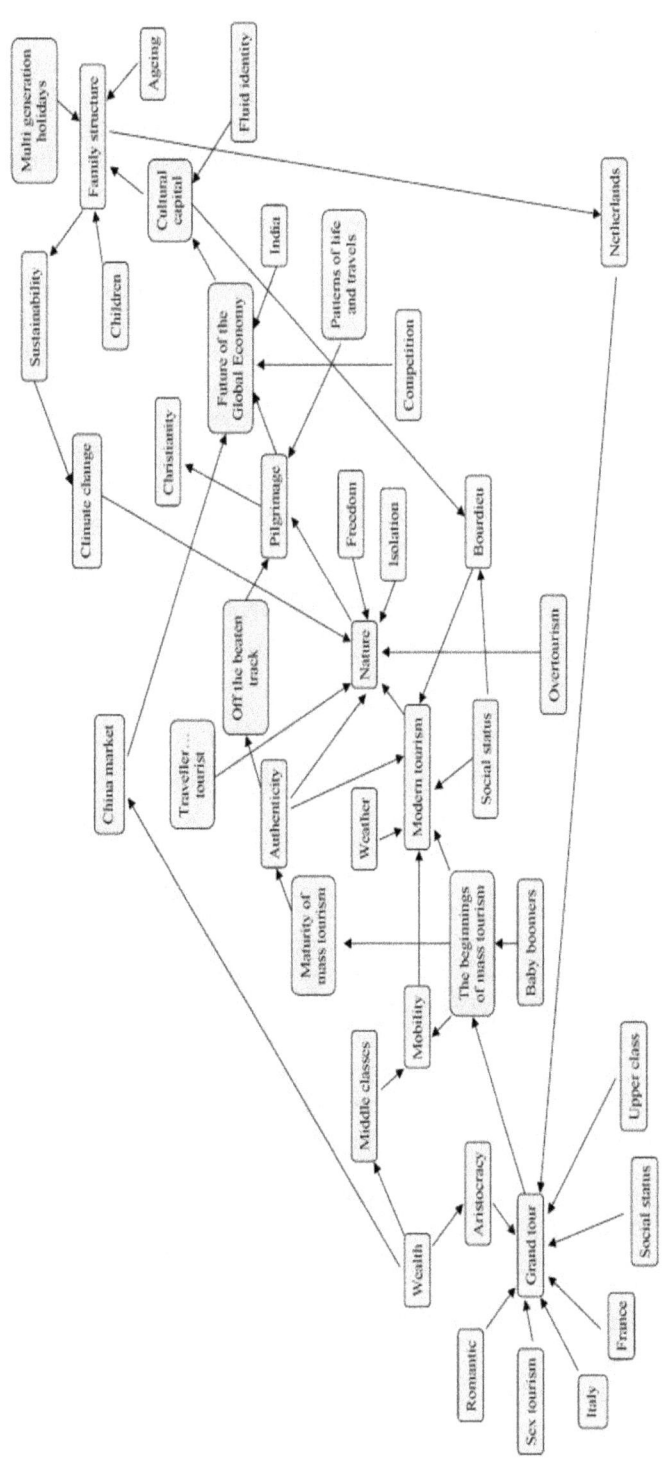

**Figure 19.1** History of the future

traveller was consciously searching for loneliness, for the modern tourist loneliness is a given, possibly without even being aware of it. It is the loneliness of 'other-directedness', of searching for safety, recognition and being valued by the like-minded. They travel through a world which has already been completely mapped out. They have become part of the (economic) system which they have been temporarily trying to break out of. They are regressing into the very existence from which they have been trying to escape.

Looking to the future, in 2035 tourism will continue to be just as important, particularly for the hardworking generation of 2035. This generation will really need its holidays because they will be working hard until their 70th birthday to bear the collective load which will have significantly increased. They will be able to do this because they will remain healthy for longer and will be well-informed about their own health. Healthy ageing will be the growth market from the 2040s onwards. Family holidays will still be seen as an escape from the emptiness of modern times, from the drifting masses of tourists who, for a few weeks a year, are served pre-cooked dishes of authenticity and instant pleasure. Tourists will strive for a Nietzschean *Umwertung aller Werte* (revaluing all values) escape.

## Chapter 3: The Development of Mass Tourism

This chapter by Butcher considers the development of mass tourism from the industrial revolution to the present day. Key concepts identified include: the beginnings of mass tourism; the Grand Tour; the culture of mass tourism; the maturity of mass tourism; and authenticity.

Butcher argues that mass tourism emerged with mass, modern society and is hence distinct in important respects from previous forms of leisure travel. He draws principally on the experience of the UK, the pre-eminent industrial power of the 19th century, but also includes examples from elsewhere, to illustrate the development of mass tourism. The key turning points are set out through a broad periodization, namely: the 19th century when industrialization created both the means for mass travel and the wealth that enabled greater numbers of people to take holidays from work

Table 19.2 Chapter 3: Historical and future turning points

| Historical turning points | Future turning points |
|---|---|
| Tourism as a mass phenomenon in the 19th century | China and India as global powers |
| Mass tourism from 1900 to 1945 | African potential |
| Post 1945 – international tourism as a mass phenomenon | Overtourism |

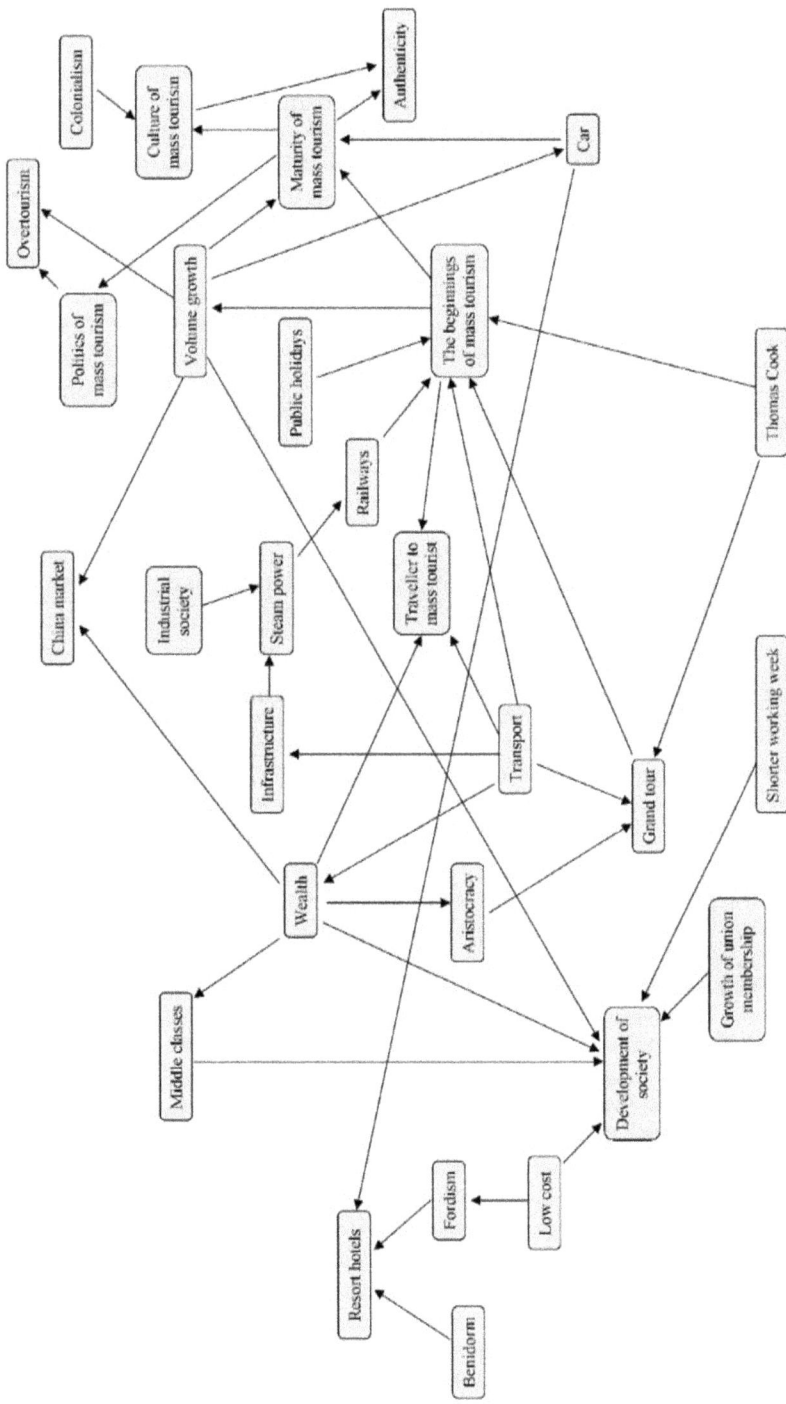

**Figure 19.2** The development of mass tourism

away from home; 1900–1945, covering a period when great technological gains were made but the tragedy of the two world wars dominated; and the period from 1945 to the present, when aviation technology and greater wealth established the package holiday abroad, and later a great variety of holiday choices for those able to pay.

The chapter establishes the historical specificity of mass tourism, rooting it in the 19th century advent of mass society. It emphasizes the importance of developments in technology at different times (broadly, steam in the 19th century, the motor car in the 1900–1945 period and the jet engine post-1945) and how the growing wealth of the masses enabled mass consumption. It examines: the application of technology to industry through Fordism, making tourism cheaper and more accessible; the resort and how mass tourism shaped destinations; and the social and cultural concerns levelled at mass tourism. Finally, it provides a reflection on the recent discussion of overtourism, suggesting that mass tourism may have reached its cultural and environmental limits.

Overall the chapter seeks to provide a short historical review through which the reader can readily establish the origins of mass tourism, and it ends with three 'turning points', highlighting what are likely to be the key future events.

## Chapter 4: The Historical Future of Tourism: The Case of Malta's Policies

Mangion examines evolutionary patterns in Malta's tourism system. This is a captivating case study not only because of Malta's distinctive combination of features, with complexities condensed into a small-territory context, but also because there are important policy-related reflections for the future for similar island destinations and the wider tourism

**Table 19.3** Chapter 4: Historical and future turning points

| Historical turning points | Future turning points |
|---|---|
| Conscious and active initiation of tourism as an economic activity due to political-economic scenario and potential economic downturn | Society's reaction and involvement |
| Tokenism | Technology, digital supremacy and competition from digitalization of virtual visiting experiences |
| Policy crises | |
| Path dependence – proactive policy support for tourism | |
| Institutional adaptation and policy leadership | |
| Air connectivity policy – independent travel and low-cost airlines | |

Does the Past Shape the Future of Tourism? 251

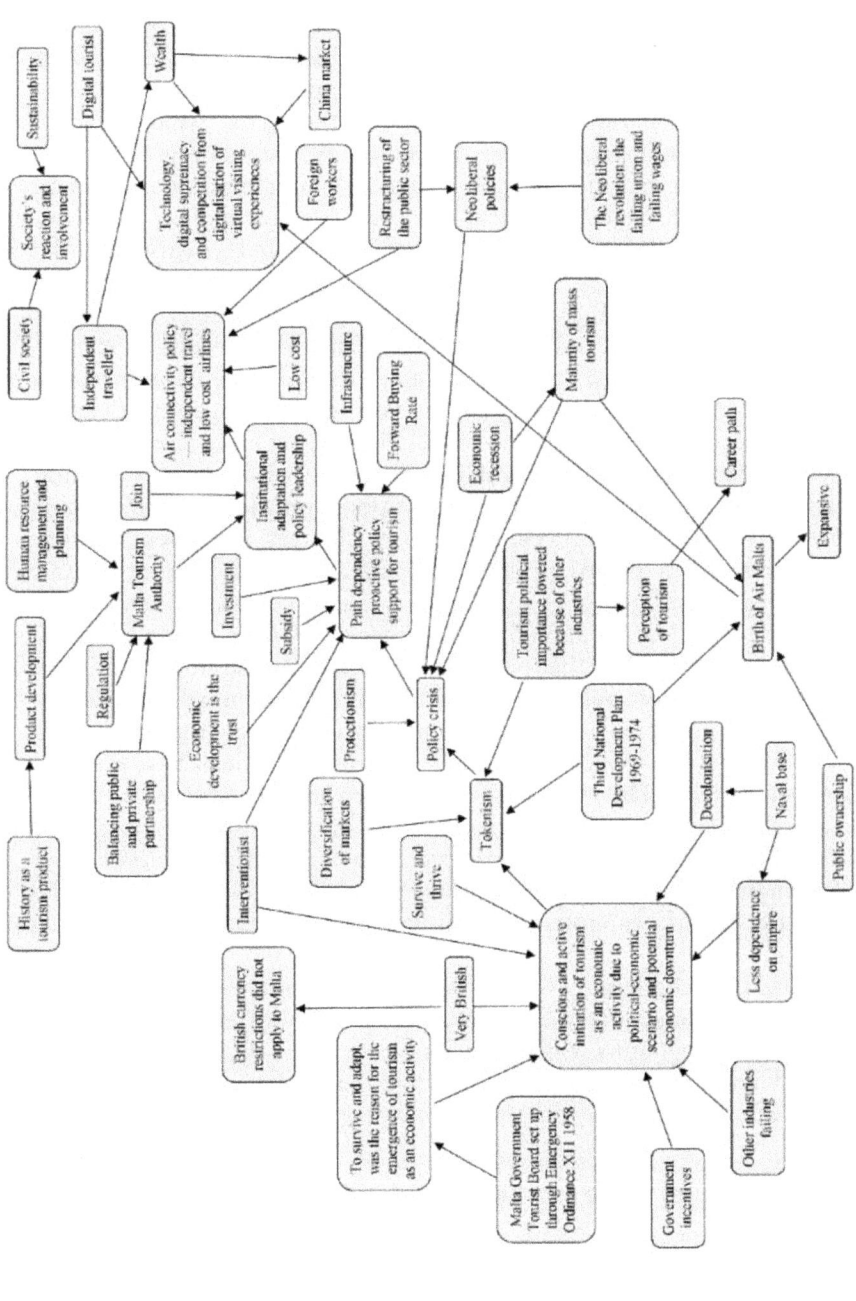

**Figure 19.3** The historical future of tourism: The case of Malta's policies

research field. Malta's island geography, its extremely limited land area and its high population density compounded by tourism activity amplify the challenges of accessibility, increased vulnerability to Mediterranean geopolitics and competing demands which need to be met by minimal resources. Taking a historical perspective on the country's tourism policy and the private sector's response over a 60-year period, the chapter identifies the transformations that took place in the sector, what triggered change and what instigated policy responses. The key concepts emerging from the chapter include: conscious and active initiative of tourism as an economic activity due to political economic downturn; air connectivity policy – independent travel and low-cost airlines; and tokenism and the birth of Air Malta.

Looking to the future, the development of Malta's tourism is expected to continue on this evolutionary path, where public policy and private sector responses play the primary shaping role. However, one expects a third player, namely civil society, in its broadening diversity and its own ongoing changes, to request and expect more direct involvement, beyond consultation, in shaping tourism policy. A second future turning point could arise from competition from the digitalization of virtual visiting experiences, instigating changes in current market dynamics and requiring private and public sector collaboration and policy leadership.

The central contribution of this chapter arises from a simple but significant point which is that policy can nudge evolution in tourism. The chapter depicts how policy responds to the broader context and to external factors, how public institutions interact with private tourism service providers and how policy leadership in specific policy fields have prodded tourism activity. The chapter also illustrates how the absence of policy leadership impacts tourism performance. Tourism in Malta, in spite of critical complexities, consequently evolved as a result of policies that led the initiation of market identification and penetration and created the right framework, support mechanisms and institutions for tourism to function.

The distinct relevance to futurism-based approaches arises primarily through the detection of common threads drawn from the identified historic turning points which, when projected onto the future, highlight critical developments that may occur and that policy makers need to be aware of.

### Chapter 5: Jules Verne as a Key to Understanding Irish Tourism

Pfatschbacher's analysis of Verne's novel *Foundling Mick* can be compared to searching for archetypes in human behaviour; it reveals constant features in Irish identity which serve to elicit inspiration for touristic strategies. Those archetypes can be summarized as a special bond

Table 19.4 Chapter 5: Historical and future turning points

| Historical turning points | Future turning points |
|---|---|
| Luxury accommodation is recommended | National identity and individualism |
| Round trips more and more relevant | |
| History guarantees rising numbers of visitors | |
| Britons are Ireland's essential target group | |
| Authenticity boosts Irish tourism | |

between the British and the Irish, an outstanding historical development, poverty and simplicity in Irish lifestyle, a variety of unusual landscapes, and a contrast in housing leading either to emanations of luxury or depressive scenarios. These archetypes represent the reoccurring themes in Figure 19.4.

Experts could apply the archetypes to the whole range of existing offers to evaluate how far Irish tourism has achieved coherence between engrained patterns of identity and touristic projects, thus paving the way for a natural way of experiencing the Emerald Isle. These activities refer, broadly speaking, to heritage tourism which a majority of tourists engage in. Several principles should be considered to make this type of tourism a success. First and foremost, the spirit, the special atmosphere of a location, must be maintained and rendered accessible to tourists; the Joyce tower in Sandycove constitutes a perfect example. It evokes the major characteristics of a very special place of literature and therefore demands investment. Another example would be the renovation of traditional seafront hotels in Bray, for instance: they represent what authentic Irish tourism at the seaside is all about, lending an elegant touch to the resort. It would be extremely regrettable to abandon those impressive buildings to decay, thereby spoiling beautiful views and vistas and damaging the reputation of the town. Instead, a renovation project should be launched, involving funds coming from the state and even the EU.

Talking about accommodation, we should of course focus on one specific expectation, which also became manifest in Verne's novel: variety should be provided in order to be in line with traditional ways of living on the Emerald Isle. In the past, accommodation was always marked by strong contrasts, with the affluent living next to people facing misery and starvation. Tourists should be given the opportunity to see both dimensions. The service industry thus not only has to offer historic hotels with specific narratives on their renowned past but also stone cottages on remote islands, campsites or just caravans to conjure up and engrain images of disadvantaged social classes. Another requirement for the future relates to the outstanding importance of Irish history. How can experts exploit this tendency which often emerges in interviews with travellers who are keen on experiencing authentic Irish history? Probably one

254 Part 6: Evolution

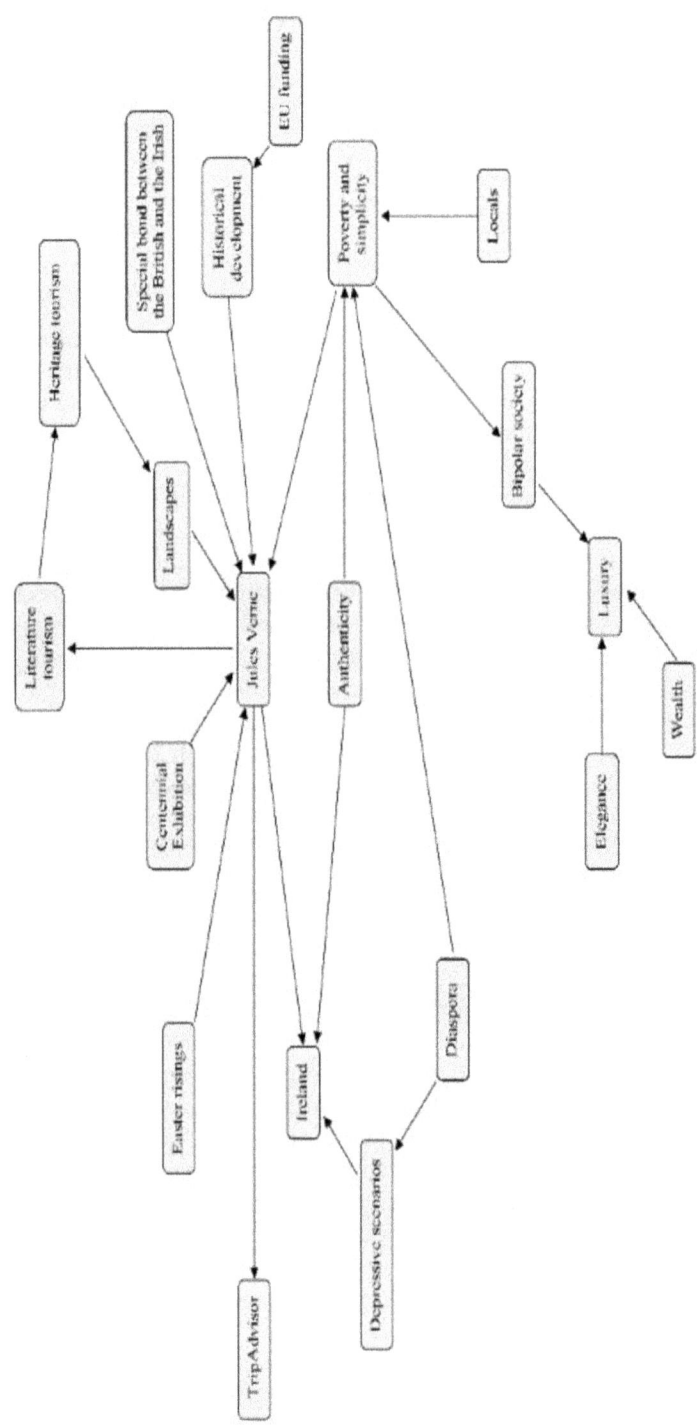

**Figure 19.4** Jules Verne as a key to understanding Irish tourism

of the best examples dates back to 2016 when Ireland commemorated the Easter Rising with the Centennial Exhibition at the GPO in the heart of Dublin. The reviews that can still be found on TripAdvisor prove to be extraordinary and reflect the public acceptance of historical highlights.

## Chapter 6: The Growth, Decline and Resurgence of the City-State

The historical turning points identified in this chapter are the central focus of Figure 19.5. The growth of city-states was initially linked to their relationship with the Church, followed by their need to develop free-trade relationships with other city states in order to sell their surplus goods. This created a central banking system, which encouraged the growth of business/political tourism. However, some city-states have declined because their boundaries were a hindrance to commercial trading. Territorial-states encouraged the development of railways, standardized time zones and common language and currency, resulting in a market for holidays.

The re-emergence of the city-states can be seen in the growth of world/global cities. This is because of their duty-free zones, sovereign funds, private military contractors and cabinet forms of government, where the rights of individuals are secondary to the needs of the State. City-to-city tourism will drive their tourism, rather than country-to-country tourism. Hay concludes by looking to the future in which illiberal democracy dominates as it places priority on an efficient and effective government. City-states will restrict tourism, through laws to govern the interaction/behaviours between citizens and tourists. New technologies will control and manage the behaviour of residents and tourists, thus directing the social activities of its tourists and citizens alike, through a daily programme of activities to minimize the impact of tourism on its citizens. The management of city-states will shift from the public to the private sector, with the development of specialized models (entertainment, sexual-orientation city-states). City-states will restrict entry to pre-approved tourists, whose social, economic and political profiles match those of the city-state. This will be used to pre-vet its future citizens.

Table 19.5 Chapter 6: Historical and future turning points

| Historical turning points | Future turning points |
|---|---|
| The growth of city-states has been driven by both their political and their economic power | Efficient governance is assured through the free election of an illiberal democracy |
| The decline of the city-state was driven by the expansion of territorial-states and their growing commercial interests | Technology will be used to control and manage the behaviour of residents and tourists |
| The emergence of world/global trading | Development models for city-states of the future |

256 Part 6: Evolution

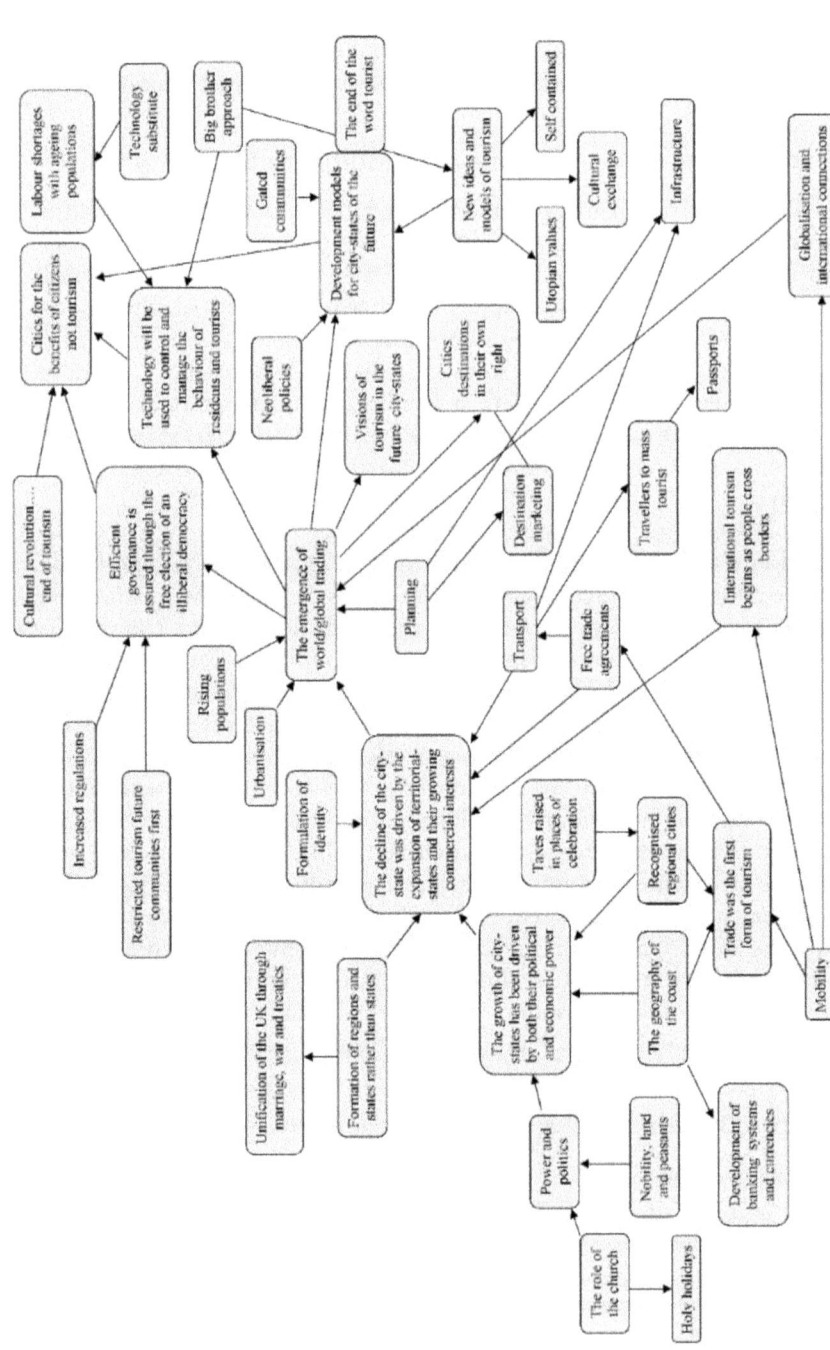

**Figure 19.5** The growth, decline and resurgence of the city-state

This chapter suggests that the term tourism will become redundant and tourists will be viewed as 'non-citizens'. DMOs will become destination monitoring organizations, whose function will be to control 'non-citizens'. It concludes that city-states are only part of a continuous cycle of change (city-states, territorial-states, global-cities/world-cities, private/commercial city-states, nano-states, city-states). Its central contribution is that future city-states will be open to tourists, but only on their own terms and under their control. They will not welcome tourists as strangers who need to be supported and guided, but rather as like-minded individuals who conform to their visions and ideals.

## Chapter 7: Geohistorical Analysis of Coastal Tourism in China (1841–2017)

In this chapter Taunay offers a geohistorical analysis of the relationship between tourism and the coast in China. The key concepts identified from Figure 19.6 include: from Europe to Chinese coasts, a global history of sea bathing; a progress and rapid opening, taking deserving workers to the beaches in large numbers (1979–2007); German development; and beaches reserved for the Chinese Communist Party (1949–1996).

Although hot baths had been around for hundreds of years, the first forms of coastal tourism in the People's Republic of China (PRC) only began to take place within foreign concessions in the second half of the 19th century. This spread of European customs in China also coincided with the arrival of Western powers in the country during the same period (1841–1911) and the relationship between the Chinese population and the coast has since been redefined many times. Taunay has identified three changes which took place, and a further two that are still underway. From the fall of the Qing Dynasty (1911) to the establishment of the PRC these activities were confined to foreign concessions. Then, up until the beginning of the Cultural Revolution (1966), it was only the leaders of the Chinese Communist Party and military veterans who used the former colonial sites. Between the time when Deng Xiaoping launched his reforms

Table 19.6 Chapter 7: Historical and future turning points

| Historical turning points | Future turning points |
| --- | --- |
| Arrival of Western powers | Controlling bodies and increasing standard setting |
| The hesitant spread of bathing activities (first half of the 20th century) | Banalization |
| Beaches reserved for the Chinese Communist Party (1949–1966) | |
| A progressive and rapid opening (1979–2007) | |

258　Part 6: Evolution

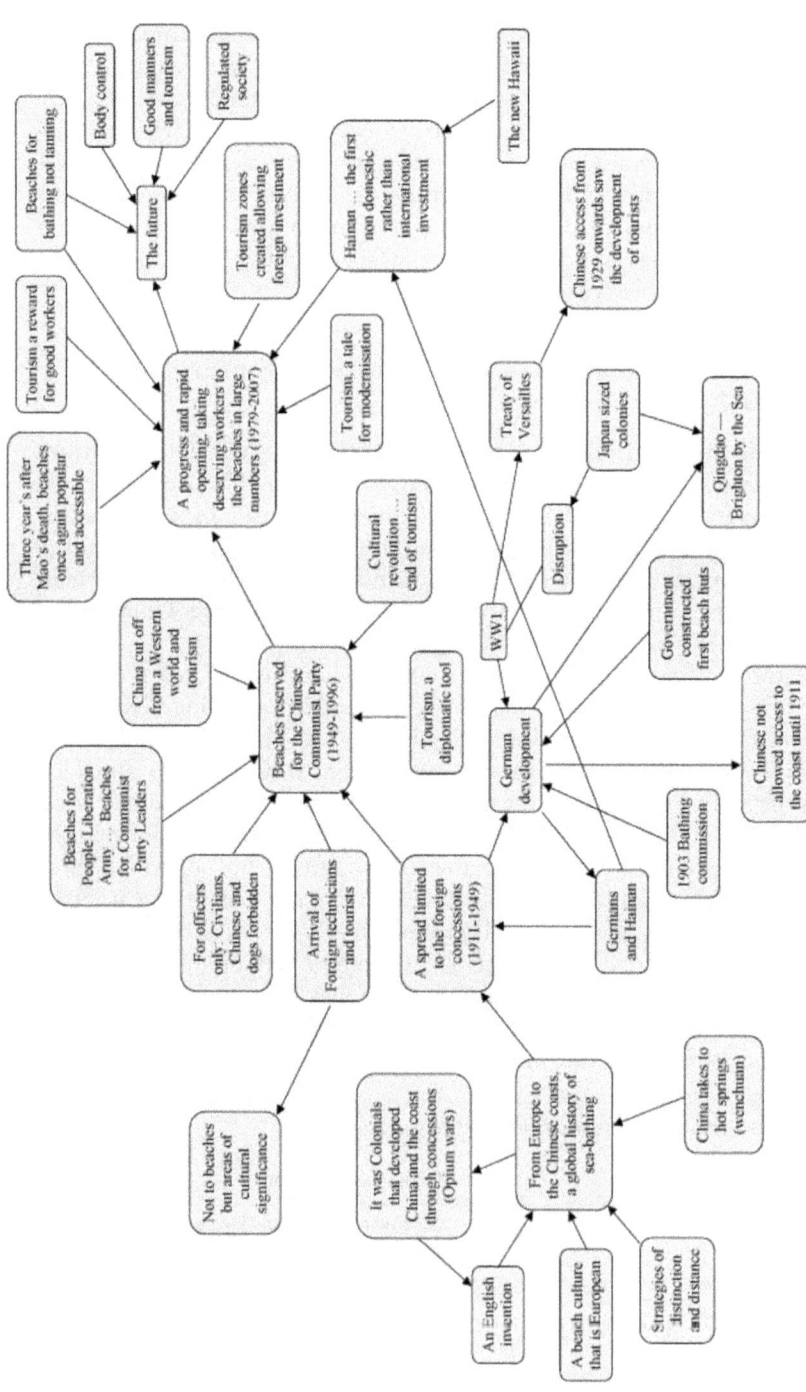

**Figure 19.6** Geohistorical analysis of coastal tourism in China

(1978) and 2007 (with a reworking of the *golden week* system) sea-bathing attracted ever more tourists each year. From the Beijing Summer Olympic Games (2008) to the end of Xi Jinping's first term (2017), central government gradually strengthened its control over seaside activities. Activities involving exposure of the body particularly reflect this political control. The ongoing individualization of seaside activities, in places reserved for certain population groups and to the detriment of others, is probably the next important step in this geohistory. Taunay observes that as society progresses in China, it is a progression based upon statehood and how central government wants its citizens to behave. This to a certain extent represents the banalization of beach practices.

This chapter thus contributes to deconstructing the idea that sea-bathing in China is a recent practice. The filiations between current and other, older practices in China or elsewhere in the world are longstanding, emphasizing that the study of tourism in a given place and period cannot avoid looking at it in a global historical context.

## Chapter 8: The Future Past of Aircraft Technology and its Impact on Stopover Destinations

Castro and colleagues trace the development of aircraft technology focusing on the turning points listed in Table 19.7 and as a linear trajectory shown in Figure 19.7.

Aircraft technology has markedly shaped the global tourism industry, shrinking the world by making travel affordable and fast. In particular, long-haul travel has significantly increased over the last 100 years, with new aircraft technology expanding the range of flights and the comfort of passengers. Using the example of the Kangaroo Route between Australia and the UK, this chapter discusses the relationship between aircraft technology development, the implications of non-stop long-haul travel, and how the role of stopover destinations has changed over the years. It highlights critical milestones in the historical development of aircraft

Table 19.7 Chapter 8: Historical and future turning points

| Historical turning points | Future turning points |
|---|---|
| Beginning of the Kangaroo Route | Future capacity |
| Introduction of the Super Constellation | Airport futures |
| Beginning of the jets operation | |
| The Super Jumbo takes over | |
| The Boeing 747-400 era | |
| The Airbus A380 | |
| Sunrise Project | |

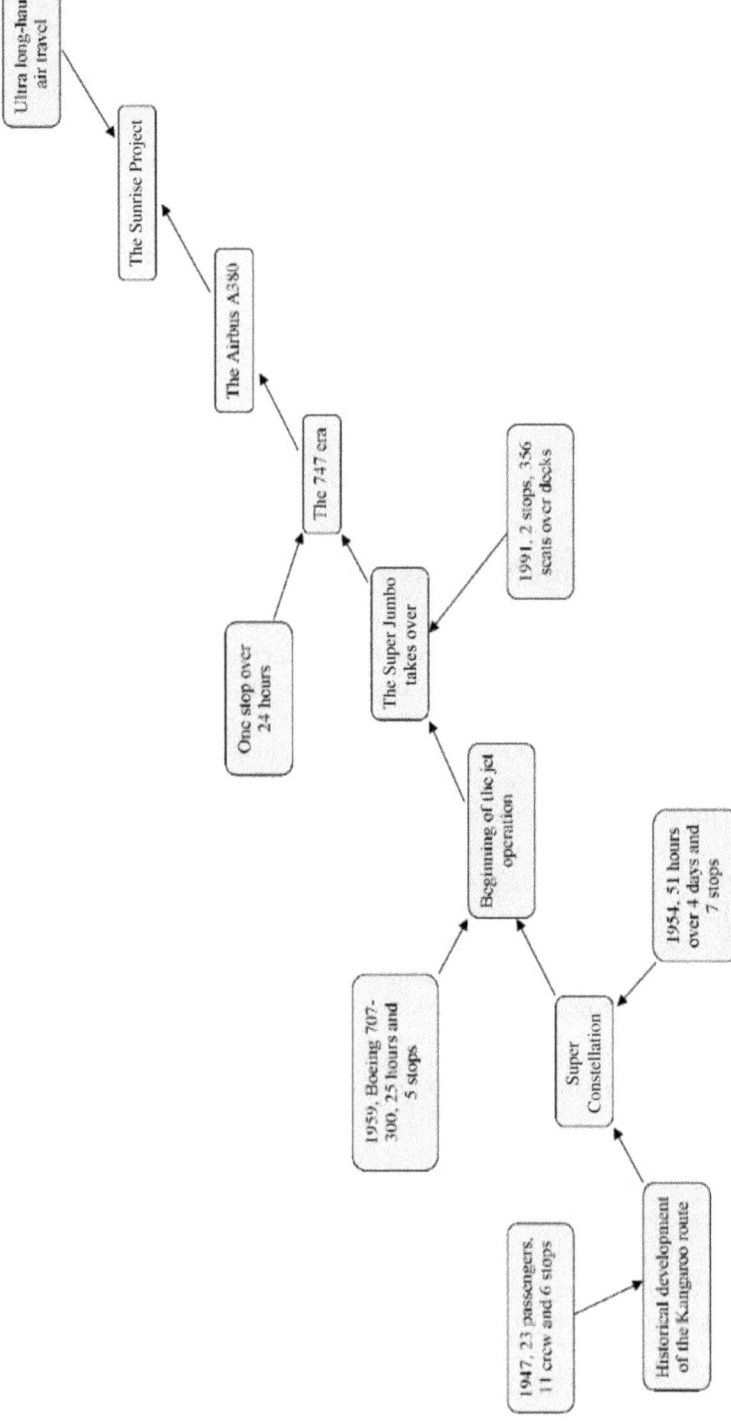

**Figure 19.7** The future past of aircraft technology and its impact on stopover destinations

technology and their effect on passengers' preferences and also on stopover destinations. It covers a number of aircraft technological developments, from the Lockheed Constellation era with multiple stopovers, to the emergence of the Boeing 707 and 747, as well as the Airbus A380 and the rise of global stopover destinations such as Singapore and Dubai. The introduction of ultra-long range (ULR) flights (e.g. Doha to Auckland and Perth to London) has already shortened flight times and created the environment where stopover destinations are no longer necessary. However, passengers' preferences on ULR flights are yet to be studied. Potential impacts on the health and discomfort of passengers and crew, as well as the advantage of adding additional travel destinations into the itinerary, might prevent a certain number of passengers from flying ULR direct routes. From a destination management perspective, some of the stopover hubs (i.e. Dubai and Singapore) have not only developed their aviation infrastructure but also invested in marketing and branding strategies, events management, built attractions, accommodation and other leisure/tourism features. These destinations have taken the advantage of the stopover role to boost their destination credentials. In many aspects, a number of these places are now established enough to support their tourism industry without the need of the stopover function. Other destinations that develop their tourism as a stopover product are urged to rethink their tourism and marketing business strategies. With the likelihood of aircraft technology improving further, the stopovers as we know them today will not be required in the future.

### Chapter 9: Forever Young and New: Cruise Tourism

Cruising continues to be the fastest growing sector within the leisure tourism industry. In Figure 19.8 London and Wallace identify the following concepts central to its development: transport innovation and industrialization; image; luxury; volume growth; the beginnings of mass

Table 19.8 Chapter 9: Historical and future turning points

| Historical turning points | Future turning points |
| --- | --- |
| The birth of leisure cruising | Climate change |
| The winds of war, waves of immigration and economic downturn | Conflict |
| The impact of air travel | Technology |
| Oil shocks and the reinvention of ocean cruising | Future growth |
| Coming of age | |
| The floating resort | |
| Terrorism and economic downturn | |
| The widening of the Panama Canal | |

262  Part 6: Evolution

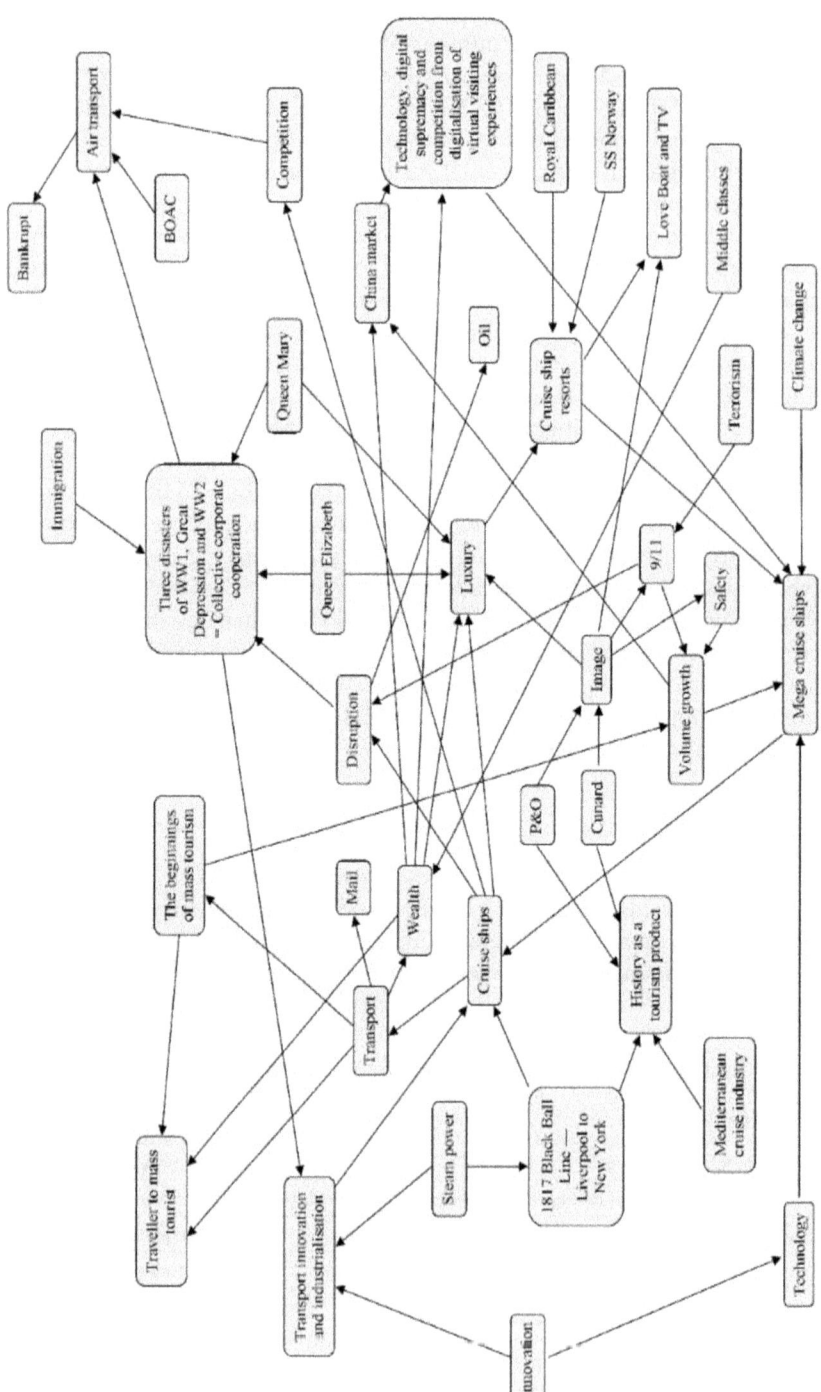

**Figure 19.8** Forever young and new: Cruise tourism

tourism; and three disasters of WWI, the Great Depression and WWII = collective, corporate, cooperation.

Undoubtedly the industry's proactivity in adopting technological innovation and its ability to reinvent itself despite an expansive catalogue of geo-political, economic and environmental challenges has contributed to its growth. Technology has allowed the industry to introduce positive marketing attributes such as high levels of comfort, speed, enhanced passenger experiences, feature-rich ship designs and environmentally sensitive systems. Reinvention has ensured that the cruise industry continues to survive despite such global events as conflict and economic downturns. Cruising is the great survivor and it does so in style. This chapter examines the confident and effective responses of an industry which is otherwise limited by the nature of its large, ocean-plying revenue-generating assets to events which might seem to be too difficult to surmount. The story told in this chapter traces a truly evolutionary lifecycle which appears to have a future grounded in growth.

The chapter identifies a number of historical milestones starting in 1822 when the first shoots of the modern leisure cruising industry could be detected when steamships were deployed for passenger transport. In 1840, passenger comfort came to the fore when Cunard's *Britannia* sailed from New York with a cow on board to provide fresh milk for passengers. In 1889, passenger comfort took another leap forward when electric lighting was installed on a ship for the first time. The world wars resulted in the appropriation of passenger ships for troop transport but after each episode the industry demonstrated its resilience and continued to grow. By 1958, air transport became the preferred method of transport across the Atlantic, resulting in a short-lived downturn in the passenger shipping industry. However, by the early 1960s, the passenger shipping industry responded by reinventing itself. Modern leisure cruising began with vacation trips to the Caribbean on cruise ships. The oil crisis in 1973 caused instability in the cruise industry, but Captain Stubing came to the rescue, generating huge interest in cruising. More recently, 11 September 2001 and the GFC of 2007–2008 counter-intuitively resulted in a stronger cruise industry, as passengers took to the seas which were considered safer than the air. With the discounted fares available during the Global Financial Crisis (GFC), more passengers and a significantly more diverse passenger cohort led to substantial growth in the industry. In 2016, the first ship transited the wider Panama Canal, making it easier and faster to reposition mega cruise ships between the Northern and Southern Hemispheres.

Looking to the future, climate change is likely to impact popular cruise destinations, thereby potentially necessitating a re-imagination of the cruise industry as it is known today. Ongoing conflict is likely to result in developing strategies to deal with the threat including changes to itineraries, and contingency plans for any future ship appropriation for troop transport. Technology is already starting to alter the face of cruising, and may perhaps be its saviour if, for example, islands disappear, and the cruise lines respond by making their ships into floating islands.

## Chapter 10: From Muscles to Electrons: A Technological Look at the Futures of Energy, Transport and Tourism

This chapter by Hui traces five key prime movers driving transportation and tourism, starting with the human body and the domestication of horses, before moving on to three inventions since the Industrial Revolution: the steam engine, the internal combustion engine and the gas turbine. Together, these five prime movers reshaped societies along different trajectories of energy mixes and economies. Evolved over millions of years, the *homo sapiens* body is bipedal, lean and optimized for long-distance treks and heat dispersal through the skin, facilitating a group hunter-gathering nomadic lifestyle. The core concepts as seen in Figure 19.9 include: the beginnings of mass tourism; steam power; gas turbines; luxury; the maturity of mass tourism; electric vehicles and energy storage; and the combustion engine.

The domestication of horses and pack animals provided access to powerful pushing and pulling companions for hauling heavy loads and extending travel. With the Industrial Revolution came the coal-powered steam engine and the development of large steamships, trains and power stations, while the internal combustion and diesel engines opened the door to gasoline-powered vehicles of many types, such as cars, motorcycles, snowmobiles and freighters. Lastly, the post-WWII design of the gas turbine made intercontinental jet travel and subsequent globalized tourism a stark reality. Through this historical examination of the link between energy and mobility, this chapter concludes by examining their futures through the overlapping trends of electrification and autonomous driving, regarding automobiles, buses and ships. This extrapolation is based upon the Paris climate goal of net-zero emissions by the second half of the century, a push that is accelerating and poised to transform a transportation sector currently contributing 40% to total greenhouse gas emissions.

Table 19.9 Chapter 10: Historical and future turning points

| Historical turning points | Future turning points |
|---|---|
| Sapiens, metabolism and movement | Electric vehicles and energy storage |
| Horse power | Cycling and scooters |
| Industrial innovations – steam power | Buses and high-speed trains |
| Industrial innovations – combustion engines | |
| Industrial innovations – gas turbines | |

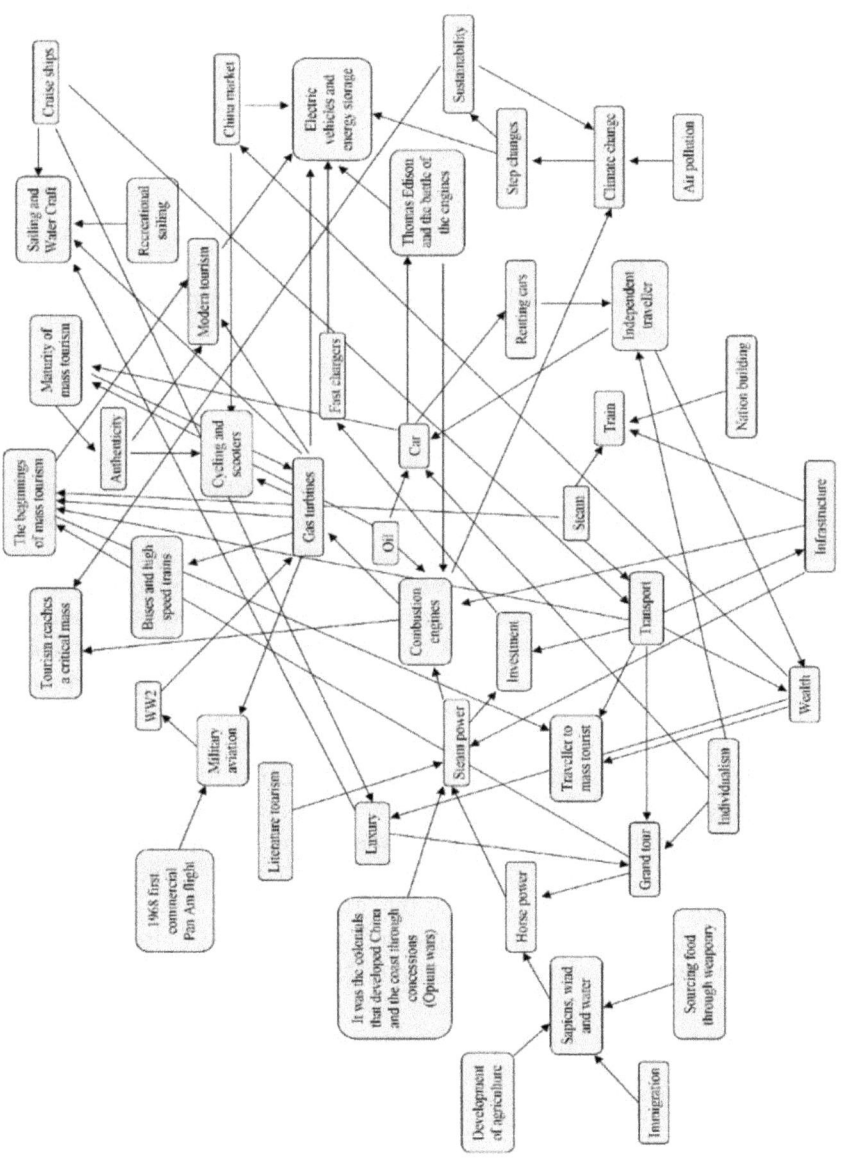

**Figure 19.9** From muscles to electrons: A technological look at the futures of energy, transport and tourism

The thrust of this piece is to examine the nexus of energy and transportation, and how their co-evolution reshapes the types of tourism that become possible. It therefore seeks to extrapolate outwards from the energy-capturing technologies that diffuse and become normalized in society. This examination is inspired by the materialist and quantitative orientation of writers like Václav Smil (2005), who popularized the idea of energy return on investment (EROI) as an attempt to quantify the amount of energy gained per unit used up in the process of extraction. Broadly, a historical and energy-based analysis helps to contextualize tourism as a mode of energy expenditure, whereas the focus on transportation brings to the surface how increasing flows of energy inputs have been harnessed in the service of travel. In light of the Anthropocene, it also provides a starting point for re-conceptualizing tourism as an object of inquiry, one that examines it as a bio-geophysical force impacting planetary feedbacks and also as a political-economic sector that will radically change as different energy infrastructure and social norms come to dominate.

## Chapter 11: Hotel History

The chapter identifies and analyzes the hotel as part of the infrastructure of travel and also as an imaginative construct. The central concepts identified in Figure 19.10 include: transport innovation and industrialization; car; hotel from French word; Grand Tour; the beginnings of mass tourism; innovations in business structure; and 'escaping' the hotel.

This chapter by James explores the milestone events and processes that influenced the position of the hotel within the travel and hospitality sectors, and also impacted on its imaginative form over two centuries. These landmarks included the expansion of the urban lodging networks associated with the Grand Tour, the development of the railway and industrial society in the West, innovations in legal frameworks and in corporate organization, the influence of the automobile over the geography and structure of lodgings, and commercial traveller preferences. It traces the historical consolidation of the 'grand hotel' as an ideal type, despite the proliferation of types of commercial lodging, and also explores the implications of the waning influence of that institution in defining the

**Table 19.10** Chapter 11: Historical and future turning points

| Historical turning points | Future turning points |
| --- | --- |
| The country house's commercialization | |
| Transport innovation and industrialization | Redefining the hotel form |
| Innovations in business structure | 'Escaping' the hotel |
| Spatial efficiencies and global expansion | |
| Automobility | |

Does the Past Shape the Future of Tourism? 267

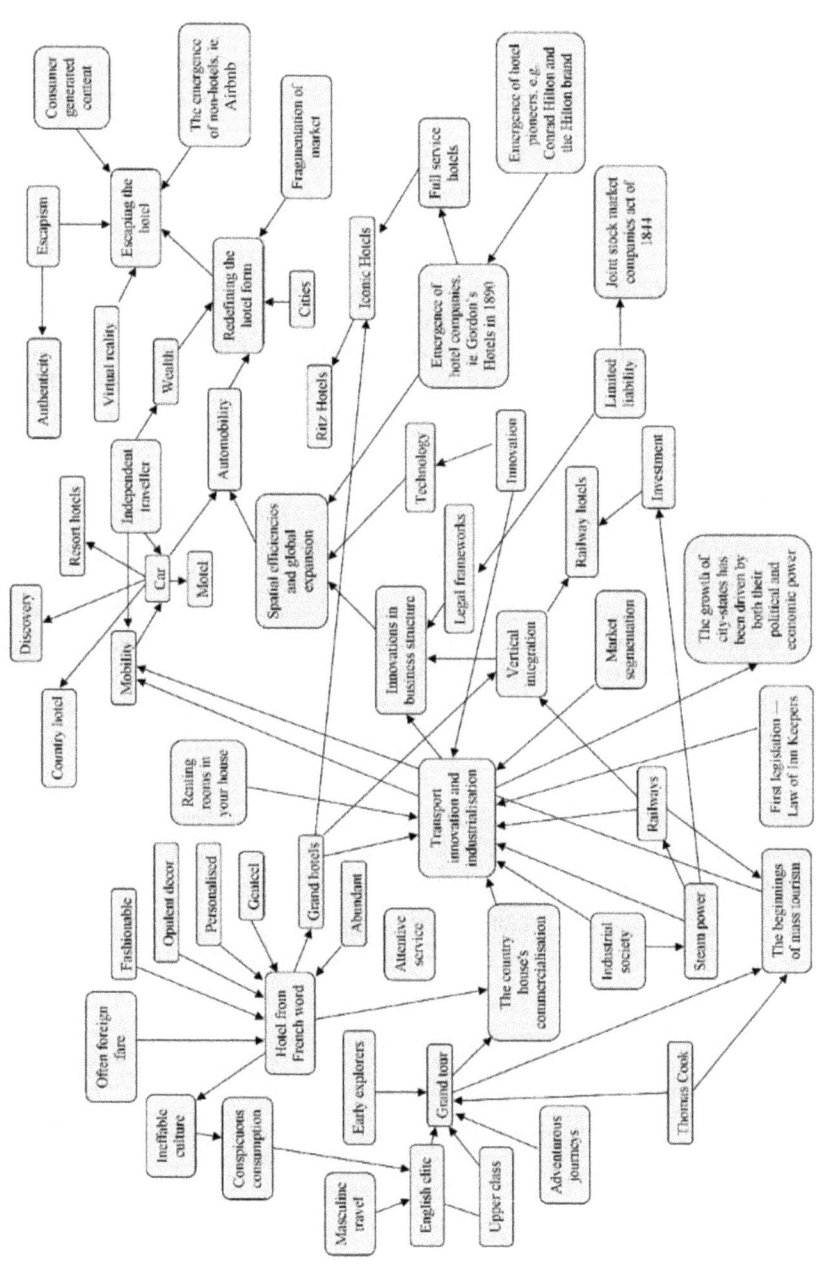

Figure 19.10 Hotel history

hotel in popular culture. It also examines the association between America and technological, managerial and organizational innovations in the sector, from the 19th century onwards. Looking forward, it evaluates the impact of the repurposing of buildings – both hotels that are adapted for other purposes and building 'conversions' into hotels. Lastly, it suggests ways in which the sector will respond to the new markets and products associated with the rise of online brokerages. As we enter an era in which the hotel no longer denotes a particular physical form, let alone a historical service culture, both anchored in a specific imaginative construct, the paid accommodation sector will continue to navigate contemporary and future challenges associated with new technologies, changing demand for products and evolving popular ideas of what forms a hotel is meant to take and what services it is meant to supply.

Overall, this chapter argues that the material form of the hotel is highly protean and adaptive, and that hotel history must account for that feature, as well as for the dominance and cultural salience of particular hotel ideas and ideals in understanding this complex institution. In so doing, it demonstrates the need to orient interpretations of the hotel's past and future towards a more holistic analysis of the hotel's evolving material and imaginative forms and functions, exploring their historical divergences and future relationships, rather than focusing largely on the hotel solely as part of the material infrastructure of travel. Adopting this integrated approach, and bridging the study of the hotel's past and future, also underscores its protean forms and meanings, and its continuous adaptations to new demands, consumer preferences, technologies, transport systems and politico-cultural environments.

## Chapter 12: Historical Employment Relations in the New Zealand Tourism Hotel Sector: From a Collective Past to an Individual Future?

As in many countries, tourism has become a major economic driver in New Zealand, overtaking the dairy sector to become the nation's largest

**Table 19.11** Chapter 12: Historical and future turning points

| Historical turning points | Future turning points |
|---|---|
| The Hotel Workers Union and rising wages | Ongoing neoliberal consensus and the 'individual' future |
| The birth of the Tourist Hotel Corporation: training, skills and careers | Disruption drives change: a collective future |
| The neoliberal revolution: the failing union and falling wages | |
| The end of the THC: the stagnation of training and development | |

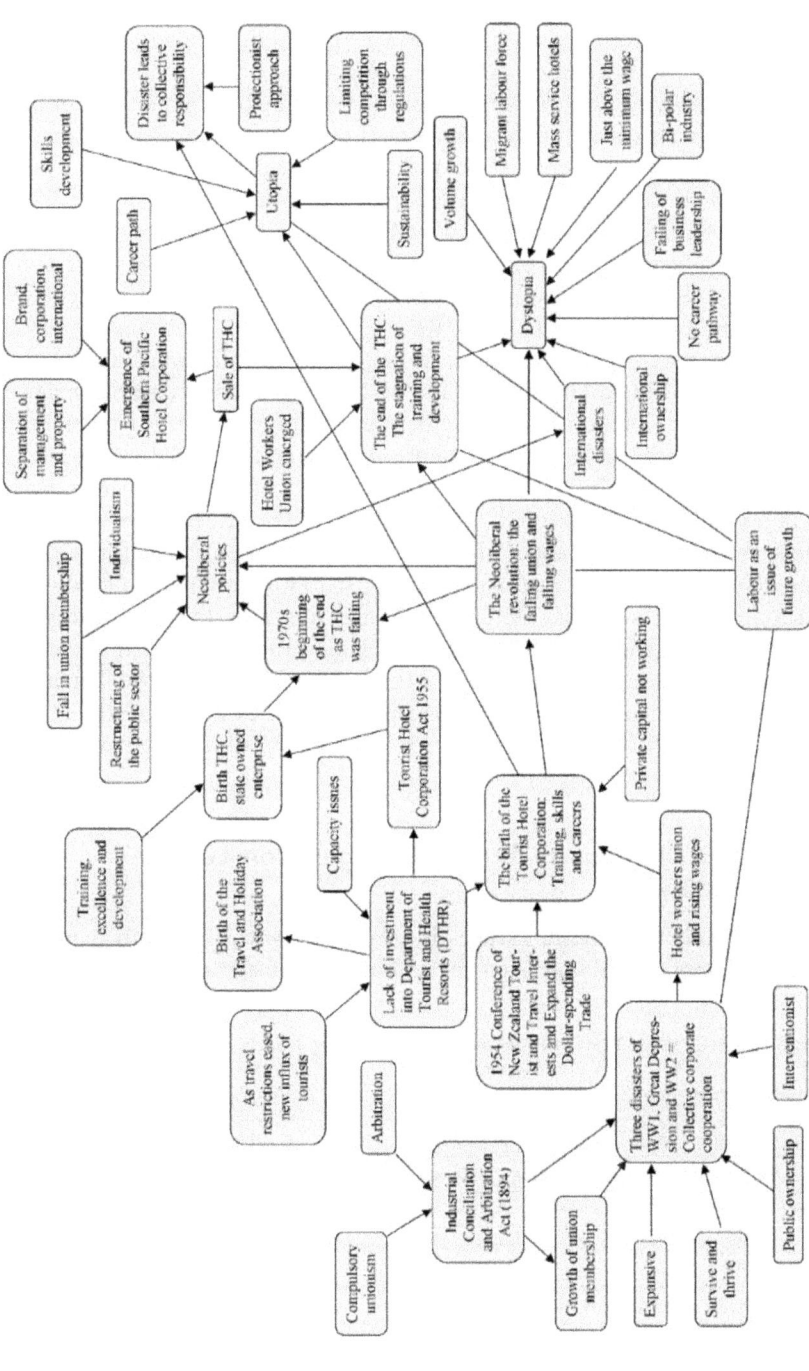

**Figure 19.11** Historical employment relations in the New Zealand tourism hotel sector: From a collective past to an individual future?

earner of export dollars. Yet despite its economic importance, labour in this sector demonstrates all of the challenges commonly associated with the international tourism workforce: low wages, high turnover, high levels of casualization, skills shortages and a dependence on migrant workers. There is a growing concern that labour issues may be a major limiter on future tourism growth and development. This chapter addresses the question: How did we get here and what does this mean for the future? Thus the central concepts identified in Figure 19.11 include: three disasters of WWI, the Great Depression and WWII = collective, corporate and cooperation; the birth of the Tourist Hotel Corporation: training, skills and careers; neoliberal policies; and lack of investment into the Department of Tourist and Health Resorts (DTHR).

The chapter takes a critical, historical employment relations approach, drawing on interviews and archival research to describe a cascading series of impacts on employment in the hotel sector. An argument is presented that the key employment relations milestones discussed are greatly influenced by, or are the direct result of, dramatic post-war changes in both international and New Zealand political, economic and social policies. The chapter highlights New Zealand's transformation from a post-war corporatist nation to a post-1980s example of neoliberal orthodoxy and suggests that this change has greatly disadvantaged the workers in the tourist hotel sector. Four milestones are discussed, covering the rise and fall of the Hotel Workers Union and the Tourism Hotel Corporation (THC). The chapter concludes by suggesting two dramatically different scenarios for the future of the tourist hotel workforce.

## Chapter 13: Film Tourism through the Ages: From Lumière to Virtual Reality

The study of tourism futures is concerned with what can happen, what will happen and how such changes affect industry and society. Bolan

**Table 19.12** Chapter 13: Historical and future turning points

| Historical turning points | Future turning points |
|---|---|
| Lumière lighting the way (1800s and early 1900s) | Hyper-reality to virtual reality |
| From *The Third Man* to *The Quiet Man* (1940s and 1950s) | Holographic technology |
| Screen impact of the 1960s | |
| *Crocodile Dundee* (1980s) and reinvigorating economies through film tourism | |
| The *Braveheart* effect and the rise of the hero (1990s) | |
| The magic of Middle Earth and Harry Potter (2000s) | |
| *Game of Thrones* (2010 onwards) | |

Does the Past Shape the Future of Tourism? 271

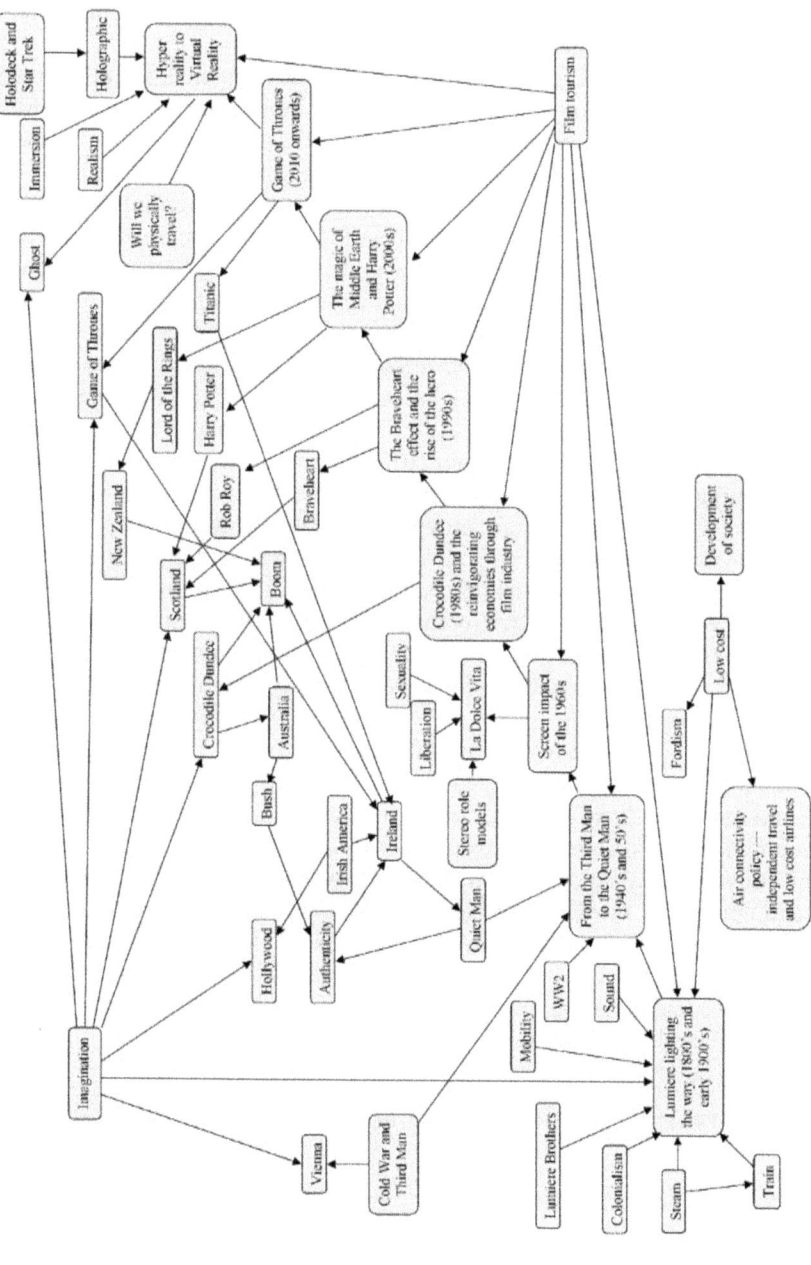

**Figure 19.12** Film tourism through the ages: From Lumière to virtual reality

and Ghisoiu's chapter addresses such issues for film tourism by examining what we can learn from key turning points in the past through the present and extrapolate these to future turning points. The core concepts identified in Figure 19.12 include: Ireland; film tourism; imagination; authenticity; New Zealand; and *Crocodile Dundee*.

Since its initial presentation to mass audiences in the late 1800s, the medium of film has undergone a myriad of adaptations across the 20th and 21st centuries impacting profoundly upon the viewers' choice of travel and their tourism experiences. The chapter examines the influences and significance of film tourism through history, from the Lumière brothers' first screening in 1895, through 1940s classics such as *The Third Man*, the appeal of Ireland to the American diaspora with *The Quiet Man* in the 1950s, through the newly found sexual liberation of the 1960s and the 'glamification' of cities such as Rome, the establishment of Australia on the tourism map through *Crocodile Dundee* in the 1980s, the effect on Scottish tourism of the rise of folk heroes in the 1990s, entering the era of magic and mystery in the 2000s via Lord of the Rings and Harry Potter, followed by *Game of Thrones* from 2010 onwards. Although initially happening organically from the 1940s to the 1980s, destinations soon started to realize the economic impact movies/series could have on tourism. The early 2000s brought renewed interest in the phenomenon with New Zealand and The Lord of the Rings, triggering a large number of studies on the phenomenon. Finally, Northern Ireland and the success of *Game of Thrones* not only showcases the huge impact this can have on a rather unassuming destination but also paves the way for future trends and insights in film tourism.

As part of post-modernist discourse it is clear that the future of film tourism is linked to advances in technology. The growing influence of digital media on how viewers, especially Generations Y and Z, consume their film/television content (through digital streaming services, on mobile devices, wearable tech and evolving elements of AR, VR and holographic technology) will fundamentally change the nature of such content itself and in turn the impact it will have on the tourist of tomorrow. As such the chapter contributes to key turning points in film tourism from a historical perspective, culminating in the future trends that have already started to impact this industry sector.

## Chapter 14: The Evolution of the Grand Tour in the Digital Society

Seeler's chapter provides an overview of the development of tourism since the times of the Grand Tour in Europe in the 17th and 18th centuries. The key concepts identified in Figure 19.13 include: the Grand Tour; luxury; authenticity; Nazi Germany; transport; wealth; and the beginnings of mass tourism.

**Table 19.13** Chapter 14: Historical and future turning points

| Historical turning points | Future turning points |
| --- | --- |
| The Grand Tour of Europe – the foundation of leisure travel | German experienced tourists – tomorrow's Grand Tourists |
| The era of Thomas Cook – the development of mass tourism | Co-creating educational travel – the future of *Studienreisen* |
| The changing image of tourism – from education to political power and consumerism | Digital narratives – the travel writings of tomorrow's Grand Tourists |
| Travelling to be educated – drowning in information and starved for knowledge | Wearable, virtual and augmented reality – new forms of knowledge creation and experience consumption |
| The evolution of the Grand Tour – daring to predict tomorrow's tourists | |

The chapter describes the early explorers and their travel motives and explains the spatial and temporal developments of the Grand Tour. It also addresses the declining significance of the Grand Tour as a result of technological advancements during the Industrial Revolution and the emergence of a leisured society in the early 19th century. Thereby the chapter brings the development of mass tourism and the changing motivation of tourists into focus. With an emphasis on the development of tourism during the Great Wars in Germany, Seeler reflects on the changing image of tourism and the utilization of tourism to shape national identity and mobilize political ideology. The chapter then provides insights into the sustained growth of tourism since the 1950s and the diversification of travel motives of modern tourists which led to travel typologies. Changing the lens from history's tourist to contemporary tourists, the chapter sheds light on the digitized travellers and their continuous desire for knowledge enhancement and self-development. The chapter describes today's knowledgeable tourists as increasingly experienced and highlights changes in travel behaviour when higher levels of prior travel experiences are achieved. In this regard, the chapter draws attention to the role of *Bildungsreisen/Studienreisen* (educational travel/study tours) in the German travel market.

Lastly, the chapter establishes a relationship between early forms of travelling when educational purposes, identity formation and personal growth were the dominant motivation to travel and today's increasingly experienced tourists who seek similar outcomes when engaging in leisure travel. The chapter shows not only a resemblance between history's Grand Tourists and today's experienced tourists in regard to their dominant travel motivation, namely educational purposes and personal growth, but it also demonstrates the critical role of travel writings throughout time. The chapter argues that despite changes in the communication channels, reach and purposes of travel narratives, they remain an essential element

274  Part 6: Evolution

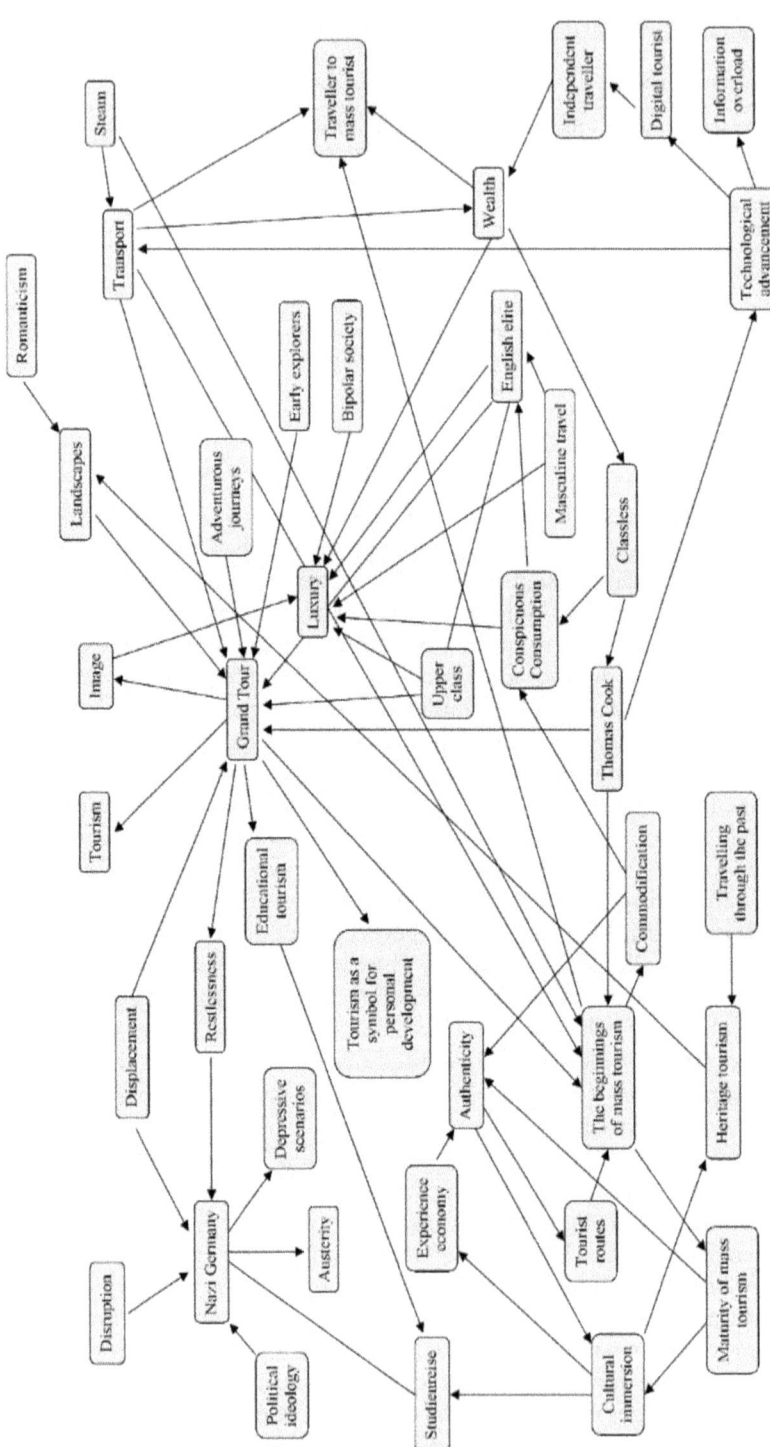

**Figure 19.13** The evolution of the Grand Tour in the digital society

in travel decision making and behaviour and subsequently have an impact on the overall development of tourism. The chapter conceptually contributes to the existing body of knowledge as it provides a comparison between travellers of different travel epochs and establishes relationships between history's Grand Tourists and today's experienced tourists. It encourages the reader to understand tourism as iterative and to consider the past to understand the future.

## Chapter 15: Shopping on the Edge: Identifying Factors Contributing to Tourist Retail Development in Heritage Villages

Tourist shopping villages (TSVs) are small towns that base their visitor appeal on heritage and retailing (Getz, 2000). As researchers have begun to focus attention on this phenomenon, some of the following issues have been identified: negative environmental and social impacts; authenticity and commodification; a failure to provide expected benefits to regional communities; and undesirable changes in the nature of the village and retail experiences offered to tourists (Murphy *et al.*, 2011). One option to address these challenges is to understand the different evolutionary pathways that have led to particular outcomes, either positive or negative, for both visitors and the local communities in which they are situated. In this chapter, Moscardo and colleagues report on historical case studies of three TSVs: Hahndorf in Australia, St Jacobs in Canada and Cheddar in England. These three villages are well-established tourist destinations with diverse tourism development histories. The chapter identifies the following concepts in Figure 19.14: local retail shifts attention from locals to tourists; disruptive, creative enhancement and/or rejuvenation; tourism reaches critical mass; and property development.

The chapter's central contribution to the study of tourism futures lies in demonstrating that a historical or evolutionary approach uncovers a wider range of underlying processes than cross-sectional analyses. Decisions made by tourism governance organizations that encouraged

Table 19.14 Chapter 15: Historical and future turning points

| Historical turning points | Future turning points |
|---|---|
| The discovery and establishment of a tourist attraction | The emergence of new Asian TSVs |
| Local retail shifts attention from locals to tourists | A return to the rural idyll |
| Arrival of external entrepreneurs and departure of local residents | |
| Tourism reaches a critical mass | |
| Disruption, creative enhancement and/or rejuvenation | |

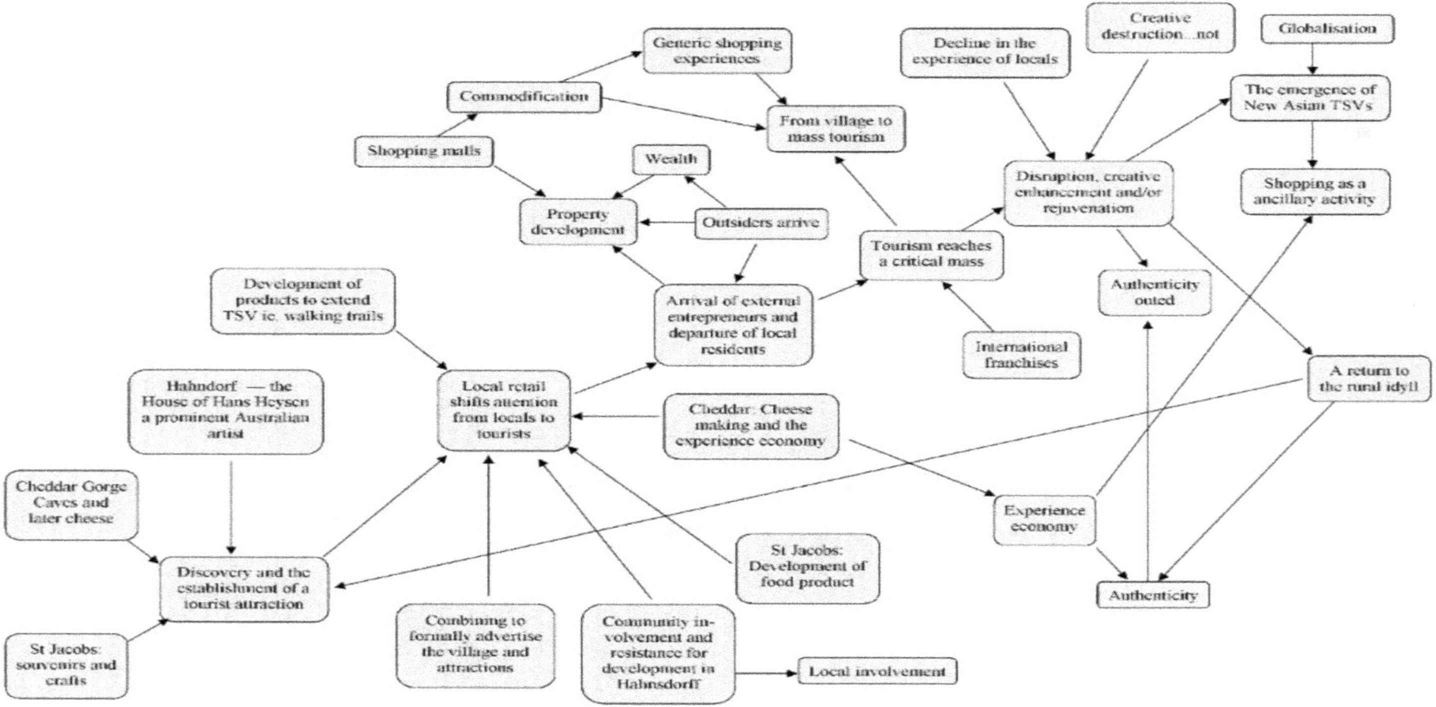

**Figure 19.14** Shopping on the edge: Identifying factors contributing to tourist retail development in heritage villages

growth without consideration of impacts, a shift in attention from locals to tourists in retail and services, and supported external investment to support growth all contributed to the turning points presented in the chapter. In addition, the chapter demonstrates the importance of looking at tourism within the wider context of economic and cultural changes. These contributions provide tourism decision makers with insights into the unexpected negative consequences of decisions made at these turning points and serve as a reminder that decisions need to prioritize the wellbeing of local resident communities. An examination of the history of TSV development is therefore critical in understanding the links between decisions and consequences and guiding the future evolution of the sector.

## Chapter 16: Tourism and Religion: Pilgrims, Tourists and Travellers – Past, Present and Future

The major contribution of this chapter by Butler and Suntikul is in clarifying the roles that the world's organized religions have played in motivating tourism and influencing its past and current patterns and practices. Figure 19.15 highlights the following concepts central to the authors' arguments: the future; maturity of mass tourism; pilgrimage; and sites of significance, e.g. Lourdes. Six of the world's major faiths are examined in terms of their roles in and influences on tourism with specific reference to major historic events and the continued relevance of these for present and future tourism. These events include seminal historical religious occurrences that continue to inspire pilgrimage and other travel and tourism in the present day, as well as the ways in which annual or cyclical events in the religious calendar determine many aspects of travel patterns and practices. The chapter then discusses the ways in which modern societal, political and technological factors are shaping traditional religious tourism practices.

The focus then moves to the identification of possible future events and turning points in these religious traditions which may be expected to have an influence on both international and domestic tourism globally in the future. Finally, the chapter reviews current trends and developments related to religion including patterns of growth and decline, demographics and geographical distributions, and related sociopolitical entanglements,

Table 19.15 Chapter 16: Historical and future turning points

| Historical turning points | Future turning points |
| --- | --- |
| Births of specific individuals | Discoveries and events |
| Division | Peace |

278 Part 6: Evolution

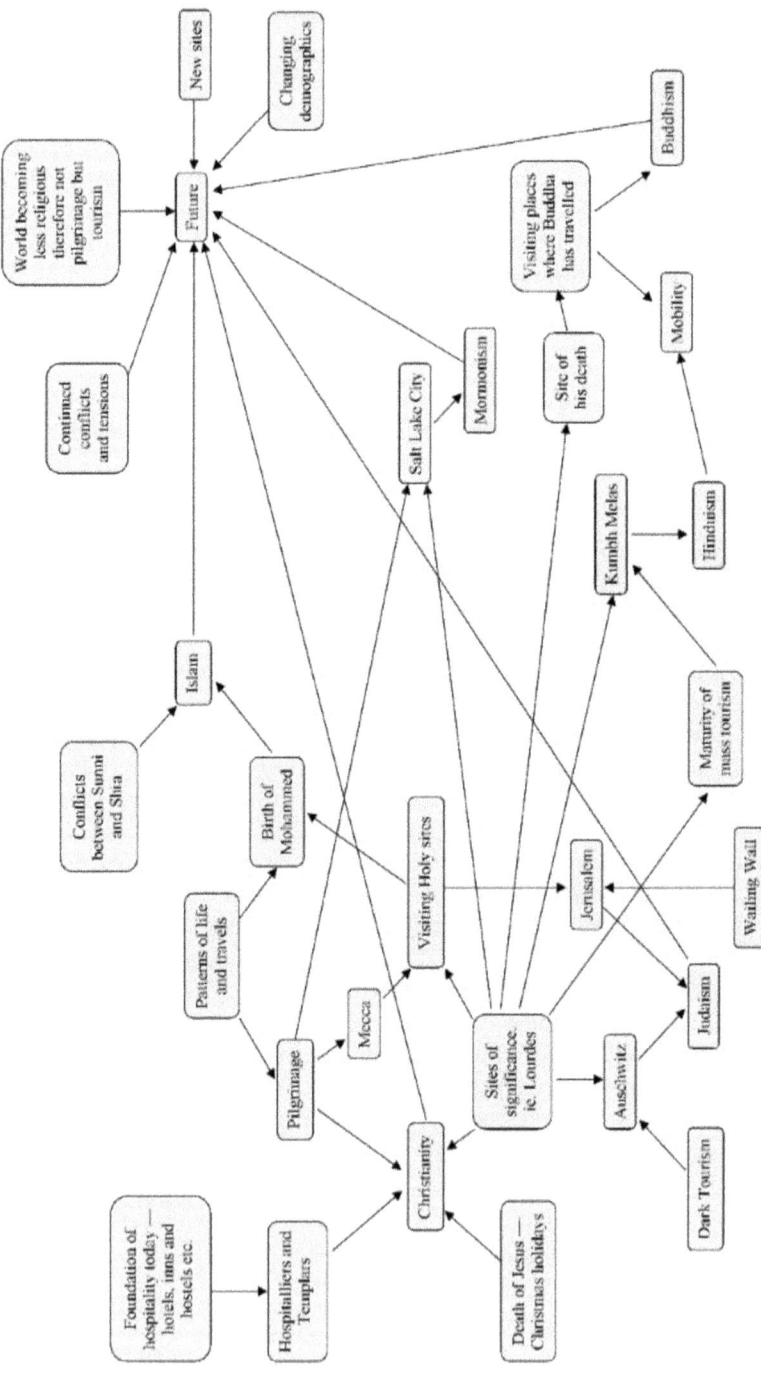

**Figure 19.15** Tourism and religion: Pilgrims, tourists and travellers – past, present and future

which are also likely to shape the future face of tourism at specific time periods in the future. It concludes that religions, and their influences on both tourism and the world in general, are inextricably linked to political realities, and that any major future religious event of global significance will almost certainly have major political ramifications, which will in turn have additional implications for religious tourism and the tourism industry more broadly.

## Chapter 17: The History and Future of Mountaineering Tourism

Researchers have devoted their efforts to conceptualizing mountaineering tourism by studying it from various aspects and perspectives such as activity, people and place, the mountain tourism experience, environments, communities and sustainability. Previous studies present mountaineering tourism in relation to geography, social and historical development, issues related to the person involved in mountaineering and environmental management. In Figure 19.16, the core concepts that emerge from this chapter by Musa and Sarker include: modern mountaineering (tourism era); the beginnings of mass tourism; 'the last frontier' of mountaineering; and mountaineering during the golden era (1854–1865).

This chapter accounts for mountaineering tourism's historical development in order to understand its future by identifying a number of turning points including the early era, the emergence of the golden era (when mountaineering emerged as an adventurous activity), mountaineering's post golden era (when the sport went global), the climbing of the last frontiers, and the modern tourist era of mountaineering. The chapter concludes with future forecasts.

The central contribution of the chapter for the future of mountaineering tourism is that despite the grim effect of global warming, mountaineering tourism activity will remain resilient. There will be entrepreneurial adaptations that witness a variation in the mountaineering activities offered in the Alpine regions. Technology will improve in terms of equipment, medicine and nutrition, allowing greater access among

Table 19.16 Chapter 17: Historical and future turning points

| Historical turning points | Future turning points |
|---|---|
| Mountain climbing in the early era (before 1854) | Climate change |
| Mountaineering during the golden era (1854–1865) | Entrepreneurial adaptations and technology |
| Mountaineering after the golden era (after 1865–early 20th century) | |

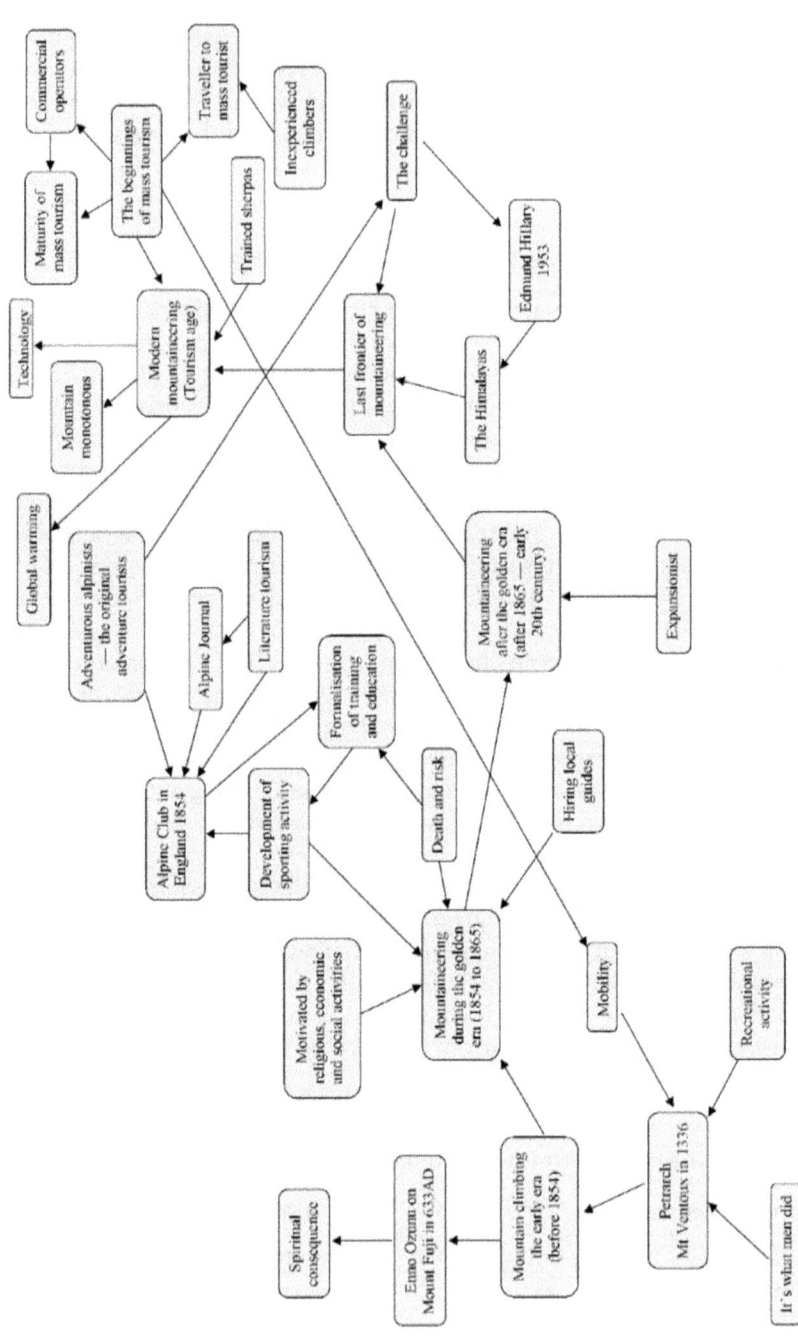

**Figure 19.16** History and future of mountaineering tourism

mountaineers to remote locations in mountain regions. In addition, various technological inventions will make mountaineering activities much safer in the future.

## Chapter 18: Sustainability, Ecotourism and Scotland: Concerns, Complaints, Conflicts and Conservation

In this chapter, Durie and colleagues examine past incidents and issues that are relevant to the debates about the future of sustainable tourism in Scotland. In Figure 19.17, the chapter identifies the following reoccurring themes: societies and trusts; hospitality and manners; culture of mass tourism; wealth; maturity of mass tourism; and the beginnings of mass tourism.

The authors use a historical analysis and archive search to identify turnings and incidents associated with Scottish tourism from an environmental and sustainable perspective. These include the abuse and degradation of Scotland's landscapes by early tourists, and the impact of tourism on the physical, natural and cultural environment focusing on incidents relating to vandalism and amateur science. The chapter continues by exploring the beginnings of organized tourism and the development of mass tourism. It examines 'Scotland' as a cultural resource and the role of game, societies and trusts. As in the past, climate change and sustainability dominate future discourses about Scottish tourism. The significance of this chapter reinforces Hobsbawm's philosophy that the future is a reoccurrence of the past. The chapter identifies several incidents in Scottish tourism that connect to the debates about communities, overtourism, conservation and access, thereby reminding us that to understand the future, you only have to look to the past.

**Table 19.17** Chapter 18: Historical and future turning points

| Historical turning points | Future turning points |
|---|---|
| Vandals, communities and tourism | Climate change |
| Tourism as a hobby and amateur science | A sustainable approach to the future |
| The beginnings of organized tourism | |
| Mass tourism | |
| Mobility | |
| Cultural gain – Scotland as a cultural oasis | |
| Cultural loss – Gaelic identity | |
| Hospitality and manners | |
| Land use | |
| Game | |
| Societies and trusts | |

282  Part 6: Evolution

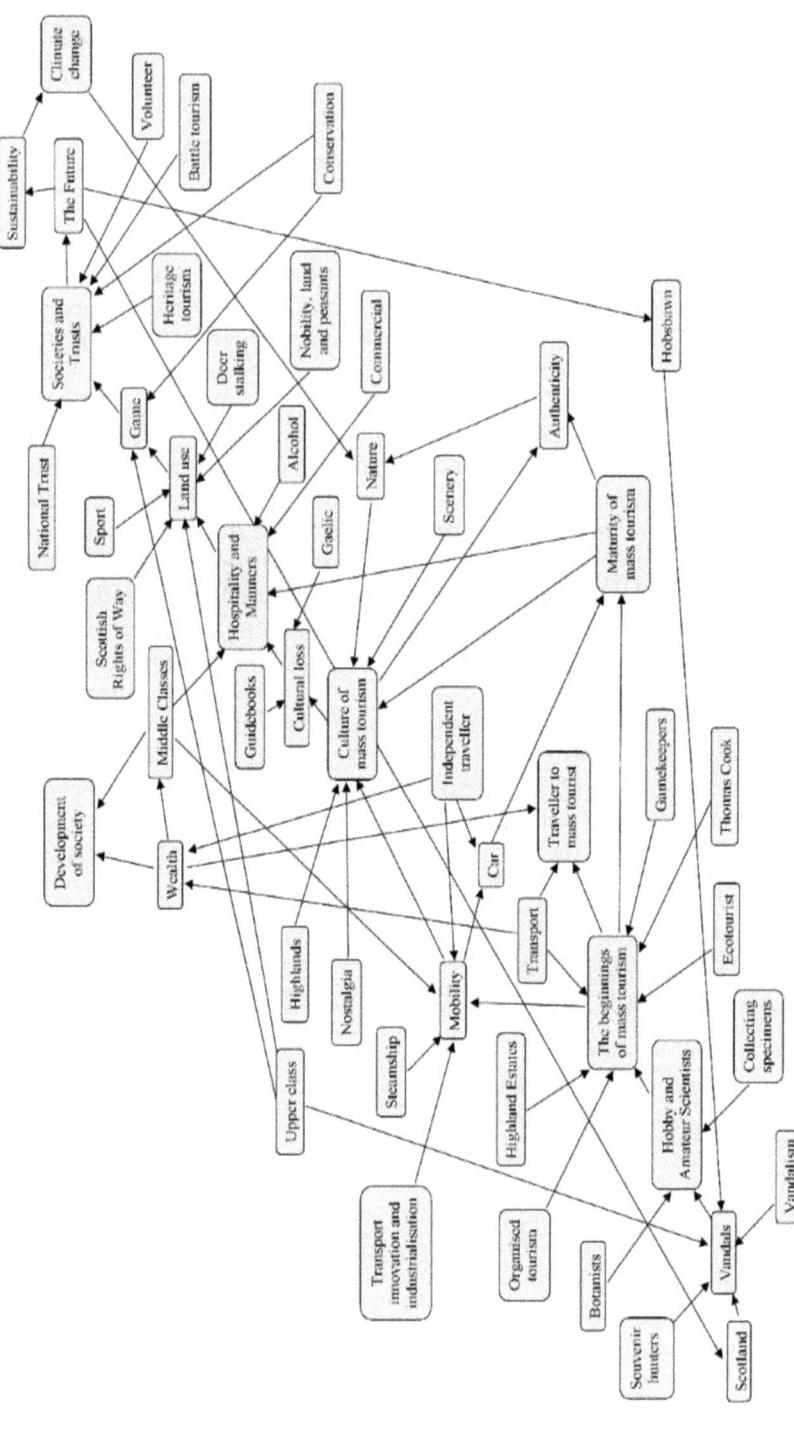

**Figure 19.17** Sustainability, ecotourism and Scotland: Concerns, complaints, conflicts and conservation

## Developing an Aggregated Map for the Future of Tourism Using a Historical Lens

The purpose of this section in the chapter is to demonstrate how the construction of the aggregate cognitive map took shape. Because of the complexity and subjectivity of the construction, the section is only an illustration of the process to guide readers' understanding of how the process happened. At this stage, all the chapters have an individual cognitive map. The merging of the individual cognitive maps into an aggregation is a process by which the researchers immerse themselves into the maps and search for concept connections driven by semantic similarity. This allows the drawing out of key concepts from each individual map and remapping

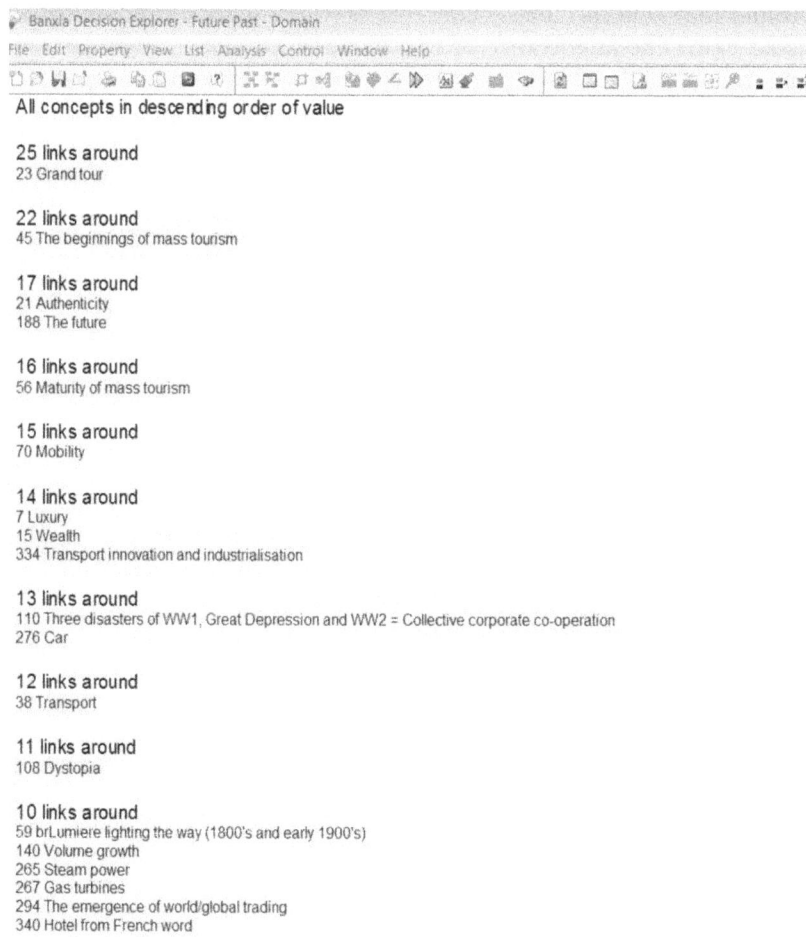

**Exhibit 19.1** Central command

the concept in Decision Explorer (DE). Once this is complete and after several iterations, an aggregate cognitive map is formed.

As this aggregate map is complex, and given that are a great number of connections, DE has a number of features which allow the breaking up of the aggregate map into viewpoints. From this, the researcher can build, explore and reflect on these maps as component parts of the total aggregate map. The 'central' command looks at specified band levels which are connected to the concepts. This allows the researchers to view the importance of the length of linkage between concepts. Each concept is weighted according to how many concepts are traversed in each band level. Fundamentally, the central command shows how many concepts are dependent upon one concept. Exhibit 19.1 demonstrates this view.

The 'domain' (Exhibit 19.2) command performs a hierarchical domain analysis which lists each concept in descending order of the linked density around that concept. Those concepts with the higher link density are listed first. The importance of the 'domain' command highlights the importance of the closeness of the local links between concepts. The researchers used both the 'central' and 'domain' commands as a means to identify the most important concepts in order to explore and construct maps. Further, both the 'central' and 'domain' commands identify a number of concepts to map in which the modeller makes a judgement to construct and explore these concepts while holding them as a central view.

Using these two commands, 'domain' and 'central', the researchers explored the concepts using DE commands such as 'show unseen links'

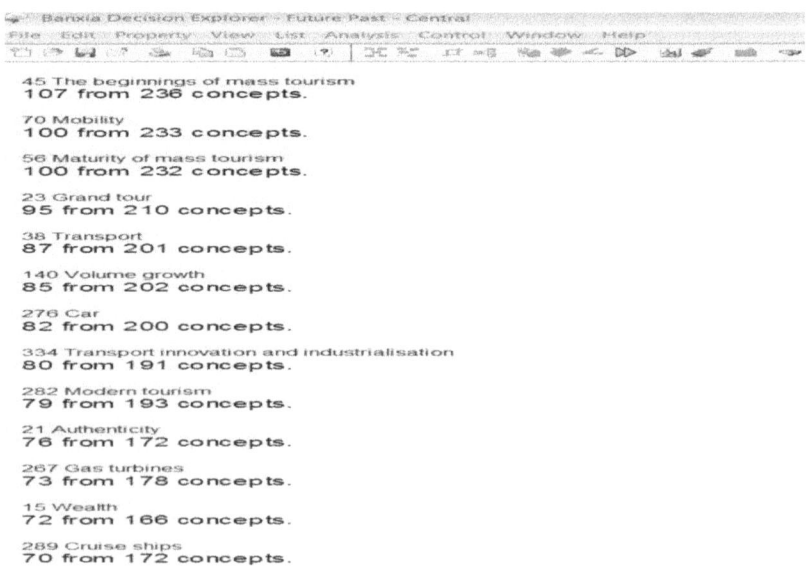

**Exhibit 19.2** Domain command

with which the modeller is able the find connections between concepts and thus start to build a cognitive map, explore links and reflect upon them. Other commands within DE can be used to recall multiple concepts that surround other concepts. This process is repeated several times until several views make sense to the researchers. Therefore, a series of concepts were identified as significant that are common to both the domain and central analysis. These concepts were 'explored' and 'mapped' using further commands in DE which resulted in four historical turning points that determine the future past of tourism: mindfulness; mobility; step changes determining mass tourism; and the leisure class of consumption.

## The Future Past of Tourism

In this section, we discuss the historical turning points identified in the previous section that determine the future past of tourism. First of all, the turning points in the evolution of tourism to the present day are: mindfulness; mobility; step changes determining mass tourism; and the leisure class of consumption. We then conceptualize and speculate on how the past will shape the future based upon the four future turning points of: *Fluid Identity; Sustainable Futures; Ubiquitous Future* and *Mass Maturity*. We then propose two scenarios: *Degradation – If Only We Had Listened to the Past* and *A Balanced Future – Learning from the Past*.

### Historical turning point 1: Mindfulness

In reaction to the always-on lifestyle, consumers are turning inward to bring more awareness to the present moment and block out any external noise and distractions. Inner wellbeing has become a new priority for health-aware consumers. Quiet and relaxation is the one of the dominant motivations and drivers for going on holiday. According to the Foresight Factory (2016b), practising quiet and mindfulness is now a mainstream method of relaxation and improving mental balance. In Figure 19.18 the core concepts that reoccur through loops are: *authenticity; nature; off the beaten track*; and *cultural immersion*. Indeed, authenticity has always been an important concept in the evolution of tourism. As Van der Tuuk in Chapter 2 notes:

> The traveller wanted to visit authentic places not affected by civil and industrial society. The dark forest, the coast, the vast sea, the quiet lake, the deep valley and the high mountain formed the new 'holy' places for the tourist.

This is a reoccurring theme in many of the chapters. Why? Authenticity is a central concept associated with mindfulness. It is one of the core concepts associated with destinations (see Chapters 3, 4, 15 and 18) and their promotion. As Yeoman *et al.* (2007) identified, there is a growing desire

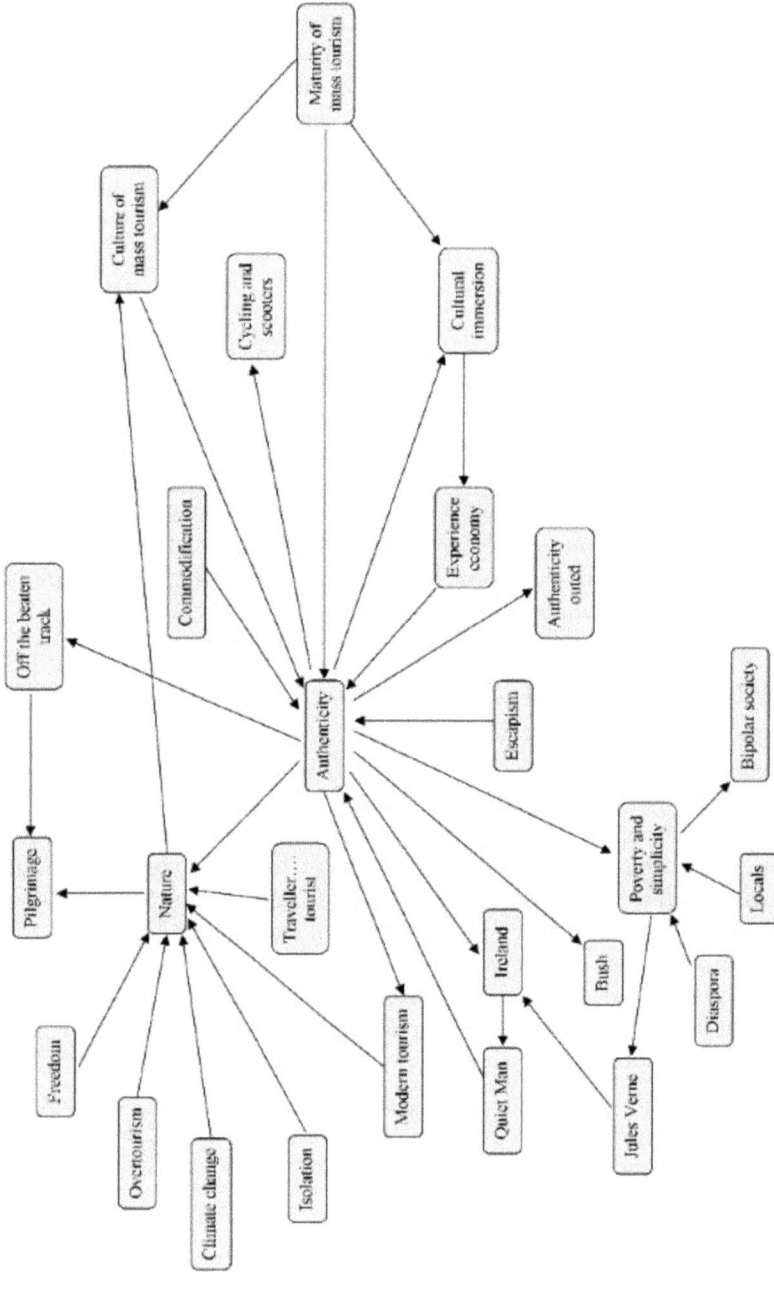

**Figure 19.18** Mindfulness

for authenticity in a complex, postmodern world and tourists will increasingly demand experiences and products that are pure, real and not associated with the homogeneity of mass tourism. The concept of authenticity is well researched in the tourism literature, with a variety of foci such as food, culture and nature (Albrecht, 2011; Boyle, 2005; Cohen & Cohen, 2012; Isaacs, 2018; MacLaren *et al.*, 2013). Yeoman *et al.* (2007) coined the term 'authentic-seeking': tourists search for authenticity in products, services and experiences or indeed within themselves (hence, mindfulness).

The trend of authenticity is about exploring the untouched and unexposed and this desire for authenticity is explored in Chapter 18 in relation to Scotland as a destination. As Durie and colleagues note:

> Scotland was one of the earliest countries to experience tourism as a large-scale activity, with a substantial part of its appeal lying in its natural resources of land and scenery, mountain and sea, air and water ...

Undeniably, nature-based tourism is concerned with the direct enjoyment of some relatively undisturbed phenomenon of nature. Those participating opt for such activities as birdwatching, kayaking and hiking, with the intention of getting in touch with nature, escaping the stresses of daily life, and seeing landscapes and wildlife (Higham *et al.*, 2016; Louv, 2008; Nyaupane *et al.*, 2004).

Looking to the future, Yeoman and McMahon-Beattie (2018) see mindfulness as symbolic of the changing nature of luxury, moving from materialism to experiences and enrichment. Right at the heart of enrichment is tranquillity, isolation and a focus on the inner self which links to the concept of spirituality (Yeoman, 2008). In Western societies spirituality is a trend as consumers search for peace of mind as an alternative to formal religions and this is alluded to by Butler and Suntikul in Chapter 16 about religions and pilgrimage.

### Historical turning point 2: Mobility

Mobility has always been a key driver of tourism, as tourism is about the movement or transportation of tourists by boat, train, car or aeroplane. As seen in Figure 19.9, mobility is identified as one of the core drivers and turning points in the evolution of tourism. This figure shows clusters centred on 'mobility', 'the beginnings of mass tourism', 'transport' and 'transport innovation and industrialization'. Certainly, Van der Tuuk (Chapter 2) and Butcher (Chapter 3) acknowledge that the beginnings of mass tourism were due to the industrial age, growing wealth and developments in transportation. Significant technological evolution in transportation can be seen in the growth of the railway industry in the 19th century, the start of the mass production of the motor car (1900–1945) and the production of jet aircraft from the 1960s onwards (Chapter 8).

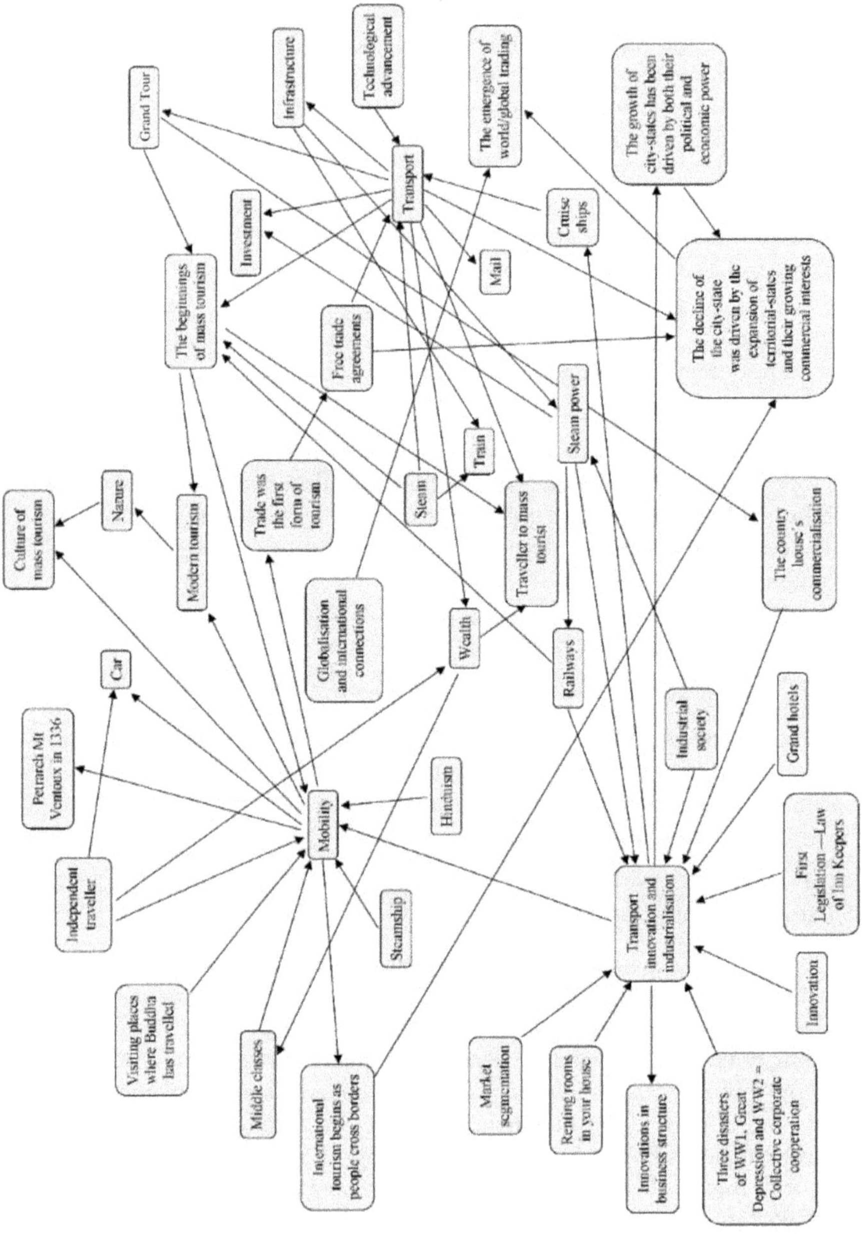

288  Part 6: Evolution

**Figure 19.19** Mobility

Even small destinations and islands like Malta (Chapter 4) have been the beneficiaries of transportation evolution. The same can be said for city-states and urban destinations. Hay highlights this in Chapter 7 and notes the importance of connectivity and integrated transport systems. In Chapter 10, Hui provides a detailed discussion about technological advancement in transportation, highlighting the correlation between the arrival of mass tourism and the growth of coal and oil economies and transport systems. He notes that tourism is an oil-dependent sector which is historically susceptible to shocks such as the continued oil crises of the 1970s. Indeed, transport as an experience is nowhere better discussed than in London and Wallace's chapter (Chapter 9) on cruise tourism. Cruising has been the beneficiary of shipbuilding innovations leading to the building of mega ships as floating hotels and resorts. In conclusion, Page et al. (2010) argue that transport is an integral part of tourism. They note that, according to the World Economic Forum, Switzerland is the most competitive tourism nation in the world with transport as the key contributing factor. What we see is that transport and mobility have been among the core drivers of the evolution of tourism and this will continue into the future.

## Historical turning point 3: Step changes determining mass tourism

Figure 19.20 identifies several clusters that tell the story of the evolution of tourism from the 'Grand Tour' through to the 'beginnings of mass tourism' and the 'maturity of mass tourism'. The earliest tourists were pilgrims (Butler & Suntikul, Chapter 16). Here the purpose of travel was religion driven by annual or cyclical events in the religious calendar, for example, the *Hajj* which is an annual Islamic pilgrimage to Mecca, Saudi Arabia. However, the beginnings of tourism for reasons of leisure, adventure or curiosity was evident in the Grand Tour which is mentioned throughout this book as a core turning point in the evolution of tourism. As Zuelow (2016) states:

> The Grand Tour is generally associated with England's so-called landed elite and with the education of young nobles, so much so that one recent historian claims 'the Grand Tour is not the Grand Tour unless it includes the following: first, a young British male patrician ...; second, a tutor who accompanies his charge throughout the journey; third, a fixed itinerary that makes Rome its principle destination; fourth, a lengthy period of absence, averaging two to three years.' While it is true that far more young Englishmen traveled around Europe than did members of other nationalities and that much of the debate that surrounded the trip centered on its pedagogical role, the reality is that this definition is too confining. (Zuelow, 2016: 16)

In support of this, Van der Tuuk (Chapter 2) identifies travel in the 17th and 18th centuries as a matter of leisure, a clear turning point away from

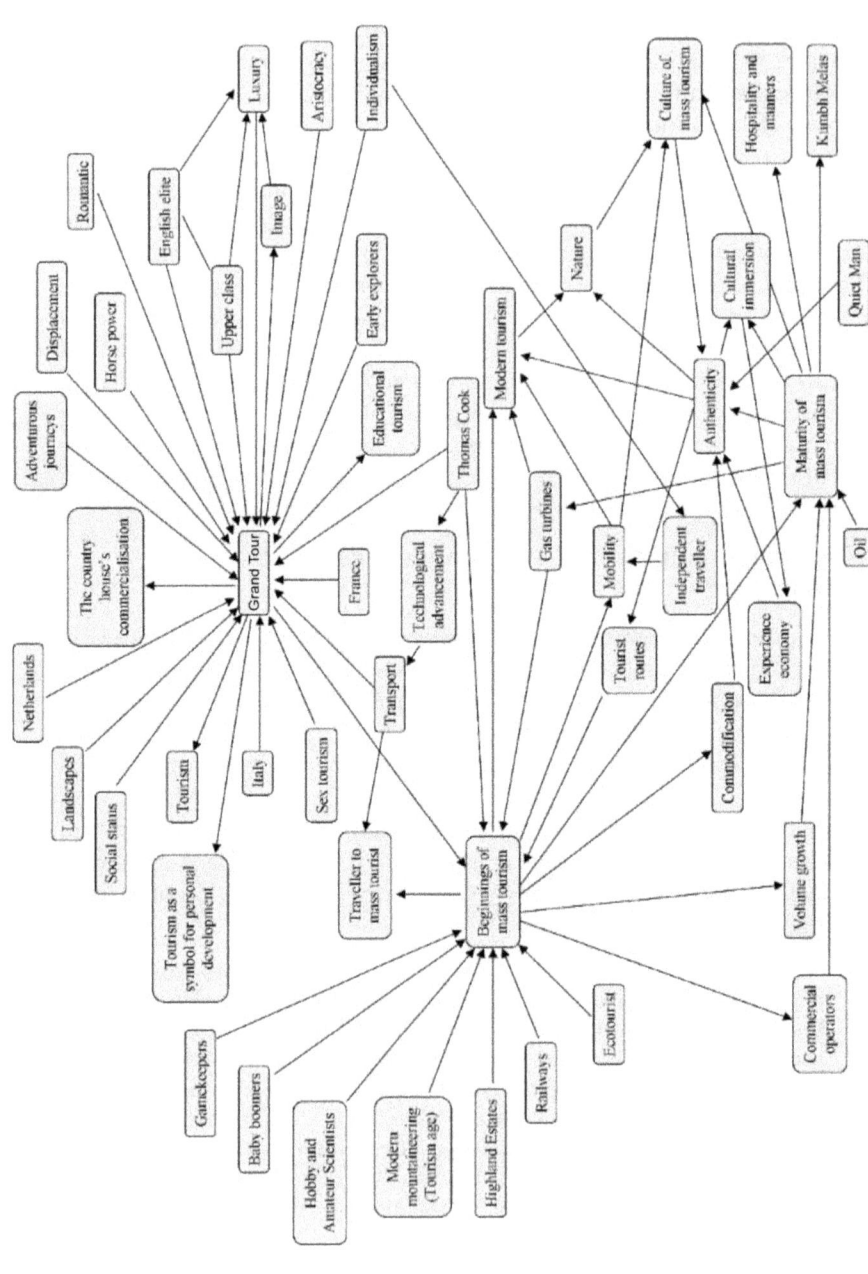

Figure 19.20 Step changes determining mass tourism

pilgrimage or trade. The Grand Tour created demand and, on the supply side, we see the development of tourist hotels (James, Chapter 11). These first tourist hotels were elaborate mansions and were seen as high-status lodgings. Certainly, many of these hotels tended to replicate the amenities and atmosphere of aristocratic abodes. Outside the cities, accommodation would often be more rustic (or authentic in modern tourism academic language). Seeler, in Chapter 14, provides an overview of the Grand Tour, drawing out the significance from a researcher perspective, as these tourists were distinguished by their social class and the route they took. Seeler goes on to identify the decline of the Grand Tour due to the emergence of rail and steamships in the 19th century and the emergence of the working and middle classes taking holidays because of the Industrial Revolution and a restructuring of society. The beginning of mass tourism was upon us, with the emergence of a tourism industry. Notably, the first travel agency was initiated by the British businessperson and explorer Thomas Cook in the 1860s. Cook advocated free trade and he claimed that travel was for everyone and needed to be enjoyed beyond the social elites (Berghoff & Korte, 2002).

As noted earlier in this chapter, transport was a core driver in the development and maturity of mass tourism. Butcher (Chapter 3) highlights the importance of the development of tourism infrastructures in the rise of intercity transport systems. The advent of the motor car post-WWII and its affordability saw mass tourism come of age, although it was the emergence of aircraft technologies that was the turning point for the maturity stage of mass tourism. By 1958, the aeroplane had overtaken cruise ships as the preferred form of transport between North America and Europe (Chapter 8). Notably the Russian émigré, Vladimir Raitz, ran the very first package holiday flight from England to Corsica in 1950. After this we saw the first package holidays and the emergence of charter flights to the Spanish Costas (Bray & Raitz, 2001; Durie, 2017; Zuelow, 2016).

The next step change came with the development of aircraft technology which supported long-haul travel. Castrol and colleagues in Chapter 8 trace the development of the Kangaroo Route between Australia and the UK, connecting the mother country with the colonies. This chapter overviews the Lockheed Constellation era with multiple stopovers, to the emergence of the Boeing 707 and 747, and then the Airbus A380. This brings us to the present day and the introduction of ultra-long range (ULR) flights such as Doha to Auckland and Perth to London. The maturity of mass tourism also represented a change in why tourists travelled, i.e. for pleasure. Additionally, Van der Tuuk (Chapter 2) notes that travelling had become the way of showing who one was and where one belonged. Mass tourism has meant that travel is safer, more convenient, faster and for the masses, whereas for the Grand Tourist the opposite was true.

### Historical turning point 4: The leisure class of consumption

The intertwined concepts of 'luxury' and 'wealth' are central to the development of leisure class consumption as depicted in Figure 19.21. Important here in terms of our understanding of these concepts is Thorsten Veblen's (1899) book, *The Theory of the Leisure Class*. Scott (2014) notes that the leisure class:

> ... is argued to be the product of the competitive struggles of modern business in industrializing America: 'absentee ownership' has isolated it from the 'instinct of workmanship' which Veblen believed was essential to continued technological development of societies. Instead, members of the class are engaged in continuous public demonstrations of their status, a process which Veblen terms 'conspicuous consumption'. This is a form of hedonism involving the ostentatious display and waste of possessions and goods. Documenting the patriarchal character of the class, he showed that women were among the 'objects' put on show as symbols of wealth. This has a specific form which Veblen termed 'conspicuous leisure'. He remarked that leisure itself, although costly, is invisible and offers no particular status advantage. In order to attract public admiration, leisure must be taken in ways that are both wasteful and highly visible – as for example casino gambling, or the use of expensive leisure products like resort clothes, sporting equipment, and the like, which signal wealth and status. (Scott, 2014: 415)

Veblen equated free time and conspicuous consumption with the decadent display of social status (Wakeman, 2012). Seeler (Chapter 14) portrays conspicuous consumption with reference to the Grand Tours of the wealthy classes travelling around the Protestant Northern countries. The Grand Tour is synonymous with the word 'hôtel' which James explores in Chapter 11. The word denotes a large style of private accommodation that was also used for large-scale public edifices and resembled the mansions and chateaus of the aristocracy such as the 'hôtel de ville' ('city hall'). By the late 18th century hotels, from a style perspective, were opulent in their décor, delivering personalized and attentive service. The fare was abundant, fashionable and often foreign.

While Grand Tours represented the very beginnings of mass tourism, the Industrial Revolution changed the image and clientele base of hotels. Luxury became democratized as a middle class emerged and hotels began to become more diverse in their standards. As an example, the growth of seaside resort hotels with comparatively modest amenities responded to the desire of a new market to reproduce elite leisure patterns at a lower cost.

London and Wallace (Chapter 9) argue that some would describe ships as hotels and/or resorts at sea. The focus on luxury and exemplary service is highlighted in Cunard's first ship, the *Britannia*, which sailed from Liverpool to New York with a cow on board to provide passengers with fresh milk. In 1880, SS *Ravenna* became the first ship to be made of a steel superstructure. Indeed advancements in marine engineering supported the

Does the Past Shape the Future of Tourism? 293

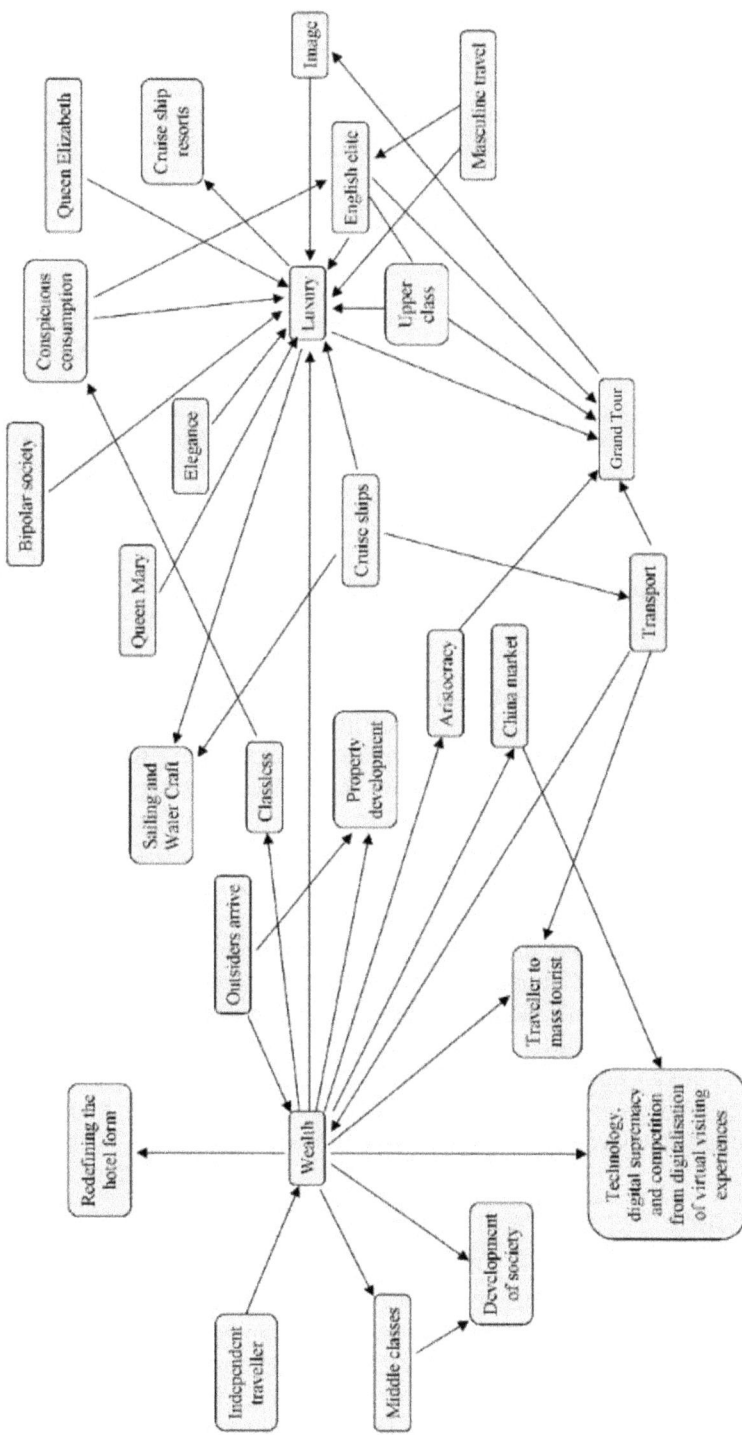

**Figure 19.21** The leisure class of consumption

development of even bigger luxury ships with swimming pools, gymnasia, opulent theatres and expansive dining rooms. Cruise ships became symbols of status and class distinction. The upper classes represented Veblen's conspicuous consumption of leisure, while the lower steerage classes carried immigrants who provided their own food and slept in the hold wherever they could find room.

In more recent years the cruise ship has become a symbol of mass tourism and the democratization of luxury. London and Wallace note in Chapter 9:

> The cruise lines came of age in the 1980s, with many new ships being built or refurbished from older ships, new cruise lines being established, and others changing their names through consolidation or mergers. Cruise ships became increasingly self-contained, capturing passenger shopping revenue with new on-board shops and other facilities. ... As a result of the expansion into non-traditional demographic groups, at least one cruise company (Royal Caribbean Cruise Lines) began to deploy increasing features and technology-rich mega cruise ships which could be mistaken for land-based resorts. New passengers continue to be attracted to cruising from the growing middle classes in India, China and Russia (Branchik, 2014), with some ships being retrofitted to cater to Asian tastes.

Today tourism can be seen to provide ample opportunities for conspicuous consumption. As consumers become wealthier and better educated, they like to boast about where they have travelled to and what they have been doing. Being on an exotic holiday in the Maldives or at an Elton John concert provides great social media content for Facebook or Twitter.

Looking to the future, Yeoman (2012) highlights the rise of the middle classes in China, India and southeast Asia, noting that that that they are dominating growth patterns in global tourism. Certainly, China is rapidly moving to become the number one inbound market for many countries. Notable here is the importance of China for the luxury-goods market globally, with huge demand from the Chinese at home and abroad on holiday (Yeoman & McMahon-Beattie, 2018). In contrast, Western tourists are seeking an Eastern lifestyle which from a luxury tourism perspective focuses on inconspicuous consumption, tranquillity and enrichment – the very opposite of Veblen's leisure consumption theories.

### The Future

So, what does the past tell us about the future? Figure 19.22 highlights the integration and connection between the concepts of tourism that have been identified through the series of cognitive maps presented in this chapter. What we see is how tourism has become democratized as a result of major changes or happenings in society. For example, the Industrial Revolution was the catalyst for the development of the steam train which

Does the Past Shape the Future of Tourism? 295

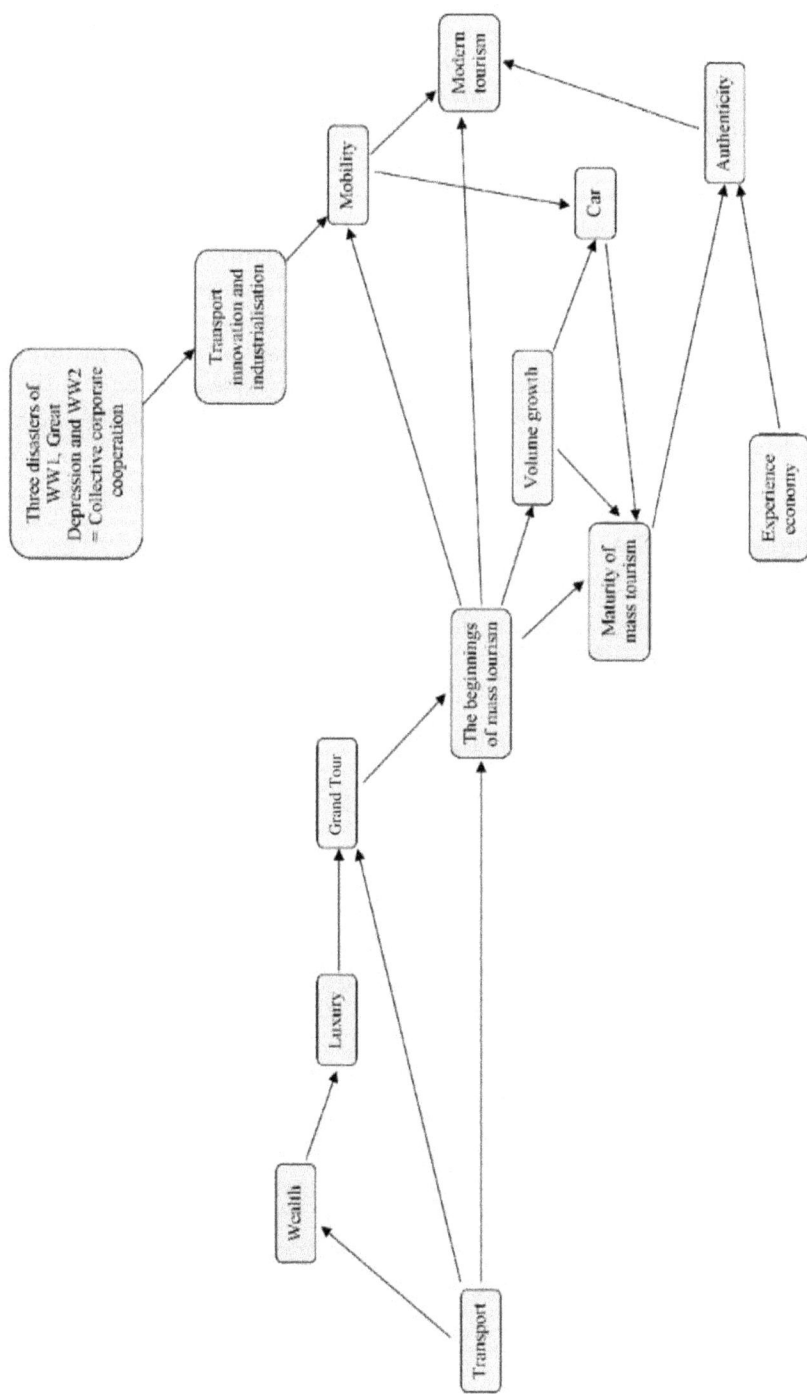

**Figure 19.22** Future past

enabled new tourists from the middle and working classes to travel further afield. Further technological developments in aviation have meant that tourists can travel long distances in relatively short time periods (e.g. the Kangaroo Route). This type of advancement in technologies was certainly a game changer as the cost of plane travel fell in real terms, making flying not a form of luxury but a commodified product. With the onset of mass tourism so was born a large infrastructure of distribution channels and supply such as travel agents, tour operators, travel intermediaries, airlines, hotels, resorts and destination management organizations. However, there are some things that do not change, that is, the purpose of travel and why we go on holiday. Tourism is about adventure, connecting with family, mindfulness, relaxation, hedonism, enjoyment and cultural engagement. The motivation to travel and tourist behaviours remain the same. It is just that, as the past moves into the future, the number of tourists has grown exponentially.

**Future Turning Points**

Figure 19.23 represents a series of future turning points and scenarios. The future turning points are: *fluid identity; sustainable futures; mass maturity;* and *ubiquitous future*.

### Future turning point 1: Fluid identity

Rising incomes and individual wealth accumulation have altered the balance of power in tourism from suppliers to consumers. As a result of the changes in distribution channels moving from the high street travel

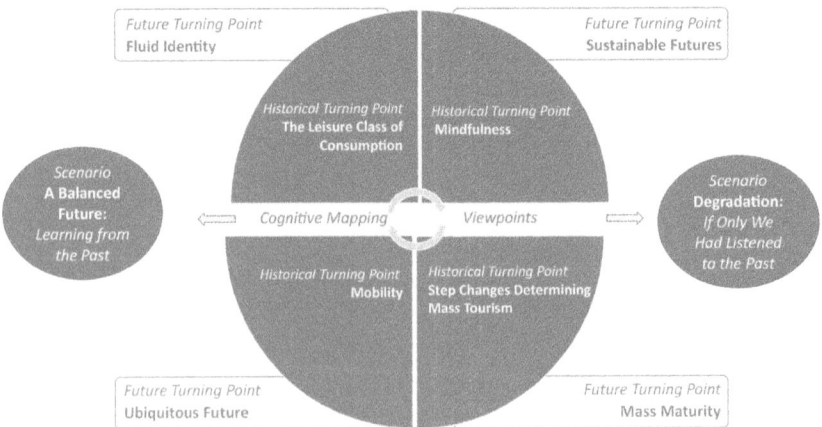

**Figure 19.23** A conceptualization of tourism's future

agent to opaque booking websites the tourist now has the power to shop around multiple land-based and online offerings looking for the best deal At the same time, today is rich for new forms of connection and association, allowing a liberated pursuit of personal identity that is fluid and much less restricted by influences of background or geography. This society of networks has in turn facilitated a mass of innovative options provided by communication channels leading to the paradox of choice. As Seeler points out in Chapter 14, the new 'Grand Tourist' will be defined by a duality of experiences:

> The increasingly connected, engaged and knowledgeable tourists aim to actively participate in the creation of their subjective meaningful experiences (Campos *et al.*, 2018). These experiences will contribute to the desired transformation of self-identity sought by tomorrow's Grand Tourists. At the same time, experienced tourists pursue personal enrichment, want to be challenged in new places, want to define boundaries and test personal limits.

In the future market place, tourists can holiday anywhere in the world. They can vacation in Afghanistan or Las Vegas, take a trip to the North or South Poles and even a day trip into outer space with *Virgin Galactic* (Yeoman, 2008). The concept of fluid identity is supported by Boztug *et al.*'s (2015) research on the hybrid tourist who challenges the concept of market segmentation. In consumer studies the hybrid consumer buys cheaper generics and low-end brands but trades up on some occasions. The hybrid consumer likes to sample and try new experiences and does not have a brand preference (Ehrnrooth & Gronroos, 2013; Silverstein *et al.*, 2003). Boztog *et al.* (2015) emphasize that hybrid tourists' purchases vary dramatically from one buying occasion to another given the different situations and circumstances.

### Future turning point 2: Sustainable futures

As Hobsbawm (1995) reminds us, debates about the future are just a recurrence of the past; it is just that the actors are different. This is the case with sustainable tourism in Scotland as noted by Durie *et al.* in Chapter 18. Tourism contributes significantly to the global gross domestic product and is forecast to grow at an annual 4%, thus outpacing many other economic sectors. In Chapter 10 Hui raises the issues of climate change and the human footprint on tourism. According to Lenzen *et al.* (2018), between 2009 and 2013 tourism's global carbon footprint has increased from 3.9 to 4.5 $GtCO_2e$ (four times more than previously estimated), accounting for about 8% of global greenhouse gas emissions. Terms such as 'overtourism' or 'tourismphobia' have also made headlines in recent times. They reflect the challenges of managing growing tourism flows into urban destinations and the impact of tourism on cities and their residents (Koens *et al.*, 2018). Tourism will only be sustainable if it is developed and managed considering both visitors and local communities. This can be achieved through community engagement, congestion management, reduction of seasonality, careful planning that respects the limits of capacity and the specificities of each destination, and product diversification. This is not a new debate. Durie *et al.* recalls (Chapter 18) in relation to the impact of mass tourism at the Scottish seaside that:

> There was ... concern over the impact of mass tourism at the seaside, in terms of the behaviour of the day-tripper and the excursionist. There were tensions over mixed bathing, over the use or non-use of the Sunday, over dress and language.

Given tourism's exponential growth, the debate about a sustainable future for tourism has become mainstream and the focus of tourism leaders globally. Most notably, destinations like Copenhagen have embraced sustainable tourism, operating a triple bottom-line approach to sustainable tourism which incorporates environmental, social and economic aspects of sustainability. This means the city's tourism strategy is focused on the UN Sustainable Development Goals (Scheyvens, 2018; Yonglong *et al.*, 2015) and takes a community-wide approach to tourism and its future.

### Future turning point 3: Ubiquitous future

It is an acknowledged fact that human beings cannot travel at the speed of light and that technological advancement in transport has a boundary (Sweeney, 2014). This is why intergalactic travel maybe confined to science fiction and our imagination (Yeoman, 2012). Indeed, advancements in transport technologies from a speed perspective may have reached a limit. For example, Concorde failed because it could not achieve scalability (Sakade, 2011) and Cole (2015) has noted that the same can be said of space tourism. This is why Hui in Chapter 10 does not focus on speed in relation to the future of transportation but rather sustainability and the end of the combustion engine. He notes:

> For example, horse-drawn carriages, steam-powered trains and muscle-powered humans all coexisted at the turn of the 20th century and the same will be true of the rising renewable-powered and existing fossil-fueled infrastructure. Having focused on the changes brought by transitioning to fossil fuels, I will now center on forecasting the trend towards electrification and different zero- and low-carbon forms of transport.

However, transformational change will come in relation to personal technologies. This is the interface between tourist and tourism which is bringing about changes in how we experience tourism, how we purchase tourism and how we seek information about tourism (Yeoman, 2012). Technologies in a modern society are highly interlinked. The concept of ubiquitous computing was first invented by the researcher Mark Weisier

(1993) who developed the idea of 'invisible computing', that is, where the computer is not intrusive. This concept is characterized by the external environment, allowing users to become more connected to their surroundings and other people. Ubiquitous computing allows the integration of devices and technology applications in a world where everything is shared (Yeoman, 2012).

With the emergence of big data, a transformational future is occurring in tourism in a ubiquitous, connected manner. Technological and scientific developments like the global positioning system (GPS), facial recognition, gestural interfaces, predictive algorithms, light computing, virtual reality (VR), augmented reality (AR), autonomous vehicles, artificial intelligence, biometrics, haptic technologies and brain computer interfaces (BCI) are converging to create ubiquitous change in the technological interface between tourist and tourism (Robertson *et al.*, 2015). As Seeler argues in Chapter 14:

> Further expansions of the actual lived experiences are generated through new technological developments related to wearables and virtual and augmented reality (Tussyadiah *et al.*, 2018). The opportunities these new technologies bring to support on-trip experiences and knowledge creation will be relevant for historical sites: where witnesses of the past have disappeared, stories can be retold with a leveraged support such as 360-degree views of the landscapes of the past, and value can be added as educational components (Seeler, 2018).

We are seeing a transformational future (Dator & Yeoman, 2015). From hotels to hostels, restaurants to visitor attractions, the tourism industry has woken up to a digital world which is data-intensive and enabled by a consumer–technology interface that is ubiquitous (Dator & Yeoman, 2015).

## Future turning point 4: Mass maturity

Worldwide, Yeoman (2012) forecasts that world tourism will be valued at US$4.7 trillion in 2050, with 4.2 billion international arrivals. This represents, a huge growth considering that, according the UNWTO, the number of international arrivals was only 25 million in 1950. Today, tourism has an economic value of US$1.3 trillion and 1.3 billion international arrivals. Tourism has come of age to the point where it has political and economic capital (Findlay & Yeoman, 2015; Power, 1999). This is a point at which political leaders are seriously engaged in tourism because of its economic benefits. Tourism is used as a mechanism to shape place and create identity. Through place marketing, destinations are presenting brands embedded in place culture and history to the mass tourist (Bornhorst *et al.*, 2010).

The evolution of tourism has predominately been a Western phenomenon. However, looking to the future, world economic growth will come from within Asia and particularly China. Indeed, rising disposable incomes, improved transport infrastructure and the spread of internet technology are leading to the rapid development of the travel industry in China. According to Passport (2018), while inbound and outbound tourism experienced steady growth, domestic tourism also posted a robust volume sales increase in 2018 which reflected the increased desire to travel among Chinese residents. This rapid growth of domestic tourism in China is a representation of the final stages of Rostow's (1952) economic theory of high mass consumption. This refers to the period of contemporary comfort afforded to consumers concentrating on durable goods but, in a modern era, the focus is on experiences rather than goods (Yan *et al.*, 2017; Yeoman & McMahon-Beattie, 2018). In line with this, international tourism/travel has been seen as the leading desire of China's emerging middle classes (Foresight Factory, 2016a). Butcher highlights in Chapter 3:

> Recent decades have witnessed the rise of China and India as global economic powers and as societies with growing wealth and a growing class of people with the disposable income that tourism – international tourism especially – relies on. Domestically, too, China is developing to cater for increasing domestic and overseas tourism, including beach and ski resorts and city-based and rural tourism.
>
> The cruise ship industry is witnessing a substantial new market among the Indian middle class. The growth of air travel is rapid in both. China's rate of infrastructure growth has been phenomenal in recent decades and the budget airline model which had a profound impact on intra-European tourism from the mid-1990s has more recently grown rapidly in India.

### Scenarios from the Past

In the preceding section we identified four turning points that we see as dominating the future discourses of tourism, notably: *fluid identity, sustainable futures, ubiquitous future and mass maturity*. Taking these turning points into consideration, we propose two scenarios about the future: *A balanced future – learning from the past* and *Degradation – if only we had listened to the past*. Both scenarios are fictional accounts based on a speech to the UN World Tourism Organization General Assembly in 2050 by Duke Trump, Secretary General.

### Scenario 1: Degradation – if only we had listened to the past

We all ignored Steffen *et al.*'s (2018) report and his predictions about a scorched planet. Looking back, he predicted the domino-like cascade of melting ice, warming seas and Earth's hothouse state. Although some destinations have benefited from climate change, notably New Zealand as the new eco-paradise, many others have collapsed. It is just too hot. The ski resorts of the Alps are no more and wine tourism in California is limited to touring hydroponic vineyards. The thriving summer destinations of the Mediterranean are too hot and tourists only go there in winter. Africa is barren. India has literally run out of water and Shanghai has sunk! Driven by the need to generate high tourist revenues, tourism's rapid growth through the 2010s and 2020s overrode environmental concerns. A 'me first' policy by presidents and prime ministers, including my dear departed Uncle Donald, meant nations could not work together in an effective manner to address the issues of climate change, and expert scientific opinion was ignored or ridiculed by political leaders. It got to the point where even if we wanted to change, it was too late.

The success of tourism has been its failure. We had overtourism in Scotland, Venice, Paris, Mount Li, Patagonia and many more destinations. It is a pity we did not listen to Dr Alastair Durie who reminded us about the vandalism caused by tourists in Victorian times, whether it was picking rare flowers, stealing eagles' eggs, trampling over preserved spaces or inscribing graffiti on historical properties. There were too many tourists in Bournemouth in 1936 which caused a typhoid outbreak due to water shortages and sewage issues; the same happened in 2029. The number of tourists in Bournemouth overwhelmed the seaside resort. The infrastructure could not cope, resulting in the first typhoid outbreak in the UK since 1964. We never learn.

A turning point in tourism's success can be attributed to the GFC of 2008, when tourism became the recovery strategy for many nations and destinations. Tourism was seen as the cash cow for economic growth when many industries where stagnant or in decline. Tourism was at the centre of the experience economy. At the same time, middle-class Chinese tourists started to travel in droves. These tourists had wealth, were brand conscious and liked to shop. For example, Chinese tourists in Dubai would spend nearly 10 times more than European visitors on retail goods. As they say, money talks. Professor Marie-Louise Manion, a prominent tourism academic, noted that the rapid growth of tourism in Europe during the 2010s represented a period of poor planning resulting in tourists overwhelming local facilities. This was due to underinvestment in tourism and destination infrastructure in previous years, under-forecasting, political malaise, a lack of legislative frameworks, corruption and an amateur approach to planning.

Some destinations had invested in a sustainable tourism strategy to mitigate this growth, but it had a utopian feel about it. Community tourism was portrayed as the saviour but this strategy could never match the revenues generated from mass tourism. Anyway, community tourism was consumed by the wealthy, educated and creative classes, not the average citizen. Technological advancements or green technologies helped a bit, but they could not achieve the economies of scale necessary to mitigate all this growth.

The real failure from a tourism perspective was a focus on the short term. By the point we realized this was not the correct strategy, it was too late.

### Scenario 2: A balanced future – learning from the past

When the city of Copenhagen launched its 'Tourism is Good' strategy in 2018, there was a realization that something had to change. Copenhagen had started a kind of movement based on a realization that tourism as we knew it was no more: a sustainable future was the only future. Simply put, tourism was not the goal in itself, but a means to a sustainable end. It is a balance between community, society, business and the environment. If one of these elements is out of balance, the impact maybe small initially but, in the long term, the consequences are huge. In 2050, tourism is a US$5 trillion industry with 4.3 billion international arrivals. This is an exponential increase compared to 100 years ago, when international arrivals were reported as 25 million according to the International Union of Official Travel Organisations (the predecessor to the UNWTO).

Building on the United Nations Sustainable Development Goals meant focusing tourism on the following principles:

- *Leading by example*: Being the voice for tourism within communities that advocates tourism as a lever to a sustainable future.
- *Extended partnerships*: Tourism is not just about hotels or restaurants; it is everyone's business. This is an industry which touches so many, from the farmer who supplies the food to the retailer where the tourist shops. Tourism is vital to many communities. Partnership means working with these extended stakeholders such as businesses, residents, students or community groups to create a balanced future.
- *Balancing tourism choices*: Tourism can be an inclusive and sustainable industry bringing substantive benefit to tourists and those that live in host communities. This means making choices that balance the social, economic, cultural and environmental benefits of tourism and managing the negative elements in an informed manner.
- *Meaningful action*: In order to create a balanced future that is meaningful, it is necessary to use a range of instruments to achieve our goal. These include advocacy, money, government actions and laws.

Over the last 30 years we have seen much change, such as the shifting flow of tourists, with Asia now the number one region for inbound and outbound tourism. China's approach to governance and doing 'tourism the right way' embellishes the soft power approach. Across the world, following Copenhagen's lead, countries and cities legislated for good sustainable practices while penalising bad practice. Some things did not change, such as the reasons why tourists go on holiday: spending time with families, relaxing on the beach or connecting with nature. The modern tourist still wants both hedonistic and authentic experiences. Green lifestyles have become the norm with eco-adventures the number one trend in holiday products. But old favourites like Monte Carlo, Las Vegas and Macau are still here, as gambling is still an important tourism driver.

Innovative technologies have transformed the way we travel. The emergence of electric planes in the late 2020s fundamentally changed short-haul aviation, making the sector emissions-neutral by 2037. Richard Branson eventually got his *Virgin Galactic* into orbit in 2021 and began the day tripper market into outer space. Interestingly, the material these space ships were made of was a new polymer hydrocarbon which used less energy and water to make and was more robust and completely recyclable. *Virgin Galactic* was launched and also hypersonic travel. On the seas, cruise ships got bigger but cleaner and more environmentally friendly.

Claytronics was a game changer in hotel design. Guests could change the configuration of their rooms using programmable matter. Information technology innovations gave the tourist more choices. Destination holograms or personal guides would appear in hot spot locations redirecting tourist flows. Technological advancements also 'blinded' destinations when they reached capacity; thus the tourist could not access that place at a particular time. In terms of tourism accessibility, exoskeletons transformed the tourist experiences of people with disabilities.

Tourism priorities in 2050 are certainly different compared to the past. Tourism has become a driver to create a sustainable future for humankind.

## Conclusion

One of the roles of futures research is to model the social development of society by spotting key milestones (Mannermaa, 1991) or turning points. The central contribution of this book is to provide a reflective continuum in a systematic manner of the past, present and future of tourism. The analysis provided through the cognitive maps is a robust approach to understanding the past and its key turning points. It tells the story of the development of tourism from the era of the Grand Tour to that of mass tourism, noting how technological developments in transport and mobility have created change. Transport was the transformational driver of tourism which, along with the rising wealth of the Industrial Revolution, brought about a new type of consumer, that is, the conspicuous leisure consumer of the Victorian era. Although authenticity is one of the most talked about trends in tourism this century (Boyle, 2005; Isaacs, 2018; Yeoman *et al.*, 2007), we have seen its origins in the past in the concept of mindfulness. Both authenticity and mindfulness are associated with the need to get away from everyday life both physically and mentally and tourism provides the opportunity to do so.

Looking to the future, we identify a series of future turning points based on our cognitive mapping analysis of the preceding chapters in this book. *Step changes of mass tourism* evolves into *mass maturity* given the rise of the Asian tourist. This is a tourist who is sophisticated, demanding and educated. This links to *fluid identity* which emerges from *the leisure class of consumption*, with a tourist who is simultaneously into conspicuous and inconspicuous consumption (Yeoman, 2008) or hedonistic and simple experiences. *Mindfulness* leads to *sustainable futures* given the global debates about sustainability, global warming and climate change. The final turning focuses on *mobility* which is transport based, to a *ubiquitous future* which is based around developments in personal technology. We conclude with two scenarios based on the decisions and behaviours linked to these turning points. *Degradation* means we have not learned from the past and *a balanced future* means we have listened to the past.

This book allows historians to predict the future, and futurists to reflect on the importance of the past. The arguments and insights of the contributing authors are paramount to our understanding of the passage of time. The book represents the order of social change in tourism, concluding with future scenarios. Thus, we have demonstrated the idea of development as a directional course of events leading to an understanding of the complexity and plurality of the future of tourism.

## References

Ackermann, F. (2011) How OR can contribute to strategy making. *Journal of the Operational Research Society* 62, 921–923.

Albrecht, M. (2011) 'When you're here, you're family': Culinary tourism and the Olive Garden Restaurant. *Tourist Studies* 11, 99–113.

Berghoff, H. and Korte, B.B. (2002) Britain and the making of modern tourism: An interdisciplinary approach. In H. Berghoff, B. Korte, R. Schneider and C. Harvie (eds) *The Making of Modern Tourism: The Cultural History of the British Experience, 1600–2000*. New York: Palgrave.

Blass, E. (2003) Researching the future: Method or madness? *Futures* 35 (10), 1041–1054.

Bornhorst, T., Ritchie, B.J.R. and Sheehan, L. (2010) Determinants of tourism success for DMOs and destinations: An empirical examination of stakeholders' perspectives. *Tourism Management* 31 (5), 572–589.

Boyle, D. (2005) *Authenticity: Brands, Fakes, Spin and the Lust for Real Life*. London: Harper.

Boztug, Y., Babakhani, N., Laesser, C. and Dolnicar, S. (2015) The hybrid tourist. *Annals of Tourism Research* 54, 190–203.

Branchik, B. (2014) Staying afloat. *Journal of Historical Research in Marketing* 6 (2), 234–257.

Bray, R. and Raitz, V. (2001) *Flight to the Sun: The Story of the Holiday Revolution*. London: Continuum.

Campos, A.C., Mendes, J., do Valle, P.O. and Scott, N. (2018) Co-creation of tourist experiences: A literature review. *Current Issues in Tourism* 21 (4), 369–400.

Cohen, E. and Cohen, S.A. (2012) Current sociological theories and issues in tourism. *Annals of Tourism Research* 39 (4), 2177–2202.

Cole, S. (2015) Space tourism: Prospects, positioning, and planning. *Journal of Tourism Futures* 1 (2), 131–140.

Corbin, J.M. (2015) *Basics of Qualitative Research: Techniques and Procedures for Developing Grounded Theory*. International Institute for Qualitative Methodology, Anselm Strauss. Los Angeles: Sage.

Dator, J. and Yeoman, I. (2015) Tourism in Hawaii 1776–2076: Futurist Jim Dator talks with Ian Yeoman. *Journal of Tourism Futures* 1 (1), 36–45.

Durie, A.J. (2017) *Scotland and Tourism: The Long View, 1700–2015*. London: Routledge.

Eden, C. and Ackerman, F. (1998) *Making Strategy: The Journey of Strategic Management*. London: Sage.

Ehrnrooth, H. and Gronroos, C. (2013) The hybrid consumer: Exploring hybrid consumption behaviour. *Management Decision* 51 (9), 1793–1820.

Elliott, G. (2010) *Hobsbawm: History and Politics*. New York: Palgrave Macmillan.

Ellis, A., Park, E., Kim, S. and Yeoman, I. (2018) What is food tourism? *Tourism Management* 68, 250–263.

Findlay, K. and Yeoman, I. (2015) Dr Spock's Food Festival. *Journal of Tourism Futures* 1 (2), 148–151.

Foresight Factory (2016a) *Demanding Consumers*. London: Foresight Factory.

Foresight Factory (2016b) *Hospitality and Tourism Trends*. London: Foresight Factory.

Getz, D. (2000) Tourist shopping villages: Development and planning strategies. In C. Ryan and S. Page (eds) *Tourism Management: Towards the New Millennium* (pp. 211–225). Oxford: Elsevier Science.

Gladwell, M. (2002) *The Tipping Point: How Little Things Can Make a Big Difference*. Boston, MA: Back Bay Books.

Higham, J., Reis, A. and Cohen, S.A. (2016) Australian climate concern and the 'attitude–behaviour gap'. *Current Issues in Tourism* 19 (4), 338–354.

Hobsbawm, E.J. (1995) *Age of Extremes: The Short Twentieth Century, 1914–1991*. London: Abacus.

Huff, A.S. and Jenkins, M. (2002) *Mapping Strategic Knowledge*. London: Sage.

Isaacs, D. (2018) Kentucky's authenticity is key to tourism success. *The Lane Report* 33, 22–25.

Jones, M. (1993) *Decision Explorer: Reference Manual (Version 3.1)*. Kendal: Banxi Software.

Kelly, G.A. (1955) *The Psychology of Personal Constructs*. New York: Norton.

Kelly, G. (1977) Personal construct theory and the psychotherapeutic interview. *Cognitive Therapy and Research* 1, 355–362.

Koens, K., Postma, A. and Papp, B. (2018) *Overtourism? Understanding and Managing Urban Tourism Growth Beyond Perceptions*. Madrid: UN World Tourism Organization.

Lenzen, M., Sun, Y.-Y., Faturay, F., Ting, Y.-P., Geschke, A. and Malik, A. (2018) The carbon footprint of global tourism. *Nature Climate Change* 8 (6), 522–528.

Louv, R. (2008) *Last Child in the Woods: Saving Our Children from Nature-Deficit Disorder*. London: Algonquin Books.

MacLaren, A., O'Gorman, K., Stringfellow, L. and Maclean, M. (2013) Conceptualizing taste: Food, culture and celebrities. *Tourism Management* 37, 77–85.

Mannermaa, M. (1991) In search of an evolutionary paradigm for futures research. *Futures* 23 (4), 349–372.

McMahon-Beattie, U., McEntee, M., McKenna, R., Yeoman, I. and Hollywood, L. (2016) Revenue management, pricing and the consumer. *Journal of Revenue and Pricing Management* 15 (3–4), 299–305.

Mingers, J. (2014) *Systems Thinking, Critical Realism and Philosophy: A Confluence of Ideas*. Florence: Routledge.

Murphy, L., Benckendorff, P., Moscardo, G. and Pearce, P.L. (2011) *Tourist Shopping Villages: Forms and Functions*. New York: Routledge.

Nyaupane, G.P., Morais, D.B. and Graefe, A.R. (2004) Nature tourism constraints: A cross-activity comparison. *Annals of Tourism Research* 31, 540–555.

Page, S., Yeoman, I., Connell, J. and Greenwood, C. (2010) Scenario planning as a tool to understand uncertainty in tourism: The example of transport and tourism in Scotland in 2025. *Current Issues in Tourism* 13 (2), 99–137.

Passport, E. (2018) *Travel in China – Country Report, September 2018*. London: Euromonitor.

Pearce, D. and Butler, R. (eds) (2010) *Tourism Research: A 20-20 Vision*. Oxford: Goodfellow.

Power, E.M. (1999) An introduction to Pierre Bourdieu's key theoretical concepts. *Journal for the Study of Food and Society* 3, 48–52.

Robertson, M., Yeoman, I., Smith, K. and McMahon-Beattie, U. (2015) Technology, society, and visioning the future of music festivals. *Event Management* 19 (4), 567–587.

Rostow, W.W. (1952) A historian's perspective on modern economic theory. *American Economic Review* 42 (2), 16–29.

Sakade, T. (2011) Trapped in a loveless marriage: The Anglo–French Concorde crisis of 1974. *Kyoto Economic Review* 80 (2), 134–147.

Schänzel, H.A. and Yeoman, I. (2014) The future of family tourism. *Tourism Recreation Research* 39 (3), 343–360.

Scheyvens, R. (2018) Linking tourism to the sustainable development goals: A geographical perspective. *Tourism Geographies* 20 (2), 341–342.

Scott, J. (2014) Leisure class. *A Dictionary of Sociology* (4th edn). Oxford: Oxford University Press.

Seeler, S. (2018) Continuum of an experienced tourist's multidimensionality – explorations of the experience levels of German and New Zealand tourists. PhD thesis, Auckland University of Technology.

Silver, C. and Lewins, A. (2014) *Using Software in Qualitative Research: A Step-by-Step Guide*. London: Sage.

Silverstein, M., Fiske, N. and Butman, J. (2003) *Trading Up: The New American Luxury*. New York: Portfolio.

Smil, V. (2005) *Creating the Twentieth Century: Technical Innovations of 1867–1914 and their Lasting Impact*. Oxford: Oxford Scholarship Online.

Steffen, W., Rockström, J., Richardson, K., *et al.* (2018) Trajectories of the Earth System in the Anthropocene. *Proceedings of the National Academy of Sciences* 115, 8252–8259.

Sweeney, J. (2014) Einstein's dreams. *Review of Metaphysics* 67 (June), 811–834.

Tussyadiah, I.P., Wang, D., Jung, T.H. and tom Dieck, M.C. (2018) Virtual reality, presence, and attitude change: Empirical evidence from tourism. *Tourism Management* 66, 140–154.

Veblen, T. (1899) *The Theory of the Leisure Class: An Economic Study of Institutions*. London: Allen & Unwin.

Wakeman, R. (2012) *Veblen Redivivus: Leisure and Excess in Europe*. In R. Wakeman (ed.) *Oxford Dictionary of Sociology*. Oxford: Oxford University Press.

Weatherford, L. (2016) The history of forecasting models in revenue management. *Journal of Revenue and Pricing Management* 15 (3), 212–221.

Weick, K.E. (1989) Theory construction as disciplined imagination. *Academy of Management Review* 14 (4), 516–531.

Weiser, M. (1993) Some computer science issues in ubiquitous computing. *Communications of the ACM* 36 (7), 75–84.

Yan, L., Sid, G. and Hiroko, O. (2017) Chinese consumers' luxury value perceptions – a conceptual model. *Qualitative Market Research: An International Journal* 20 (2), 247–262.

Yeoman, I. (2004) The development of a conceptual map of soft operational research practice. PhD thesis, Napier University.

Yeoman, I. (2008) *Tomorrow's Tourist: Scenarios and Trends*. London: Elsevier Science.
Yeoman, I. (2012) *2050: Tomorrow's Tourism*. Bristol: Channel View Publications.
Yeoman, I. and McMahon-Beattie, U. (2016) An ontological classification of tourism futures. In M. Scerri and L.K. Hui (eds) *CAUTHE 2016: The Changing Landscape of Tourism and Hospitality: The Impact of Emerging Markets and Emerging Destinations*. Sydney: Blue Mountains International Hotel Management School.
Yeoman, I. and McMahon-Beattie, U. (2018) The future of luxury: Mega drivers, new faces and scenarios. *Journal of Revenue and Pricing Management* 17 (4), 204–217.
Yeoman, I. and Watson, S. (2011) Cognitive maps of tourism and demography: Contributions, themes and further research. In I. Yeoman, C.H.C. Hsu, K.A. Smith and S. Watson (eds) *Tourism and Demography*. Oxford: Goodfellow.
Yeoman, I., Brass, D. and McMahon-Beattie, U. (2007) Current issue in tourism: The authentic tourist. *Tourism Management* 28 (4), 1128–1138.
Yeoman, I., McMahon-Beattie, U., Backer, E., Robertson, M. and Smith, K. (eds) (2014) *The Future of Events and Festivals*. London: Routledge.
Yeoman, I., McMahon-Beattie, U. and Wheatley, C. (2015) The future of food tourism: A cognitive map(s) perspective. In I. Yeoman, U. McMahon-Beattie, K. Fields, J. Albrecht and K. Meethan (eds) *The Future of Food Tourism: Foodies, Experiences, Exclusivity, Visions and Political Capital*. Bristol: Channel View Publications.
Yonglong, L., Nebojsa, N., Martin, V. and Stevance, A.-S. (2015) Policy: Five priorities for the UN Sustainable Development Goals. *Nature* 520 (7548), 432–433.
Zuelow, E.G.E. (2016) *The History of Modern Tourism*. London: Palgrave.

# Index

Accessibility, 3, 39–40, 46, 51, 61, 94, 102, 135, 183, 184, 224, 252, 303
Accommodation, 4–5, 15, 23, 27, 33, 42–44, 58, 61–63, 73, 101, 134–138, 140–144, 149, 154–156, 175, 178, 193–194, 201–202, 235, 253, 261, 268, 291–292
Active initiation, 41–42
Adventure tourism, 215–219, 223, 227
Ageing population, 198
Air connectivity, 45–48, 180, 252
Air Transport, 5, 49, 70, 93, 109, 263
Airbnb, 33, 58, 143
Airbus, 5, 93, 95, 96–98, 100, 261, 291
Airline travel, 45–48, 180, 252
Airplane technology, 5, 26, 27, 93–103
Amenity migration, 196
Anthropocene, 128, 266
Architecture, 20, 198, 237
Armchair mountaineering, 114, 226
Asian tourists, 197–198, 304
Augmented reality, 114, 170, 184–185, 299
Australia, 5–7, 69, 80, 93–97, 100–101, 106, 112, 153, 161, 165, 188, 190–192, 196, 208, 220, 259, 272, 275, 291
Authenticity, 15, 18, 31, 62, 114, 163, 169, 179, 188, 190, 196, 248, 272, 275, 285, 287, 303
Aviation, 5, 25, 26, 28, 45, 93, 96, 98, 100, 103, 123, 250, 261, 296, 303

Beach practices, 79–89, 233–234, 257, 259, 302
Boeing, 5, 93, 95, 96–98, 100, 109, 259, 261, 291
Britons, 61
Buddhism, 202, 206, 210

Capitalism, 21, 137, 177
Cheddar, 7, 188, 191–193, 195, 275
China, 3, 4, 16, 32–33, 49, 66, 78–89, 100, 101, 112, 126–127, 189, 195, 198, 206, 220–221, 251, 257–259, 294, 300, 302–303
Chinese characteristics, 79
Christianity, 120, 202, 203–204, 205, 208
Cinema, 162–167
Circulation of ideas, 2, 4, 21, 31, 63, 73, 75, 76, 120, 126, 128, 244–246, 268
City-state, 4, 65–76, 255–257, 289
Civil society, 48, 71, 252
Climate change, 5, 7, 17, 112–113, 128, 224–225, 238, 243, 263, 281, 297, 300–301, 304
Coastal tourism, 33, 78–89, 251, 257–259,
Co-creation, 183, 184
Cognitive maps/mapping, 3, 8, 243–246, 283–285, 294, 303–304
Collaboration, 45, 46, 50, 51, 98, 252
Commercial lodging, 133–137, 140, 142–143, 266
Commodification, 179, 188–189, 218, 223, 275
Competing demands, 3, 231, 252
Conspicuous consumption, 179, 292, 294, 304
Corporatism, 147
Creative destruction, 189, 190, 192, 194, 195
Cruise passengers, 110, 112–115
Cruise ships, 5, 108–115, 263, 291, 294, 303
Cruise tourism, 5, 105–115, 261–264, 289
Cuisine, 17, 191,
Cultural impacts, 32

Culture/cultural, 2–7, 11–12, 14, 16–17, 20–21, 23–33, 47, 50, 53–54, 60–62, 67, 69–73, 79–81, 85, 88, 94, 99, 120, 122–125, 133–136, 141, 143, 144, 163, 168, 175–182, 185, 195–198, 203, 207, 215, 224, 230–238, 246, 248, 250, 257, 268, 277, 281, 287, 296

Decision Explorer, 245–246, 284
Decline of democracy, 71
Destination image, 189
Destination management organisations, 74, 296
Destination marketing, 189
Destinations, 3–5, 7, 14–17, 22, 39–50, 53–57, 62–63, 70, 74–75, 85–88, 93–103, 109, 111–115, 123–125, 162–167, 170, 174, 176, 177, 182, 185, 189, 204–205, 210–212, 215–216, 229, 238, 250, 257, 259–263, 272, 275, 285, 287, 289, 296
Digital,
  Digital media, 161, 272
  Digital narratives, 183–185
  Digital society, 174–186, 272–275
Digitalization, 49–51, 252
Dubai, 5, 70, 75, 94–98, 100–102, 261, 301

Economic development, 33, 44, 65, 88
Economic downturn, 41, 105, 108, 111–112, 252, 263
Economic power, 22, 32, 67–68, 300
Educational tourism, 6, 135, 156, 175–176, 179–181, 183
Electricity, 122, 124–126, 128
Employment relations, 146–157, 268–270
Energy, 5, 44, 115, 118–128, 226, 264–266, 303
Entrepreneurs, 23, 47, 149, 192, 193–194, 196, 198
Environmental tourism, 7, 20, 29, 32, 33, 51, 114, 115, 127, 128, 216, 224, 230
Evolution, 1–3, 5, 7–8, 47–49, 50, 78, 83, 100, 101, 105, 118–119, 141, 174–186, 188–189, 192–195, 198, 208, 225, 243–304
Evolutionary paradigm, 2, 47–49, 243
Experienced tourists, 6, 181, 182–185, 273, 275, 297

Experiences, 3, 6, 12, 14, 15, 17, 40, 49–51, 55, 57, 58, 65, 73, 74, 103, 106, 114, 115, 118, 123, 124, 161, 167, 169, 175–179, 181–185, 188, 196, 198, 218, 252, 263, 272–275, 287, 297, 299, 300, 303
Experiential tourism, 5, 142

Family
  Family structure, 17
  Family in 2035, 18–19
Film
  Film fans, 167–168
  Film tourism, 6, 161–171, 270–272
  Film history, 162–168
Floating resorts, 111
Fluid identity, 296–297, 304
Fordism, 24, 26–28, 250
Forecasting, 124, 128, 243, 298, 301
Foreign concessions, 78, 79, 81–84, 86–87, 257

Geohistory, 79, 259
German tourists, 178–179, 180
Global cities, 70, 72, 75, 255, 257
Global history, 79, 257
Global warming, 17, 224, 279, 304
Globalization, 2, 3, 11–19, 20–34, 70, 122, 195, 209
Golden Era in mountaineering, 17, 215–227, 279–281
Grand hotel, 24, 136–144, 266
Grand Tour, 1, 3, 6, 12–13, 20–23, 53–55, 67, 120, 122, 134–135, 174–186, 246, 248, 266, 272–275, 289, 291–292, 297
Group travel, 174, 182, 183, 185
Guided mountaineering, 221–224

Hahndorf, 7, 188, 191–195, 275
Heritage, 17, 29, 30, 53, 57, 61, 63, 141–142, 166, 169, 176, 188–198, 203, 207, 208, 211, 237–238, 253, 275–277
Hidden workforce, 75, 146, 153, 156
Hinduism, 202, 206–207, 210
History, 2, 3–8, 11–18, 31, 33, 39, 41, 46, 47, 53, 60–63, 79–80, 87, 118–121, 124, 126, 133–144, 174–175, 178, 181, 183, 185, 191–193, 198, 215–227, 246–248

Hologram, 170, 303
Holographic technology, 161, 169–170, 272
Hospitality, 47, 128, 134, 136, 140, 144, 151, 152, 155, 156, 201, 204, 234, 235–236, 266, 281
Hotel(s), 5–6, 133–144, 146–157, 266–270
  Hotel chains, 140
Hubs, 5, 94, 99, 101, 123, 261
Hunting, 11, 30, 119, 127, 231, 236–237
Hyper-reality, 168–169

Independent travel, 45–47, 180, 185, 207, 252
Industrial revolution, 5, 6, 21–22, 67, 118, 121–122, 127, 176, 248, 264, 273, 291, 292, 294, 303
Industrial/Industrialization, 20, 21, 23, 107, 111, 121, 123, 136–138, 234, 238, 261, 266, 287
Inn, 136, 137, 139, 141
Innovation, 27, 29, 57–62, 105, 111, 118, 121–122, 127, 136–140, 144, 146, 195, 225, 261, 263, 266–268, 287, 289, 303
Institutional adaptation, 44–45
Ireland, 4, 6, 53–62, 69, 161, 163, 165–168, 211, 255, 272
Islam, 120, 134, 202, 204–205, 208–211, 289
Island destinations, 41, 112, 250

Judaism, 202, 205
Jules Verne, 4, 53–64, 252–255

Kangaroo Route, 5, 93–98, 259, 291, 296
Knowledge enhancement, 6, 174, 179–185, 273

Last Frontier of mountaineering, 220–223, 279
Leisure Class, 8, 244, 285, 292–294
Literature, 57–62, 93, 122, 123, 162, 167, 174, 180, 196, 201, 245, 253, 287
Loss of individual freedom, 62–64
Luxury, 11, 16, 26, 58, 61, 63, 106–108, 137, 155, 179, 194, 253, 261, 264, 272, 287, 292, 294, 296

Malta, 3, 39–51, 250–252
Managed development, 50, 65, 69
Mass maturity, 299–300
Mindfulness, 285–287
Mobility, 287–289
Modernity, 20, 133, 139
Mormonism, 202, 205–206
Mountain guiding, 218, 224
Mountaineering
  Mountaineering entrepreneurial adaptation, 224–227
  Mountaineering history, 216–223
  Mountaineering Tourism, 215–227

New generation, 18, 112
New Zealand, 1, 6, 106, 112, 146–157, 161, 166, 167, 190, 220, 268–270
Niche tourism, 3, 6–7, 27, 161–171, 174–186, 188–198, 201–212, 215–227, 229–238

Ocean liners, 106–109
Other-directed society, 18
Overtourism, 33–34, 197, 238, 250, 281, 297, 301

Paradox of tourism, 14
Path dependence, 43–44
Pilgrimage, 15, 66, 120, 134, 201–207, 212, 277, 287, 289, 291
Policy
  Policy crises, 42–43
  Policy leadership, 44–45
  Policy responses, 40, 252
Political control, 79, 259
Political power, 66, 79, 178–179
Politics, 12, 30, 32, 88, 100, 156, 246, 252
Population growth, 73, 195
Ports, 24, 60, 81, 82, 109, 111, 114, 115, 127
Post industrial, 65, 127–128
Poverty, 4, 25, 54, 56, 59, 60, 62, 253
Private sector, 3, 39, 44–45, 48–50, 73, 252, 255

Qantas, 94–98, 100, 101, 103

Railway, 22, 24, 55, 67, 69, 120–123, 127, 136, 138, 177, 191, 234, 236, 255, 266

Rejuvenation, 194–195, 275
Religion, 7, 67, 120, 201–212, 215, 277–279
Religious heritage, 211
Residents, 29, 43, 47, 49, 51, 71, 72–74, 170, 188–190, 192, 193–194, 197–198, 238, 255, 298, 300–302
Resort, 16, 22, 25, 28–30, 32–34, 63, 67, 80, 82, 85, 86, 110–111, 134–135, 137, 142, 149, 178, 234, 238, 250, 253, 289
Romantic traveller, 13–15
Round trips, 59–60, 63, 123
Rural idyll, 189, 193, 197, 198

Safety and Security, 110, 226
Scotland, 7, 22, 57, 69, 165, 229–238, 281–282, 287, 297, 301
Screen tourism, 161
Sea bathing, 4, 78–80, 82–84, 86–87, 257, 259
Seamless futures, 8
Sharing, 18, 125–127, 175, 184
Sherpas, 222
Shopping, 101, 110, 111, 188–198, 275–277
Singapore, 4, 5, 70, 71, 94–98, 100–102, 261
Small island, 3
Souvenirs, 191, 193
St. Jacobs, 192
Stopover destinations, 5, 93–103, 259–261
Studienreisen, 180–184
Sunrise Project, 97–98
Sustainability, 18, 195, 197, 198, 229–230, 281–282
Sustainable futures, 297–298

Target group, 61, 63
Technology

Growth of technology, 24–27, 49–51, 93–103, 105, 114–115
Territorial states, 65–66, 68–70, 73–76
Terrorism, 33, 105, 111–112, 113, 203, 209–212
The Foundling, 59
Thomas Cook, 21–23, 27, 28, 67, 123, 176–178, 233, 273, 291
Tokenism, 42, 252
Tourism
  Mass tourism, 2, 3, 6, 8, 13, 20–34, 84, 123, 176–180, 185, 215, 230, 233–234, 244, 248–250, 264, 266, 272–273, 277, 279, 281, 285, 287, 289–291
  Modern tourism, 13–14, 26, 62, 120, 122, 123, 149, 201, 291
Tourist attractions, 24, 30, 191, 208
Tourist Hotel Corporation, 149–150, 157, 270
Tourist motivation, 48
Transformations, 4, 39, 47, 118, 141, 244, 252
Transportation, 5, 61, 107, 118–124, 127–128, 175–177, 180, 204, 264, 266, 287, 289, 298
Travel, 3, 5–6, 11–12, 15, 20, 22-33
Travel writings, 175, 183–184
Turning points, 2, 3, 5–7, 32–34

Ultra Long-Haul Flights, 93
Unions, 25, 123, 148, 151, 152, 154–157

Virtual reality (VR), 114, 161–171, 226, 270–272

War, 5, 25, 32, 60, 66, 84, 105, 108, 110, 119, 122, 176, 203, 212

For Product Safety Concerns and Information please contact our EU Authorised Representative:

Easy Access System Europe

Mustamäe tee 50

10621 Tallinn

Estonia

gpsr.requests@easproject.com

www.ingramcontent.com/pod-product-compliance
Ingram Content Group UK Ltd.
Pitfield, Milton Keynes, MK11 3LW, UK
UKHW021542050825
2593IPUK00005B/77